Essentials of Mathematica

Nino Boccara

Essentials of Mathematica

With Applications to Mathematics and Physics

 Springer

University of Illinois at Chicago
Department of Physics (M/C 273)
845 West Taylor Street
Chicago, IL 60607
USA
boccara@uic.edu

Additional material to this book can be downloaded from http://extras.springer.com

ISBN 1-4939-0242-3   ISBN 0-387-49514-2 (eBook)
ISBN 978-1-4939-0242-2   ISBN 978-0-387-49514-9 (eBook)

Printed on acid-free paper.

9 8 7 6 5 4 3 2 1

springer.com

Preface

This book consists of two parts. Part I describes the essential *Mathematica* commands illustrated with many examples and Part II presents a variety of applications to mathematics and physics showing how *Mathematica* could be systematically used to teach these two disciplines.

The book is based on an introductory course taught at the University of Illinois at Chicago to advanced undergraduate and graduate students of the physics department who were not supposed to have any prior knowledge of *Mathematica*.

Mathematica is a huge mathematical software developed by Wolfram Research Inc. It is an interactive high-level programming language that has all the mathematics one is likely to need already built-in. Moreover, its interactivity allows testing built-in and user-defined functions without difficulty thanks to numerical, symbolic and graphic capabilities. All these features should encourage students to look at a problem in a computational way, and discover the many benefits of this manner of thinking. For instance, when studying a new problem, *Mathematica* makes it easy to test many examples that might reveal unsuspected patterns.

The reader is advised to first study Chapter 1 of Part I entitled A Panorama of *Mathematica* which presents an overview of the most frequently used commands. The following chapters—dealing with Numbers, Algebra, Analysis, Lists, Graphics, Statistics and Programming—go into more details. The reader would probably make the most of the book browsing, as soon as possible, Part II, devoted to Applications to Mathematics and Physics, coming back to Part I to go deeper into specific commands and their various options.

This book is intended for beginners who want to be able to write a small efficient *Mathematica* program in order to solve a given problem. Having this in mind, we made every effort to follow the same technique: first the problem is broken up into its different component parts, then each part of the problem is solved using either a built-in or a user-defined *Mathematica* function, checking carefully that this function does exactly what it was supposed to do, and the program is finally built up by grouping together all these functions using a standard structure.

Note concerning the figures

Most figures have been generated using colors as indicated by their *Mathematica* code but are represented in the book using only various shades of grey. However all the figures can be found in color in the accompanying CD-ROM which also contains all the *Mathematica* cells that appear in the book.

Nino Boccara

Contents

Part I Essential Commands

Part II Applications

List of Figures

Essential Commands

This first part describes the essential *Mathematica* commands.

Chapter 1 gives a detailed overview of the most frequently used *Mathematica* commands, starting from the most elementary and culminating in an introduction to *Mathematica* programming with a detailed application to the Collatz conjecture and possible generalizations. After studying this chapter, the reader should be able to tackle the applications presented in Part II coming back to a specific chapter of Part I to study more closely a particular command and its various options to better understand how a user-defined function solving a specific problem is built up.

Chaper 2 is dedicated to numbers. *Mathematica* distinguishes integer—odd, even, prime, and Gaussian—, and rational, real, and complex numbers. *Mathematica* can manipulate these numbers in different bases with any precision. The chapter ends with a discussion of positional number systems, the Zeckendorf representation, and calendars which, as a matter of fact, are multibase positional number systems.

Chapter 3 deals with algebra. It examines algebraic and trigonometric expressions, how to solve equations either exactly or numerically, and describes a few built-in *Mathematica* functions related to linear algebra.

Chapter 4 is devoted to calculus. It studies differentiation and integration, differential equations, sums and products, power series and limits, complex functions, Fourier transforms and Fourier series, Laplace and Z transforms, and in conclusion shows how *Mathematica* can help solve recurrence equations and partial differential equations.

Chapter 5 studies lists that provide an efficient way of manipulating groups of expressions as a whole. It shows how to create lists; extract or add elements to lists; and find, group, rearrange, and count elements. Many built-in *Mathematica* functions are listable indicating that the function should automatically be threaded over lists that appear as its arguments. User-defined functions can also be made listable.

Chapter 6 explains how to generate graphics that are important components of many applications. *Mathematica* provides powerful graphics capabilities. We can plot two- and tridimensional graphics, using different coordinate systems. We can also plot lists of data and use a lot of options dealing with colors, text, labels, and legends, Graphics can be grouped in arrays. Using specific packages, we can produce special plots such as log-log plots, bar charts, pie charts, and histograms. Manipulating graphics primitives, that is, points, line, polygons, and circle, we can draw a variety of figures. We can animate graphics, draw vector fields, gradient fields, contour plots, and density plots.

Chapter 7 is dedicated to probability and statistics. *Mathematica* can generate various types of random numbers: integers or reals uniformly distributed in a given interval. *Mathematica* can also generate random numbers distributed according to most discrete and continuous probability distributions such as Bernoulli, binomial, Poisson, normal, Cauchy, gamma, Pareto, and so on. To analyze data we have at our disposal a variety of statistical tools with the possibility of drawing graphics illustrating our results.

Chapter 8 explains how to write simple and efficient basic programs. After a brief review of the *Mathematica* language, we examine functional programming which is characteristic of *Mathematica* although other types can also be used. In order to build up a function generating, for example, the Fibonacci number of a given order, we study different programming methods and show that the CPU time necessary to generate such a Fibonacci number may vary by many orders of magnitude ranging from hours to a fraction of a second depending on how efficient the program is.

1

A Panorama of *Mathematica*

This rather long chapter presents an overview of the most frequently used *Mathematica* commands.

1.1 Notebooks and Cells

Mathematica consists of two separate programs: the *kernel* and the *front end*. The kernel is the computational engine, whereas the front end is the user interface. The user sends commands to the kernel through the front end. The kernel sends back a postscript code that is displayed in the front end.

A *Mathematica* notebook is an interactive document combining text, graphics, and calculations. Notebooks are platform independent. The present document is a notebook.

A notebook is organized in cells. On a computer screen, a cell is defined by a square bracket on the right-hand side. There are three types of cells: text, input, and output cells.

Commands sent to the kernel are entered in input cells. The cell below is an input cell:

```
23 + 14
```

When an input cell is evaluated by pressing $\boxed{\text{SHIFT}}$ $\boxed{\text{RETURN}}$, the result of the kernel computation is sent back to the front end and displayed in an output cell. The cell below is the output cell resulting from sending the previous command to the kernel. This cell is a text cell.

1.2 Basic Syntax

All built-in function names have an initial capital letter. Most function names are explicitly spelled out (`Integrate`, `Plot`, ...) except a few abbreviations of common use (`Sin`, `Det`, ...). If a name consists of more than one word, the first letter of each word is capitalized, and no spaces separate the words (`ListPlot`, `FindRoot`, ...). The number of built-in functions is extremely large. *Mathematica* is case sensitive: `x` and `X` are two different symbols.

It is good practice to name variables and functions as explicitly as possible and avoid using an initial capital letter when naming user-defined functions.

Mathematica uses different types of bracketing. Parentheses ( ) are used to explicitly group terms and force the correct order of evaluation as in `x - (y-x)`. Square brackets [ ] are used for functions; for example, the sine of x is denoted `Sin[x]` and not `sin(x)`. Curly braces {...} are used to group the elements of a list as {a,b,c}. Double square brackets are used for indexing: `v[[n]]`, for instance, represents part n of v. *Mathematica* gets confused when the wrong bracket type is used.

Commas are used to separate the elements of a list or the arguments of a function. A semicolon at the end of an input tells *Mathematica* to perform the operation but not display the output. Semicolons are also used to separate different expressions written on the same line. For example,

```
a = 5; b = 3; c = 7;
```

tells *Mathematica* to assign the values 5, 3, and 7 to the symbols a, b, and c respectively without, however, displaying an output.

A space between two expressions is understood by *Mathematica* as multiplication: `final result` is not an acceptable name, it is understood as the product of `final` and `result`. An acceptable name would be `finalResult`.

1.3 Basic Operations

Mathematica can be used as a pocket calculator but arithmetic operations can be done with any number of significant digits.

```
(4536784519876453286 - 443217654393562751 +
7659432176587356289 - 321736482582441593) / 5467821
```

$$\frac{11431262559487805231}{5467821}$$

```
3429854 67532098
```

231625236453692

```
3²¹
```

10460353203

```
(2.67 + 5.72 / 3.4) / 1.58
```

2.75465

Using parentheses (...), the expression above could also have been written:

```
(2.67 + (5.72 / 3.4)) / 1.58
```

2.75465

Numbers can be manipulated using an arbitrary base whose maximum value is 36. BaseForm[number, b] displays number in base b. If b > 10, Mathematica uses letters.

```
{BaseForm[137, 7], BaseForm[379, 27]}
```

$\{254_7, \text{e1}_{27}\}$

Using the notation b^^digits, where each of the digits is less than b, the number digits in base b is displayed in base 10.

```
{7^^254, 27^^e1}
```

{137, 379}

In the following cell, the first term has infinite precision and the second one has machine precision. The command N[expression] gives the numerical value of expression. N[expression, n] attempts to give a result with n-digit precision.

Sqrt[x] or $\sqrt{x}$ gives the square root of x. $\sqrt{x}$ is entered using the BasicInput submenu of File → Palettes.

```
{Sqrt[5], √5, N[Sqrt[5]]}
```

{Sqrt[5], Sqrt[5], 2.23607}

We can see the precision from the InputForm:

```
{Sqrt[5], N[Sqrt[5]] // InputForm}
```

{Sqrt[5], 2.23606797749979}

By definition, the Precision of x is equal to minus the decimal logarithm of the ratio $\Delta x/x$, where Δx is the uncertainty on x. The machine precision is 15.9546. Precision is different from Accuracy which is equal to minus the decimal logarithm of the uncertainty Δx. That is,

$$\textbf{relative error} = 10^{-\texttt{Precision}} \text{ and } \textbf{absolute error} = 10^{-\texttt{Accuracy}}$$

Irrational numbers can be manipulated with any chosen precision.

```
N[Pi, 100]
```

3.14159265358979323846264338327950288\
4197169399375105820974944592307816640\
62862089986280348253421170680

The \ indicates that the output is continuing on the next line. The function Precision[] gives the number of significant digits.

```
Precision[N[Pi, 100]]
```

100

To avoid printing a long output, end the input expression with a semicolon (;).

```
Timing[N[Pi, 100000];]
```

{1.10356 Second, Null}

Timing[expression] evaluates expression and gives the CPU time in seconds spent in the *Mathematica* kernel. Null is returned when no output is printed.

1.4 Mathematica as a Functional Language

In *Mathematica* everything is an expression. A *Mathematica* expression is any string of symbols of the form

$$■[■,■,\ldots]$$

where ■ is a placeholder in which we can write either pure symbols or other expressions. At the front of the square bracket is the head of the expression, inside the square bracket are the elements of the expression. In *Mathematica*, this is the internal form of everything. For example, gd[x,ab] is an expression whose head is gd and x and ab are elements. This expression may also be viewed as the function gd of x and ab. Variable names can consist of attached letters and numbers, but the first symbol cannot be a number; v3 is an accepted variable name but 3v will be understood by *Mathematica* as 3 times v.

```
Head[gd[x, ab]]
```

gd

Inputs in operator notation are transformed into internal forms that are combinations of expressions. These internal forms could be used instead of traditional arithmetic notations. This is, however, too cumbersome and not recommended.

```
{3 + 6, Plus[3,6]}
```

{9, 9}

```
{4 * 5, 4 5, Times[4, 5]}
```

{20, 20, 20}

Instead of the symbol * to multiply two numbers, it is simpler to leave an empty space between the two numbers.

```
{2⁴, Power[2, 4]}
```

{16, 16}

```
{5 + 2 I, Complex[5, 2]}
```

{5 + 2 I, 5 + 2 I}

In the full forms of the last four examples, the Head of the expressions is explicit.

```
{Head[Plus[x + y]], Head[x + y]}
```

{Plus, Plus}

To find internal forms, use the function FullForm.

```
{FullForm[x + y], FullForm[x y], FullForm[xʸ],
FullForm[x - y], FullForm[x + y I], FullForm[a, b, c]}
```

{Plus[x, y], Times[x, y],

Power[x, y], Plus[x, Times[-1, y]],

Plus[x, Times[Complex[0, 1], y]],

List[a, b, c]}

1.5 Getting Help

To access the *Mathematica* help system in the notebook environment we just have to go to the Help menu and click on Help Browser. Entering a command,

say, Plot and clicking the Go button we have to choose among various types
of plots such as 2D Plots, 3D Plots, Contour Plots, and so on. Selecting 2D
Plots and clicking, for example, on ListPlot, a window appears with detailed
information on how to enter the command. This information is completed with
Further Examples illustrating how to use the ListPlot command. The Help
Browser gives also access to Wolfram's *Mathematica* book [68]. Also worth
consulting when looking for help are Ruskeepää's *Mathematica Navigator* [48]
and the very detailed four-volume *Mathematica Guidebooks* by M. Trott [?].

It is also possible to get information about a specific *Mathematica* command
by entering the symbol ? followed by the command name. For example:

```
?Plot
```

```
Plot[f, {x, xmin, xmax}] generates a plot of f as a function
of x from xmin to xmax. Plot[f1, f2, ... , x, xmin, xmax]
plots several functions fi.
```

The double question mark ?? adds information about attributes and options.
For example:

```
??Plot
```

```
Plot[f, {x, xmin, xmax}] generates a plot of f as a function
of x from xmin to xmax. Plot[{f1, f2, ... }, {x, xmin, xmax}]
plots several functions fi.
Attributes[Plot] = {HoldAll, Protected}
Options[Plot] = {AspectRatio → 1/GoldenRatio, Axes → Automatic,
AxesLabel → None, AxesOrigin → Automatic,
AxesStyle → Automatic, Background → Automatic,
ColorOutput → Automatic, Compiled → True,
DefaultColor → Automatic, DefaultFont :→ $DefaultFont,
DisplayFunction :→ $DisplayFunction,
Epilog → { }, FormatType :→ $FormatType,
Frame → False, FrameLabel → None,
FrameStyle → Automatic, FrameTicks → Automatic,
GridLines → None, ImageSize → Automatic,
```

MaxBend → 10., PlotDivision → 30.,

PlotLabel → None, PlotPoints → 25,

PlotRange → Automatic, PlotRegion → Automatic,

PlotStyle → Automatic, Prolog → { }, RotateLabel→ True,

TextStyle :→ $TextStyle, Ticks → Automatic}

If we want to list all function names containing the word `Plot` we can use the wild card * as shown below. We can then obtain information on a specific function by clicking on its name.

```
?*Plot*
```

System'

ContourPlot,	ListPlot3D,	Plot3Matrix,	PlotRange,
DensityPlot,	ParametricPlot,	PlotDivision,	PlotRegion,
ListContourPlot,	ParametricPlot3D,	PlotJoined,	PlotStyle,
ListDensityPlot,	Plot,	PlotLabel,	
ListPlot,	Plot3D,	PlotPoints,	

1.6 Logical Operators

We can use logical operators to compare two expressions. These commands return either **True** or **False**.

The symbol != is equivalent to **Unequal**, that is, ≠.

```
{3 > 4, 5 == 3 + 2, 3 != 2, 4 ≤ 6, 3 ≤ 3,
5 ≥ 7, 3 ≠ 1 + 2}
```

{False, True, True, True, True,
False, False}

Function names:

```
{Head[x > y], Head[x == ]y], Head[x ≤ y], Head[x ≠ y]}
```

{Greater, Equal, LessEqual, Unequal}

&& is equivalent to **And** and || to **Or**.

{Head[x && y], Head[x || y]}

{And, Or}

Here are more examples combining several logical operators.

(!(4 == 3))

True

4 == 2 + 2 && 3 ≠ 5

True

4 ≤ 2 || 3 ≤ 5

True

Xor is the exclusive Or, that is, Xor[expression1, expression2, . . .] gives True if an odd number of expressionk are True, and the rest are False, it gives False if an even number of expressionk are True, and the rest are False.

Xor[3 == 2 + 1, 2 == 4 - 2

False

Xor[3 == 2 + 1, 2 == 0]

True

((4 > 2) || (5 < 1)) && ((4 < 9) || (2 < 1))

True

Be careful, do not mix up = and ==. The command **a = 6** means that the symbol a is given the value 6 (this could also be written **Set[a, 6]**), while the command **a == 6** yields **True** if a is equal to 6 and **False** otherwise.

1.7 Elementary Functions

{Sin[Pi/4], Sin[0.785], Cos[Pi/6]}

$\{\dfrac{1}{\text{Sqrt}[2]}, \; 0.706825, \; \dfrac{\text{Sqrt}[3]}{2}\}$

Because $\sin(\pi/4)$ and $\cos(\pi/6)$ can be evaluated exactly, the outputs are given with infinite precision. This is not the case for $\sin(0.785)$.

We could also have used the symbol π which, as all other Greek letters, can be entered as \[LetterName] where LetterName stands for Pi. We can also either use the command Palettes in the File menu (π is found in the palette BasicInput), or type ESC p ESC.

Tan[Pi/4]

1

{ArcSin[1], ArcCos[1], ArcTan[1]}

$\left\{\dfrac{\text{Pi}}{4}, \; 0, \; \dfrac{\text{Pi}}{4}\right\}$

If not specified, the argument unit is **Radian**. **Degree** can also be used.

{Sin[45 Degree], Cos[45 Degree]}

$\left\{\dfrac{1}{\text{Sqrt}[2]}, \; \dfrac{\text{Sqrt}[3]}{2}\right\}$

Actually **Degree** gives the value in radians of one degree. Its numerical value is $\pi/180$.

```
{Tan[22 Degree], Cot[67 Degree]}
```

{Tan[22 Degree], Cot[67 Degree]}

```
{N[Tan[22 Degree], Cot[67 Degree]]}
```

{0.404026, 0.424475}

Here are other elementary functions.

```
{Sinh[1.3], Cosh[1.3], Tanh[1.3], Coth[1.3]}
```

{1.69838, 1.97091, 0.861723, 1.16047}

```
{Log[3.78], Exp[-0.67]}
```

{1.32972, 0.511709}

1.8 User-Defined Functions

A delayed assignment is made with the **SetDelayed** function also noted :=. When the **SetDelayed** function is used, the right-hand side is not evaluated whereas in the case of the **Set function**, noted =, the right-hand side is immediately evaluated, as illustrated below, where we have used the built-in function **Random[]** that gives a uniformly distributed pseudorandom real in the interval $[0, 1]$.

```
a = Random[]
{a, a, a}
b := Random[]
{b, b, b}
```

0.218807

{0.218807, 0.218807, 0.218807}

{0.244376, 0.716337, 0.850842}

If we try to define a function by entering:

```
f[x] = x^3 - 3 x^2 + 5 x - 7;
f[x]
```

$-7 + 5x - 3\ x^2 + x^3$

it does not work because by doing so we are just assigning the expression x^3 - 3 x^2 + 5 x - 7 to f[x], and entering f[a] will not replace x by a and give $a^3 - 3a^2 + 5a - 7$ as shown below.

```
Clear[a] (* Clearing the value of a *)
f[a])
```

f[a]

Any input placed between (* and *) is ignored. Useful comments can be inserted anywhere into a *Mathematica* code using the notation (* comment *).

To define a function, we have to use a pattern object which can stand for any *Mathematica* expression. _ (short form of Blank[]) is such a pattern object, and we should use x_ to denote the formal argument of the function f[]. Moreover we have to use SetDelayed and not Set

```
f[x_] := x^3 -3 x^2 + 5 x - 7
f[a]
```

$-7 + 5\ a - 3\ a^2 + a^3$

And we can check the definition of the function f[].

```
?f
```

```
Global'f
```

$$f[x] = -7 + 5 \ x - 3 \ x^2 + x^3$$

$$f[x\_] := -7 + 5 \ x - 3 \ x^2 + x^3$$

Observe the difference when we use = (Set) instead of := (SetDelayed)

```
f1[x_] = Expand[x^3];
f2[x_] := Expand[x^3]
f1[a + 1]
f2[a + 1]
```

$$(1 + a)^3$$

$$1 + 3 \ a + 3 \ a^2 + a^3$$

The argument type may be specified.

```
g[n_Integer] := n (n+1) (n+2)
g[5]
```

210

```
g[5.4]
```

g[5.4]

_Integer (short form of Blank[Integer]) stands for any expression with head Integer. Because 5.4 is not an integer, g[5.4] cannot be evaluated.

```
h[x_Real] := (x^5 + 3 x^2) / (x-1)
h[3.7]
```

272.04

```
h[2]
```

h[2]

Because 2 is not a real number, h[2] cannot be evaluated but

```
h[2.]
```

44.

is evaluated.

```
gp[n_Integer?Positive] := (n+1)^2
gp[-2]
```

gp[-2]

gp[-2] is not evaluated, -2 is an integer but not a positive integer.

Functions can also be defined as pure functions, that is, not giving them a specific name.

```
f[x_] := x^2
{f[a], #^2 &[a], Function[x, x^2][a]}
```

$\{a^2, a^2, a^2\}$

Another example:

```
{#1^#2 &[a, b], Function[{x, y}, x^y][a, b]}
```

$\{a^b, a^b\}$

1.9 Rules and Delayed Rules

If, for instance, we define an expression and assign a value to one of the symbols in the expression, the expression is lost as shown below.

```
(2 x + 5)^2
x = 3
(2 x + 5)^2
```

$(5 + 2 x)^2$

x = 3

121

We can surmount this problem by using a *replacement rule*.

```
Clear[x]
(2 x + 5)^2 /. x -> 3
(2 x + 5)^2
```

121

$(5 + 2 x)^2$

The arrow → can be entered in many different ways. We can type ->, or ESC -> ESC, or \[RightArrow].

```
HoldForm[FullForm[(2 x + 5)^2 /. x -> 3]]
```

ReplaceAll[Power[Plus[Times[2, x], 5], 2], Rule[x, 3]]

The expression on the left of /., which is the short form of ReplaceAll, is evaluated before the replacement is made as shown in the following example.

```
(a + 3a) /. 3 a -> A
```

4 a

Once the expression has been evaluated, there is no 3a to be replaced by A. Compare the following outputs.

```
x + y /. x -> y /. y -> a
x + y /. {x -> y, y -> a}
```

2 a

a + y

Entering **expression //.rules** repeatedly performs replacements until **expression** no longer changes. The same result is obtained with **ReplaceRepeated**.

```
x + y //. {x -> y, y -> a}
```

2 a

```
ReplaceRepeated[x + y, {x -> y, y -> a}]
```

2 a

In the command above the number of iterations is supposed to be infinite (not really, it is given by the value of the option **MaxIterations**). We can, however, limit the number of iterations.

```
ReplaceRepeated[a, a -> 1 + a, MaxIterations -> 10]
```

RepaceRepeated : : rrlim :

Exiting after a scanned 10 times. More...

10+a

The maximum number of times a rule will be applied is given by the value of the **MaxIterations** option.

```
Options[ReplaceRepeated]
```

{MaxIterations → 65536}

If we wish to have the right-hand side of the rule evaluated only at the time the rule is applied, we have to use **:>** (short form of **RuleDelayed[]**), instead of **->** (short form of **Rule[]**).

```
{a, a, a} /. a -> Random[]
{a, a, a} /. a :> Random[]
```

{0.229831, 0.229831, 0.229831}
{0.849633, 0.922095, 0.566822}

1.10 Built-In Nonelementary Functions

There exist a huge number of built-in nonelementary functions. Here are a very few examples.

```
LegendreP[n, x]
```

Legendre polynomials of degree n

```
ChebyshevT[n,x]
ChebyshevU[n,x]
```

Chebyshev polynomials of degree n of the first and second kinds

```
LaguerreL[n, x]
```

Laguerre polynomials of degree n

```
BesselJ[n, z]
BesselY[n,z]
```

Bessel functions of order n of the first and second kinds

```
Gamma[z]
```

Euler gamma function

Probably the most complete collection of formulas and graphics about mathematical functions can be found at http://functions.wolfram.com.

1.11 Plotting

1.11.1 2D plots

The command Plot[f[x], {x, x1, x2}] generates a two-dimensional plot of f[x] for x varying from x1 to x2.

```
Plot[Exp[x], {x, -2, 2 }];
```

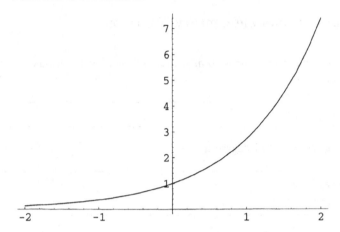

Fig. 1.1. *Graph of e^x for $x \in [-2, 2]$.*

```
Plot[BesselJ[0, x], {x, 0, 10}];
```

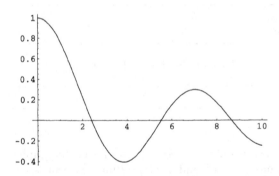

Fig. 1.2. *Graph of the Bessel function of the first kind $J_0(x)$ for $x \in [0, 10]$.*

1.11.2 3D plots

The command `Plot3D[f[x, y], {x, x1, x2}, {y, y1, y2}]` generates a three-dimensional plot of $f(x, y)$ for $x, y \in [x1, x2] \times [y1, y2]$.

```
Plot3D[Sin[x] Cos[2 y], {x, - 2, 2}, {y, - 2, 2}];
```

In the chapter dedicated to graphics, we learn how to use various options to change fonts and color, label the axes, include text, and so on.

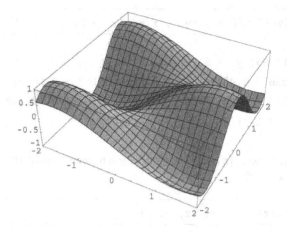

Fig. 1.3. *Graph of* $\sin(x)\cos(2y)$ *for* $\{x, y\} \in [-2, 2] \times [-2, 2]$.

1.12 Solving Equations

1.12.1 Exact Solutions

The command Solve [equations, variables] attempts to solve an equation or set of equations for the variables.

```
Solve[x^3 - 2 x^2 + 3 x - 2 == 0, x]
```

$$\left\{ \{x \to 1\}, \left\{x \to \frac{1}{2}(1 - I\sqrt{7})\right\}, \left\{x \to \frac{1}{2}(1 + I\sqrt{7})\right\} \right\}$$

```
Solve[{2 x - 4 y == 3, x + 5 y == - 2}, {x, y}]
```

$$\left\{ \left\{x \to \frac{1}{2}\right\}, \left\{x \to -\frac{1}{2}\right\} \right\}$$

Note the use of the == sign because the two sides of the equation should have identical values once the unknowns are replaced by their values.

1.12.2 Numerical Solutions

The command NSolve [equation1, equation2 ,...,variable1, variable2, ...] gives a list of numerical approximations to the roots of a system of polynomial equations.

```
Solve[{NSolve[2 x^2 - y == 1, x + y^2 == 2, {x, y}]]
```

$\{\{x \rightarrow -1.17965, y \rightarrow 1.78316\},$
$\{x \rightarrow 0.089826 - 0.451507\ I, y \rightarrow -1.39158 - 0.162228\ I\}.$
$\{x \rightarrow 0.089826 + 0.451507\ I, y \rightarrow -1.39158 + 0.162228\ I\}.$
$\{x \rightarrow 1., y \rightarrow 1.\}\}$

FindRoot[equation, {x, x0}] searches for a numerical solution to equation, starting with x = x0.

```
FindRoot[Cos[x] == 2 x, {x, 0. 3}]
```

$\{x \rightarrow 0.450184\}$

```
FindRoot[Sin[x] == 3.2, {x, 1 + I}]
```

$\{x \rightarrow 1.5708 + 1.83094\ i\}$

1.13 Derivatives and Integrals

1.13.1 Exact Results

The commands f'[x], D[f[x], x], and Derivative[1][f][x] are equivalent. They all denote the first derivative of f[x] with respect to x. Similarly, the commands f''[x], D[f[x], {x, 2}], and Derivative[2][f][x] all denote the second derivative of f[x] with respect to x. . f'''[x] is a valid command for the third derivative but $f^{(3)}[x]$ (entered using the BasicInput submenu of the Palettes menu), is understood as the cube of f[x].

```
Clear[f]
f[x_] := x^3 + 2 Cos[x]
f'[x]
f''[x]
{f'''[x], f^{(3)}[x]}
```

$3\ x^2 - 2\ Sin[x]$

$6\ x - 2\ Cos[x]$

$\{6 + 2\ Sin[x] + (f^3)[x]\}$

```
D[f[x], x]
D[f[x], {x, 2}]
Derivative[1][f][x]
Derivative[2][f][x]
```

$3 x^2 - 2 \sin[x]$

$6 x - 2 \cos[x]$

$3 x^2 - 2 \sin[x]$

$6 + 2 \cos[x]$

```
D[x^4 Cos[y^2], {x, 3}, {y, 2}]
```

$24 x (- 4 y^2 \cos[y^2] - 2 \sin[y^2])$

We have to be careful if we want to define a function as the derivative of another one.

```
Clear[f]
f[x_] := D[x Sin[x], x]
f[x]
```

$x \cos[x] + \sin[x]$

The result looks correct; but let us try to find the value of f[x] for a specific value of x.

```
f[Pi]
```

General : : ivar : Pi is not a valid variable. More...

D[0, Pi]

Because f[x] has been defined using :=, the right-hand side is kept in an unevaluated form. When we enter f[Pi], *Mathematica* tries actually to evaluate D[Pi Sin[Pi], Pi] which has no meaning. If, on the contrary, we define f[x] using = and not :=, *Mathematica* immediately evaluates f[x] and replacing x by Pi gives the correct result.

```
Clear[f]
f[x_] = D[x Sin[x], x]
f[x]
f[Pi]
```

x Cos[x] + Sin[x]

x Cos[x] + Sin[x]

-Pi

The command Integrate[f, x] gives the indefinite integral $\int f(x)\,dx$; Integrate[f, {x, a, b}] gives the definite integral $\int_a^b f(x)\,dx$. The symbol $\int$ can be entered using the Palettes submenu BasicInput.

```
Clear[f]
f[x_] := x^3 + x Cos[x]
Integrate[f[x], x]
```

$$\frac{x^4}{4} + Cos[x] + x\ Sin[x]$$

```
Integrate[f[x], {x, 0, Pi/2}]
```

$$\frac{-\ 64 + 32\ Pi + Pi^4}{64}$$

1.13.2 Numerical Integration

If Integrate does not work, NIntegrate will work if the integral is defined.

```
Plot[Tan[Sin[x]], {x, 0, Pi}];
```

The plot above shows that the definite integral is finite and we can evaluate its numerical value using NIntegrate. *Mathematica* cannot, however, find its exact value.

```
Integrate[Tan[Sin[x]], {x, 0, Pi}]
```

$\int_0^\pi Tan[Sin[x]]\,dx$

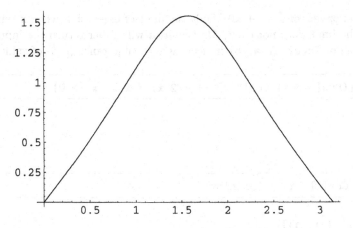

Fig. 1.4. *Graph of* $\tan(\sin x)$ *for* $x \in [0, \pi]$.

```
NIntegrate[Tan[Sin[x]], {x, 0, Pi}]
```

2.66428

1.14 Series Expansions and Limits

`Series[f[x], {x, x0, n}]` gives the power series expansion of `f[x]` about
x = x0 up to order n. It indicates that the first neglected term is of the order
$(x - x0)^{n+1}$.

```
Series[Exp[-2 x] Cos[3 x], {x, 0, 5}]
```

$$1 - 2x - \frac{5\,x^2}{2} + \frac{23\,x^3}{3} - \frac{119\,x^4}{24} - \frac{61\,x^5}{25} + O[x]^6$$

We can get rid of the term $(x - x0)^{n+1}$ and obtain a polynomial using the
command `Normal`.

```
Normal[Series[Exp[-2 x] Cos[3 x], {x, 0, 5}]]
```

$$1 - 2x - \frac{5\,x^2}{2} + \frac{23\,x^3}{3} - \frac{119\,x^4}{24} - \frac{61\,x^5}{25}$$

Limit[expression, x → x0] finds the limit of expression when x tends to x0. If the limit does not exist, *Mathematica* will either return the input un-evaluated or Interval[a, b] indicating a possible range (a, b) of values.

```
Limit[(Exp[- 2 x] Cos[3 x] -1 + 2 x) / x^2, x -> 0]
```

$$-\frac{5}{2}$$

```
Limit[Cos[x], x -> Infinity]
```

Interval[{-1, 1}]

Sometimes, the command Limit gives a wrong result!

```
Limit[Abs[x] / x, x -> 0]
```

1

The result is not correct. If x approaches 0 from the right, the limit is indeed 1 but from the left the limit is -1. From a purely mathematical point of view, the limit does not exist. Actually the wrong limit was obtained because *Mathematica* assumed that 0 was approached from the right. To obtain the limit when 0 is approached from the left, we have to specify the Direction.

```
Limit[Abs[x] / x, x -> 0, Direction -> 1]
```

- 1

The option Direction -> 1 means that we have to approach 0 going in the direction of 1 that is from the left of 0.

Here is another example. The function Sign[x] gives -1, 0, or 1 depending on whether x is negative, zero, or positive.

```
Plot[Sign[x], {x, -1, 1}];
```

```
{Limit[Sign[x], x -> 0, Direction -> 1], Sign[0],
Limit[Sign[x], x -> 0, Direction -> -1]}
```

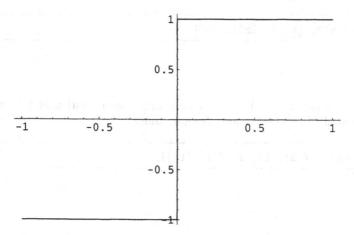

Fig. 1.5. *Graph of* sign (x) *for* $x \in [-1, 1]$.

{-1, 0, 1}

1.15 Discrete Sums

Sum[f[n], {n, n1, n2}] evaluates the sum of the values of f[n] when n varies from n = n1 to n = n2.

For example,

Sum[n, {n, 1, k}]

$$\frac{k\ (1 + k)}{2}$$

Sum[n^2, {n, 1, k}]

$$\frac{k(1 + k)(1 + 2k)}{6}$$

Sum[n^2, {n, 1, 25}]

$$\frac{23485971550561141649}{14626411683380640000}$$

```
Sum[1 / n^2, {n, 1, Infinity}]
```

$$\frac{Pi^2}{6}$$

NSum[f[n], {n, n1, n2}] gives a numerical approximation of the sum of the values of f[n] when n varies from n1 to n2.

```
NSum[Log[n] / n!, {n, 1, Infinity}]
```

0.603783

1.16 Ordinary Differential Equations

1.16.1 Symbolic Solutions

DSolve[equation, y, x] tries to symbolically solve the differential equation for the function y, with independent variable x. As for all types of equations, note the use of the sign ==. C[1] represents an arbitrary constant.

```
DSolve[y'[x] + 2 y[x] == 2, y[x], x]
```

$\{y[x] \rightarrow 1 + e^{-2x} C[1]\}$

The equation below is a Cauchy–Euler equation. It has an obvious solution of the form $y(x) = x^a$, where a is either real or complex. C[1] and C[2] are arbitrary constants.

```
DSolve[x^2 y''[x] + 2 x y'[x] - 2 y[x] == 0, y[x], x]
```

$$\{\{y[x] \rightarrow x\ C[1] + \frac{C[2]}{x^2}\}\}$$

Here is a pair of simultaneous differential equations.

```
DSolve[ {x'[t] == y[t], y'[t] == - x[t]}, {x[t], y[t]}, t ]
```

$\{\{x[t] \rightarrow C[1]\ Cos[t] + C[2]\ Sin[t],$

y[t] → C[2] Cos[t] - C[1] Sin[t]}}

As in the example above, C[1] and C[2] are arbitrary constants.

1.16.2 Numerical Solutions

NDSolve[equation, y, {x, a, b}] finds a numerical solution to the differential equation for the function y, with independent variable x in the interval (a, b).

```
solution = NDSolve[{y'[x] - x^2 y[x] == 0, y[0] == 1},y[x],
{x, 0, 1}]
```

{{y[x] -> InterpolatingFunction[0., 1., <>] [x]}}

The result is an InterpolatingFunction that represents the approximate numerical solution $y(x)$ for x in the interval (a, b). The result can be used in the following way. Define the function f[x] by

```
Clear[f]
f[x_] = y[x] /. solution[[1]]
```

InterpolatingFunction[0., 1., <>] [x]

where, as mentioned above, we use the sign = and not :=. And we can calculate the value of f[x] for a specific value of the variable x,

```
{f[0], f[0.5], f[1]}
```

{1., 1.04255, 1.39561}

plot a graph of the function,

```
Plot[f[x], {x, 0, 1}];
```

determine the derivative as an InterpolatingFunction, and calculate its value for a given value of x.

```
f'[x]
```

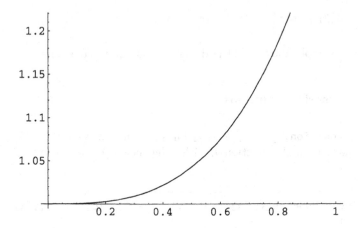

Fig. 1.6. *Graph of f(x) defined as an interpolating function for x ∈ [0, 1].*

```
InterpolatingFunction[0., 1., <>] [x]
```

```
{f'[0.3], f'[0.7]}
```

```
{0.0908133, 0.549349}
```

1.17 Lists

Lists are extremely important objects. They provide a way to group any kind of expression.

Here is a list of five elements and different operations on this list.

```
l = {a, f, b, e, c, d};
{Part[l,3], l[[3]]}
Length[l]
l^2
Exp[l]
(1+x) / l (* note this result and the next one *)
l / (1+x)
Sort[l]
Partition[l,2]
```

```
{b, b}
```

6

$\{a^2, f^2, b^2, e^2, c^2, d^2\}$

$\{E^a, E^f, E^b, E^e, E^c, E^d\}$

$\{\dfrac{a+x}{a}, \dfrac{f+x}{f}, \dfrac{b+x}{b}, \dfrac{e+x}{e}, \dfrac{c+x}{c}, \dfrac{d+x}{d}\}$

$\{\dfrac{a}{a+x}, \dfrac{f}{f+x}, \dfrac{b}{b+x}, \dfrac{e}{e+x}, \dfrac{c}{c+x}, \dfrac{d}{d+x}\}$

$\{a, b, c, d, e, f\}$

$\{\{a, f\}, \{b, e\}, \{c, d\}\}$

Partition[list, n] partitions a list into nonoverlapping sublists of length n.

Table[expression, k] generates a list of k copies of expression.

```
Table[Random[], {10}]
```

$\{0.902025, 0.539515, 0.599448, 0.880183, 0.610582, 0.989756, 0.431891,$
$0.656339, 0.798266, 0.0958655\}$

Table[expression, k, kmin, kmax, Δk] generates a list of the values of expression when k runs from kmin to kmax using steps Δk.

```
Table[Cos[x], {x, 0, Pi/4, Pi/16}]
```

$\{1, Cos[\dfrac{Pi}{16}], Cos[\dfrac{Pi}{8}], Cos[\dfrac{3Pi}{16}], \dfrac{1}{Sqrt[2]}\}$

Range[n] generates the list $\{1, 2, , n\}$, Range[n1, n2] generates the list $\{n1, n1 + 1, \ldots, n2\}$, and Range[n1, n2, Δn] uses steps Δn.

```
{Range[6], Range[3,12], Range[3,12, 3]}
```

$\{\{1, 2, 3, 4, 5, 6\}, \{3, 4, 5, 6, 7, 8, 9, 10, 11, 12\}, \{3, 6, 9, 12\}\}$

Functions with attribute `Listable` are automatically "threaded" over lists, so that they act separately on each list element. Most built-in mathematical functions are `Listable`.

```
lis = {a, b, c};
Sin[lis]
Log[lis]
```

{Sin[a], Sin[b], Sin[c]}

{Log[a], Log[b], Log[c]}

A user-defined function can be made `Listable`.

```
f[lis]
```

f[{a, b, c}]

```
SetAttributes[f, Listable]
f[lis]
```

{f[a], f[b], f[c]}]

```
ClearAttributes[f, Listable]
f[lis]
```

f[{a, b, c}]

We could also, without making a function `Listable`, evaluate the function at various points x if the function has a specified expression in terms of the variable. Look at the different behaviors of the functions g and h below.

```
ClearAll[g, h] (* clear all attributes, see below *)
g[x_] := x^2
{g[{a, b, c}], h[{a, b, c}]}
```

$\{\{a^2, b^2, c^2\}, h[\{a, b, c\}]\}$

```
{g[x] /. x → {a, b, c}, h[x] /. x → {a, b, c}}
```

$\{\{a^2, b^2, c^2\}, h[\{a, b, c\}]\}$

It is often necessary to plot a list of data. In this case, we can use ListPlot.

```
data = Table[{x, Cos[x] + 0.25 Random[]}, {x, 0, 2Pi, 0.2}]
```

$\{\{0, 1.13292\}, \{0.2, 0.999822\}, \{0.4, 0.932795\}, \{0.6, 0.993941\},$
$\{0.8, 0.711319\}, \{1., 0.618637\}, \{1.2, 0.536814\}, \{1.4, 0.385192\}$
$\{1.6, 0.129086\}, \{1.8, -0.154444\}, \{2., -0.184746\},$
$\{2.2, -0.456455\}, \{2.4, -0.662042\}, \{2.6, -0.721847\},$
$\{2.8, -0.73873\}, \{3., -0.844616\}, \{3.2, -0.932651\},$
$\{3.4, -0.943502\}, \{3.6, -0.679174\}, \{3.8, -0.63348\},$
$\{4., -0.487377\}, \{4.2, -0.343637\}, \{4.4, -0.121386\},$
$\{4.6, 0.0293136\}, \{4.8, 0.12085\}, \{5., 0.410531\}, \{5.2, 0.64273\},$
$\{5.4, 0.857554\}, \{5.6, 0.794305\}, \{5.8, 0.934053\}, \{6., 1.20993\},$
$\{6.2, 1.00418\}\}$

The data represent the variations of $\cos(x)$ with some added noise (see output in Figure 1.7).

```
pl = ListPlot[data];
```

Fit[data, functions, variables] finds the least-squares fit to a list of data as a linear combination of functions of variables.

```
s = Fit[data,{1, x, x^2}, x]
```

$1.51558 - 1.35589 x + 0.217323 x^2$

In order to visualize how good the fit is, we plot the data and the least-squares fit on the same graph using the command Show.

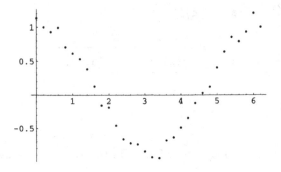

Fig. 1.7. *Plot of a list of data.*

```
pls = Plot[s, {x, 0, 2Pi}, DisplayFunction -> Identity];
Show[pl, pls, DisplayFunction -> $DisplayFunction];
```

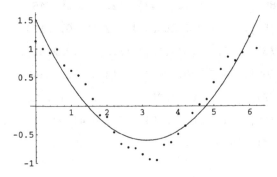

Fig. 1.8. *Data and least-square fit plots.*

The option `DisplayFunction -> Identity` suppresses the output to the screen and `$DisplayFunction` is the default setting for the option `DisplayFunction`. The command `Show[graphics, options]` displays graphics using the specified `options`.

1.18 Vectors and Matrices

An n-dimensional vector is a list of n elements that are not lists themselves. Here is a five-dimensional vector.

```
vect = {x1, x2, x3, x4, x5}
```

A vector can be multiplied by a scalar.

```
a vect
```

{a x1, a x2, a x3, a x4, a x5}

We can add two vectors having the same dimension.

```
{x1, x2, x3, x4, x5} + {y1, y2, y3, y4, y5}
```

{x1 + y1, x2 + y2, x3 + y3, x4 + y4, x5 + y5}

The Dot and Cross products of two tridimensional vectors are given by

```
{x1, x2, x 3} . {y1, y2, y3}  (* or, equivalently, *)
Dot[{x1, x2, x 3} , {y1, y2, y3}]
```

x1 y1 + x2 y2 + x3 y3

x1 y1 + x2 y2 + x3 y3

```
Cross[{x1, x2, x3}, {y1, y2, y3}]
```

{- x3 y2 + x2 y3, x3 y1 - x1 y3, - x2 y1 + x1 y2}

In general, Cross[v1, v2, ..., vn] is a totally antisymmetric product that takes vectors of length n+1 and yields a vector of length n+1 that is orthogonal to all the n vectors v1, v2, ..., vn. Here is an example for n = 3.

```
cp3 = Cross[{x1, x2, x3, x4}, {y1, y2, y3, y4}, {z1, z2, z3,
z4}]
```

{x4 y3 z2 - x3 y4 z2 - x4 y2 z3 + x2 y4 z3 + x3 y2 z4 - x2 y3 z4,
-(x4 y3 z1) + x3 y4 z1 + x4 y1 z3 - x1 y4 z3 - x3 y1 z4 + x1 y3 z4,
x4 y2 z1 - x2 y4 z1 - x4 y1 z2 + x1 y4 z2 + x2 y1 z4 - x1 y2 z4,
-(x3 y2 z1) + x2 y3 z1 + x3 y1 z2 - x1 y3 z2 - x2 y1 z3 + x1 y2 z3}

```
Length[cp3]
```

4

```
{cp3.{x1, x2, x3, x4}, cp3.{y1, y2, y3, y4}, cp3.{z1, z2, z3,
z4}}//Simplify
```

{0, 0, 0}

There is a more general product: the so-called Outer product.

```
?Outer
```

Outer[f, list1, list2, ...] gives the generalized outer
product of the listi, forming all possible combinations
of the lowest-level elements in each of them, and feeding
them as arguments to f. Outer[f, list1, list2, ... , n]
treats as separate elements only sublists at level n
in the listi. Outer[f, list1, list2, ... , n1, n2, ...]
treats as separate elements only sublists at
level ni in the corresponding listi. More ...

```
Outer[f, {a, b}, {x, y}]
```

{{f[a, x], f[a, y]},{f[b, x], f[b, y]}}

A matrix is a list of lists having the same length. Here is a 3×2 matrix.

```
MatrixForm[{{a1, a2, a3}, {b1, b2, b3}}]
```

a1   a2   a3
b1   b2   b3

Its transpose is the 2×3 matrix:

```
Transpose[{a1, a2, a3}, {b1, b2, b3}}]
```

{{a1, b1}, {a2, b2}, {a3, b3}}

Mathematica can find the inverse of an $n \times n$ matrix,

```
mat = {{1/2, 1/3, 1/4}, {1/3, 1/4, 1/5}, {1/4, 1/5, 1/6}};
TableForm[mat]
invMat = Inverse[mat]
```

$\frac{1}{2}, \frac{1}{3}, \frac{1}{4}$

$\frac{1}{3}, \frac{1}{4}, \frac{1}{5}$

$\frac{1}{4}, \frac{1}{5}, \frac{1}{6}$

{{72, - 240, 180}, {- 240, 900, - 720}, {180, - 720, 600}}

a result that can be verified:

```
mat . invMat // TableForm
```

1   0   0

0   1   0

0   0   1

Mathematica can also find the eigenvalues of a square matrix.

```
N[Eigenvalues[mat]]
```

{0.875115, 0. 0409049, 0.000646659}

1.19 Clear, ClearAll, and Remove

The command **Clear** clears values and definitions but not attributes.

```
a = 5
```

5

```
a
Clear[a]
a
```

5

a

To clear the value assigned to a symbol we can also use the command a =.
which is a shorthand notation for Unset[a].

```
b = 6
```

6

```
b
b =.
b
```

6

b

```
f[x_] := x^2
SetAttributes[f, Listable]
f[a]
Clear[f]
f[a]
f[a, b]
```

a^2

f[a]

{f[a], f[b]}

Clear does not clear attributes. Using the command ClearAll we can clear values, definitions, and attribute values.

```
f[x_] := x^2
SetAttributes[f, Listable]
f[{a, b}]
ClearAll[f]
f[a]
f[{a, b}]
```

$\{a^2, b^2\}$

f[a]

f[{a, b}]

When starting a new problem, in order to avoid interference with previous variable values or function definitions, it is a good idea to execute the command: ClearAll["Global'*"].

ExpectedValue is a built-in function defined in the package Statistics'DescriptiveStatistics'. Using the command Total[list] which gives the sum of the elements in list, we define the function ExpectedValue by

```
ExpectedValue[lis_List ] := Total[lis] / Length[lis]
```

As expected we find

```
ExpectedValue[{1, 2, 3, 4, 5}]
```

3

Now, if we load the package Statistics'DescriptiveStatistics' in order to use the *Mathematica* built-in function, we have to remove our definition using Remove.

```
Remove[ExpectedValue]
```

```
<<Statistics'DescriptiveStatistics'
```

```
ExpectedValue[{1,2,3,4,5}]
```

```
ExpectedValue[{1,2,3,4,5}]
```

The result above shows that ExpectedValue does not work as the function we defined. We can ask *Mathematica* to tell us why.

```
?ExpectedValue
```

```
ExpectedValue[f, list] gives the expected value of the pure

function f with respect to the sample distribution of list.

ExpectedValue[f, list, x] gives the expected value of the

function f of x with respect to the sample distribution

of list. More ...
```

Because ExpectedValue[f, data] gives the expected value of the function f with respect to the sample distribution of the data, we have, therefore, to specify the function f which is represented below by a pure function.

```
ExpectedValue[(#)&, {1, 2, 3, 4, 5}]
```

3

Here is another example.

```
ExpectedValue[(#^3)&, {1, 2, 3, 4, 5}]
```

45

1.20 Packages

When working in a particular area, we may need functions that are not built into *Mathematica* but that are defined in a *Mathematica* package we need to load. For instance, the following command loads the package PhysicalConstants.

```
<<Miscellaneous'PhysicalConstants'
```

The list of all the commands defined in this package is found entering the command

```
?Miscellaneous'PhysicalConstants'*
```

Here are a few examples.

```
AccelerationDueToGravity
```

$$\frac{9.80665 \text{ Meter}}{\text{Second}^2}$$

```
PlanckConstant
```

$6.62606876 \times 10^{-34}$ Joule Second

```
ElectronMass
```

$9.10938188 \times 10^{-31}$ Kilogram

1.21 Programming

1.21.1 Block and Module

The elaboration of a program is usually done in several steps, and intermediate results have to be kept. The command Module[variables, expression] is a very convenient construct to achieve this goal. One important feature of this structure is that variables are treated as local when they appear in expression. Block[variables, expression] is a similar structure that, however, behaves differently in the way it handles variables as illustrated below.

```
Clear[x, y, u]
x = Pi/4; u := Sin[x] + Cos[y]^2;
Block[{x = Pi/2, y = Pi/4}, u+1]
```

$$\frac{5}{2}$$

Replacing `Block` by `Module` yields

```
Clear[x, y, u]
x = Pi/4; u := Sin[x] + Cos[y]^2;
Module[{x = Pi/2, y = Pi/4}, u+1]
```

$$1 + \frac{1}{\text{Sqrt}[2]} + \text{Cos}[y]^2;$$

In `Block`, u is replaced by its expression $\sin(x)+\cos^2(y)$ and $u+1$ is evaluated using the local values $x = \pi/2$ and $y = \pi/4$. In `Module`, because in $u + 1$ the symbols x and y do not appear explicitly, they are not replaced by their local values $x = \pi/2$ and $y = \pi/4$, but when returning $u + 1$, `Module` replaces x by its value $\pi/4$ and leave y unevaluated since no value had been assigned to y. As shown above, variables can be assigned values inside the `Block` and `Module` structures.

The following example shows the local character of a variable value defined inside a `Block`.

```
Clear[x]
Block[{x = Pi}, Cos[x]]
Cos[x]
```

```
-1
```

```
Cos[x]
```

Here is a more interesting example.

```
Block[{$DisplayFunction = Identity},
p1 = Plot[Sin[x], {x, 0, 2 Pi}];
p2 = Plot[Cos[x], x, 0, 2 Pi];]
```

We already used DisplayFunction as an option for graphics functions that specifies the function to apply to graphics in order to display them. The local value Identity for DisplayFunction suppresses the output to the screen. To see the output to the screen, we have to use Show.

```
Show[p1, p2];
```

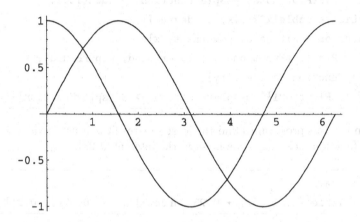

Fig. 1.9. *Graphs of* sin *x and* cos *x for x* ∈ [0, 2π].

If we did not use Block[], we should have entered

```
p1 = Plot[Sin[x], {x, 0, 2 Pi}, DisplayFunction -> Identity];
p2 = Plot[Cos[x], {x, 0, 2 Pi}, DisplayFunction -> Identity];
```

and then used the option DisplayFunction -> $DisplayFunction, which was not necessary when using Block because $DisplayFunction was a local variable. Because $DisplayFunction gives the default setting for the option DisplayFunction, the graphs are then displayed on the screen using the input:

```
Show[{p1, p2}, DisplayFunction -> $DisplayFunction];
```

The command Module has the same property. To see how to use it, it is probably better to build up a simple program that exhibits its essential features.

In the section dedicated to lists, we have obtained step by step the least-squares fit of data represented by a list, using a combination of functions. Given a data list, the function dataFitPlot, defined below, groups all the various steps together to find the best least-squares fit by a polynomial of degree n in an interval defined by its lower and upper bounds and to plot on the same graph the data points and the fitting curve.

```
dataFitPlot[data_List, degree_Integer, lowerBound_Real,
upperBound_Real] :=
Module[{lisPl, powerList, fitFunction, fitPl,x},
lisPl = ListPlot[data, DisplayFunction -> Identity];
powerList = Table[x^k, {k, 0, degree}];
fitFunction = Fit[data, powerList, x];
fitPl = Plot[fitFunction, {x, lowerBound, upperBound},
DisplayFunction -> Identity];
Show[{lisPl, fitPl}, DisplayFunction -> $DisplayFunction]]
```

We can use this program to find the least-squares fit of a list of values of the cosine function including some noise in the interval $[0, 2\pi]$.

```
Clear[data]
data = Table[{x, Cos[x] + 0.25 Random[]}, {x, 0, 2Pi, 0.2}]
```

{{0, 1.03372}, {0.2, 0.984004}, {0.4, 1.10815},

{0.6, 0.841229}, {0.8, 0.918472}, {1., 0.642946},

{1.2, 0.378395}, {1.4, 0.254044}, {1.6, 0.0212607},

{1.8, -0.00621556}, {2., -0.352155}, {2.2, -0.361068},

{2.4, -0.502693}, {2.6, -0.838658}, {2.8, -0.926133},

{3., -0.774244}, {3.2, -0.943265}, {3.4, -0.78986},

{3.6, -0.871707}, {3.8, -0.564668}, {4., -0.445388},

{4.2, -0.463964}, {4.4, -0.127786}, {4.6, -0.0343247},

{4.8, 0.262035}, {5., 0.306022}, {5.2, 0.710974},

{5.4, 0.696627}, {5.6, 0.978336}, {5.8, 1.05524},

{6., 1.18659}, {6.2, 1.2244}}

Note that because we have chosen to represent the lowerBound and upperBound variables as real numbers, they should be written 0. and N[2Pi] and not simply 0 and 2Pi.

Using the Module construct to define a function F of the variables x, y, ...
that needs to first determine an expression e1 involving x, y, ..., then an
expression e2 involving e1 and possibly x, y, ..., we observe that we have
to use a structure of the form F[x_, y_,...] := Module[{e1, e2, ...},
body] where in body are defined the intermediate expressions e1, e2, in terms
of x, y, ..., and previously defined expressions.

The definitions of each intermediate expression must end with a semicolon
(;), and all intermediate expressions have to be listed in the first argument of
the module. When defining the intermediate expressions, built-in functions or
user-defined functions (previously defined) can be used. Note that there is no
semicolon after the last command which represents the required final output.

```
dataFitPlot[data, 3, 0., N[2Pi]];
```

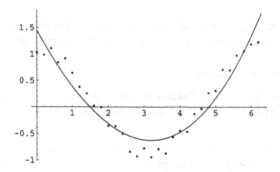

Fig. 1.10. *Least-squares fit of the data above.*

1.21.2 Collatz Problem

We first build up a few functions to study the so-called *Collatz problem*. Also
known as the $3x + 1$ problem, the Collatz problem is the following conjecture.
Starting from any positive integer n, if this integer is even, divide it by 2, if it is
odd multiply it by 3, add 1 and then divide the result by 2. Iterating this process, the sequence of integers thus obtained falls into the cycle $1, 2, 1, 2, 1, \ldots$
For a very detailed review of the Collatz problem, consult Lagarias [29].

Mathematically, this means that the sequence of iterations of the *Mathematica*
function:

```
Collatz[n_?OddQ] := (3n + 1) / 2
Collatz[n_?EvenQ] := n / 2
```

falls into the cycle 1, 2, 1, 2, 1, ...

Using an If statement we can also define the Collatz function a bit differently by

```
Collatz[n_Integer?Positive]  := If[EvenQ[n], n/2, (3n + 1)/2]
```

The statement If[condition, t, f] gives t if condition evaluates to True, and f if it evaluates to False. Consider an example.

```
NestList[Collatz, 67, 25]
```

{67, 101, 152, 76, 38, 19, 29, 44, 22, 11, 17, 26, 13, 20, 10,

5, 8, 4, 2, 1, 2, 1, 2, 1, 2, 1}

Nest[f,expression,n] gives the result of applying f n times to expression. NestList[f, expression, n] gives a list of the results of applying f to expression from 0 through n times.

If we define the length of a Collatz sequence as the smallest integer i such that the ith iterate equals 1, the following function may be used to determine this length.

```
CollatzSequenceLength[n_Integer]  :=
Module[{k = n, l = 1},
While[ k != 1, k = Collatz[k]; l = l + 1]; l]
```

```
CollatzSequenceLength[67]
```

20

```
CollatzSequenceLength[2^50 - 1]
```

384

We can also find the Collatz sequence as the sequence of iterates ending with the first iterate equal to 1.

```
CollatzSequence[n_Integer?Positive] :=
Module[{n1 = n, l = {n}},
While[ n1 != 1, n1 = Collatz[n1]; l = Append[l,n1]]; l]
```

Append[expression, element] gives expression with element appended.

```
CollatzSequence[45]
```

$\{45, 68, 34, 17, 26, 13, 20, 10, 5, 8, 4, 2, 1\}$

```
CollatzSequence[1453]
```

$\{1453, 2180, 1090, 545, 818, 409, 614, 307, 461, 692, 346, 173,$

$260, 130, 65, 98, 49, 74, 37, 56, 28, 14, 7, 11, 17, 26, 13, 20,$

$10, 5, 8, 4, 2, 1\}$

We can test the Collatz conjecture on all integers between 1 and 10 000 000 with the following function. We verify that the tested list has actually the correct length.

```
CollatzSequenceLengthList =
Table[CollatzSequenceLength[k], {k, 1, 10000000}]; // Timing
Length[CollatzSequenceLengthList]
```

$\{11.0539Second, Null\}$

10000000

1.21.3 Generalizing the Collatz Problem

In this section the reader will discover how *Mathematica* is particularly help-ful for suggesting conjectures while we try to generalize the famous Collatz conjecture.

For an initial value n, instead of considering if n is either odd or even (that is, if $n \mod 2$ is either equal to 0 or 1), we consider if $n \mod 3$ is equal to 0 or 1 or 2. The simplest generalization is then the following Collatz-type function.

```
Collatz3[n_Integer?Positive] := Which[
Mod[n, 3] == 0, n / 3,
Mod[n, 3] == 1, (4 n + 2) / 3,
Mod[n, 3] == 2, (4 n + 1) / 3]
```

Let us first study a few examples.

```
Table[NestList[Collatz3, k, 12], {k, 1,10}]//TableForm
```

1	2	3	1	2	3	1	2	3	1	2	3	1
2	3	1	2	3	1	2	3	1	2	3	1	2
3	1	2	3	1	2	3	1	2	3	1	2	3
4	6	2	3	1	2	3	1	2	3	1	2	3
5	7	10	14	19	26	35	47	63	21	7	10	14
6	2	3	1	2	3	1	2	3	1	2	3	1
7	10	14	19	26	35	47	63	21	7	10	14	19
8	11	15	5	7	10	14	19	26	35	47	63	21
9	3	1	2	3	1	2	3	1	2	3	1	2
10	14	19	26	35	47	63	21	7	10	14	19	26

On these few examples, we find that iterating the function Collatz3 we find either 1 and then the periodic sequence 1,2,3,1,2,3,1, ... or 7 and then the periodic sequence 7,10,14,19,26,35,47,63,21,7,10,14, ... If we conjecture that these are the only possibilities, we define a generalized Collatz3Sequence as a sequence of iterates ending either with 1 or by 7

```
Collatz3Sequence[n_Integer?Positive] :=
Module[{n1 = n, 1 = {n}},
While[ (n1 != 1) && (n1 != 7),
n1 = Collatz3[n1]; 1 = Append[1, n1]]; 1]
```

```
Collatz3Sequence[572]
```

{572, 763, 1018, 1358, 1811, 2415, 805, 1074, 358, 478, 638, 851,

1135, 1514, 2019, 673, 898, 1198, 1598, 2131, 2842, 3790, 5054,

6739, 8986, 11982, 3994, 5326, 7102, 9470, 12627, 4209, 1403, 1871,

2495, 3327, 1109, 1479, 493, 658, 878, 1171, 1562, 2083, 2778, 926,

1235, 1647, 549, 183, 61, 82, 110, 147, 49, 66, 22, 30, 10, 14, 19,

26, 35, 47, 63, 21, 7}

```
Collatz3Sequence[327]
```

{327, 109, 146, 195, 65, 87, 29, 39, 13, 18, 6, 2, 3, 1}

These two examples do not prove anything. Let us define, as for the standard Collatz problem, the function Collatz3SequenceLength and test many positive integers to check that the sequence does only end with 1 or 7.

```
Collatz3SequenceLength[n_Integer?Positive] :=
Module[{n1 = n, l = 1},
While[ (n1 != 1) && (n1 != 7),
n1 = Collatz3[n1]; l = l+1]; l]
```

```
Collatz3SequenceLengthList =
Table[Collatz3SequenceLength[k], {k, 1, 100000}]; //Timing
Length[Collatz3SequenceLengthList]
```

{67.5151 Second, Null}

100000

Here is another generalized Collatz conjecture.

Starting from any positive integer, by applying repeatedly the function Collatz4[n_Integer?Positive] defined by

```
Collatz4[n_Integer?Positive] :=
Which[Mod[n,4] == 0, n / 4,
Mod[n,4] == 1, (5 n + 3) / 4,
Mod[n,4] == 2, (5 n + 2) / 4,
Mod[n,4] == 3, (5 n + 1) / 4 ]
```

we find either 1 and then the periodic sequence 1, 2, 3, 4, 1, 2, 3, 4, 1, ... or 23 and then the periodic sequence 23, 29, 37, 47, 59, 74, 93, 117, 147, 184, 46, 58, 73, 92, 23, 29, 37, ...

If we assume that these are the only possibilities, we can define a generalized Collatz4Sequence as a sequence of iterates ending either with 1 or 23.

```
Collatz4Sequence[n_Integer?Positive] :=
Module[{n1 = n, l = {n}},
While[ (n1 != 1) && (n1 != 23), n1 = Collatz4[n1];
l = Append[l,n1]]; l]
```

```
Collatz4Sequence[3879]
```

{3879, 4849, 6062, 7578, 9473, 11842, 14803, 18504, 4626, 5783,

7229, 9037, 11297, 14122, 17653, 22067, 27584, 6896, 1724, 431,

539, 674, 843, 1054, 1318, 1648, 412, 103, 129, 162, 203, 254, 318,

398, 498, 623, 779, 974, 1218, 1523, 1904, 476, 119, 149, 187, 234,

293, 367, 459, 574, 718, 898, 1123, 1404, 351, 439, 549, 687, 859,

1074, 1343, 1679, 2099, 2624, 656, 164, 41, 52, 13, 17, 22, 28, 7,

9, 12, 3, 4, 1}

```
Collatz4SequenceLength[n_Integer?Positive] :=
Module[{n1 = n, l = 1}, While[ (n1 != 1) && (n1 != 23),
n1 = Collatz4[n1]; l = l+1]; l]
```

```
Collatz4SequenceLength[3879]
```

78

```
Collatz4SequenceLengthList =
Table[Collatz4SequenceLength[k], {k, 1, 100000}];//Timing
Length[Collatz4SequenceLengthList]
```

{78.7616 Second, Null}
100000

Conjecture: Starting from any positive integer, iterating the generalized Collatz function:

```
CollatzK[k_Integer, n_Integer?Positive] :=
If[Mod[n, k] == 0, n/k, ((k+1) n + k - Mod[n, k])/k]
```

we find either 1 and then the periodic sequence 1, 2, . . . , k, 1, 2,. . . , k, 1, 2, . . . or other periodic sequences.

The standard Collatz function is Collatz[2,n].

```
NestList[CollatzK[2, #] &, 45, 12]
```

{45, 68, 34, 17, 26, 13, 20, 10, 5, 8, 4, 2, 1}

We can verify that the Collatz3[n] and Collatz4[n] are, respectively, the CollatzK[3, n] and CollatzK[4, n] functions.

Let us consider a new one such as, for example, the CollatzK[6, n] function. It can be shown that, starting from any positive integer, we obtain one of the following periodic sequences 1, 2, 3, 4, 5, 6, 1, 2, 3, . . . , or 23, 27, 32, 38, 45, 53, 62, 73, 86, 101, 118, 138, 23, 27, 32, . . . , or 88, 103, 121, 142, 166, 194, 227, 265, 310, 362, 423, 494, 577, 674, 787, 919, 1073, 1252, 1461, 1705, 1990, 2322, 387, 452, 528, 88, 103, 121, ...

And we could study sequences of iterates of CollatzK[6, n] using the function CollatzKSequence[6, n] defined by

```
CollatzKSequence[6, n_Integer?Positive] :=
Module[{n1 = n, l = {n}},
While[ (n1  1) && (n1 != 23) && (n1 != 88),
n1 = CollatzK[6, n1]; l = Append[l,n1]]; l]
```

Here are three illustrative examples.

```
CollatzKSequence[6, 2154]
```

{2154, 359, 419, 489, 571, 667, 779, 909, 1061, 1238, 1445, 1686, 281, 328, 383, 447, 522, 87, 102, 17, 20, 24, 4, 5, 6, 1}

```
CollatzKSequence[6, 569]
```

{569, 664, 775, 905, 1056, 176, 206, 241, 282, 47, 55, 65, 76, 89, 104, 122, 143, 167, 195, 228, 38, 45, 53, 62, 73, 86, 101, 118, 138, 23}

CollatzKSequence[6, 5714]

{5714, 6667, 7779, 9076, 10589, 12354, 2059, 2403, 2804, 3272,
3818, 4455, 5198, 6065, 7076, 8256, 1376, 1606, 1874, 2187, 2552,
2978, 3475, 4055, 4731, 5520, 920, 1074, 179, 209, 244, 285, 333,
389, 454, 530, 619, 723, 844, 985, 1150, 1342, 1566, 261, 305,
356, 416, 486, 81, 95, 111, 130, 152, 178, 208, 243, 284, 332,
388, 453, 529, 618, 103, 121, 142, 166, 194, 227, 265, 310,
362, 423, 494, 577, 674, 787, 919, 1073, 1252, 1461, 1705, 1990,
2322, 387, 452, 528, 88}

2

Numbers

2.1 Characterizing Numbers

The command NumberQ[expression] gives True if expression is a number, and False otherwise

```
{NumberQ[5], NumberQ[2/3], NumberQ[3.7], NumberQ[Hello]}
```

{True, True, True, False}

General Remark: Many commands ending with capital Q test if an expression is of a specific type such as, for example, IntegerQ[expression], EvenQ [expression], OddQ[expression], PrimeQ[expression], and so on, which test, respectively, if expression is an integer, an even integer, an odd integer, a prime number, and so on.

```
{IntegerQ[7629], IntegerQ[2/3], IntegerQ[5.78]}
```

{True, False, False}

```
{EvenQ[46],EvenQ[51]}
```

{True, False}

```
{OddQ[46], OddQ[51]}
```

{False, True}

{PrimeQ[11], PrimeQ[14]}

{True, False}

Mathematica manipulates four types of numbers: reals, integers, rationals, and complex numbers.

2.2 Real Numbers

When, for example, we cannot find the exact solution to an equation, we have to use numerical routines to obtain an approximate numerical solution. In the first chapter we defined **Precision** and **Accuracy** which are respectively related to the relative error and absolute error by the formulas

$$\text{relative error} = 10^{-\text{Precision}} \text{ and absolute error} = 10^{-\text{Accuracy}}.$$

In general, real numbers use machine precision. This precision is given by the command:

$MachinePrecision

15.9546

Hence adding 10^{-16} to a machine-precision real number does not change this number value!

{N[Sqrt[2]]//FullForm, N[Sqrt[2]+10^(-16)]
FullForm, N[Sqrt[2]+10^(-15)]//FullForm}

{1.4142135623730951', 1.4142135623730951', 1.4142135623730963'}

Instead of using $MachinePrecision, users can manipulate numbers with any number of significant digits.

{Precision[N[Pi]], Precision[N[Pi,100]]}

{MachinePrecision, 100.}

The command N[number, n] allows us to carry out computations using number with precision n.

```
Precision[N[Sqrt[5], 30]]
Precision[(N[Sqrt[5], 30])^100]
```

30.

28.

Note that the precision on the approximate value of $(\sqrt{5})^{100}$ is less than the precision on the approximate numerical value of $\sqrt{5}$.

The command N[number, n] can also be entered as number'n.

```
{0.7'30, 0.7'30 + 10^(-30)}
```

{0.700000000000000000000000000000,

0.700000000000000000000000000001}

If we have to manipulate numbers whose values lie in a certain interval, such as experimental results, we can use the command Interval[{minimum, maximum}].

```
result = Interval[{1.9, 2.1}] + Interval[{1.4, 1.6}]
```

Interval[{3.3, 3.7}]

Min[{x1, x2, ...}] yields the numerically smallest of the list {x1, x2, ...} and Max[{x1, x2, ...}] yields the numerically largest of the list {x1, x2, ...}.

```
{Min[result], Max[result]}
```

{3.3, 3.7}

Real numbers are displayed with six digits. But we can use NumberForm[expression, n] to ask *Mathematica* to display more or fewer digits.

```
{N[Sqrt[2]], NumberForm[N[Sqrt[2]], 4],
NumberForm[N[Sqrt[2]], 12]}
```

{1.41421, 1.414, 1.41421356237}

2.3 Integers

There exist three functions that convert real numbers into integers. Floor[x] gives the greatest integer less than or equal to x. Ceiling[x] gives the smallest integer greater than or equal to x. Round[x] gives the integer closest to x.

```
{Floor[4.49], Ceiling[4.49], Round[4.49]}
```

{4, 5, 4}

```
{Floor[4.51], Ceiling[4.51], Round[4.51]}
```

{4, 5, 5}

Every integer can be written as a product of its prime factors in a unique way. FactorInteger[n] gives a list of the prime factors of the integer n, together with their exponents,

```
FactorInteger[83406151]
```

{{31, 2}, {229, 1}, {379, 1}}

a result which can be verified:

```
83406151 == 31^2 229 379
```

True

In 1832 Carl Friedrich Gauss (1777–1855) considered algebraic integers of the form $a+bi$, where a and b are rational integers, called *Gaussian integers*. They share many properties with ordinary real integers. For example, the sum, difference, and product of two Gaussian integers are Gaussian integers. Gauss proved that the Gaussian integers satisfy a generalized version of the factorization theorem. Although by default FactorInteger[n] allows only real integers, the option GaussianIntegers → True handles Gaussian integers.

```
{FactorInteger[4],
FactorInteger[4, GaussianIntegers → True]}
```

{{2, 2}}, {{-1, 1}, {1 + I, 4}}

hence

```
4 == - (1 + I)^4
```

True

Divisors[n] gives a list of the integers that divide n, including 1 and n.

```
Divisors[1364]
```

{1, 2, 4, 11, 22, 31, 44, 62, 124, 341, 682, 1364}

Here are a few more integer functions:

Mod[k, n] gives the remainder from dividing k by n. Mod[k, n, d] uses an offset d.

```
{Mod[3,3], Mod[3,3,1]}
```

{0, 3}

Instead of typing commands one can use the Palettes submenus BasicCalculation and AlgebraicManipulation (go to File → Palettes).

GCD[n1, n2, ...] gives the greatest common divisor of the integers n1, n2, ... and LCM[n1, n2, ...] gives the least common multiple of these integers.

```
GCD[72, 42, 18]
```

6

```
LCM[22, 14, 8]
```

616

These functions apply also to rationals and Gaussian integers.

```
GCD[1/3, 2/7, 5/4]
```

$$\frac{1}{84}$$

```
LCM[4 + 3 I, 2 + I, 3 - I]
```

15 + 5 I

IntegerDigits[n] gives a list of the decimal digits in the integer n.
IntegerDigits[n, b] gives a list of digits in the integer n in base b.

```
IntegerDigits[27634]
```

{2, 7, 6, 3, 4}

```
IntegerDigits[23, 2]
```

{1, 0, 1, 1, 1}

Numbers can be displayed breaking the digits into blocks of a given length separated by a specific string. For example, using the options DigitBlock -> 5 and NumberSeparator -> " " for the command NumberForm, displays numbers breaking the digits into blocks of length 5 separated by a space.

```
NumberForm[23!, DigitBlock -> 5, NumberSeparator -> " "]
```

258 52016 73888 49766 40000

2.4 Prime Numbers

Prime[n] gives the nth prime number. One is not a prime number; two is therefore the first prime.

```
Table[Prime[n], {n, 1, 10}]
```

{2, 3, 5, 7, 11, 13, 17, 19, 23, 29}

or simply

```
Prime[Range[10]]
```

{2, 3, 5, 7, 11, 13, 17, 19, 23, 29}

PrimePi[x] gives the number of primes $\pi(x)$ less than or equal to x. $\pi(x)$ is well approximated by the logarithmic integral function LogIntegral[x] defined by $\mathrm{li}(x) = \int_0^x 1/\log(t)\, dt$, where the principal value of the integral (singular for $x = 1$) is taken.

```
Plot[{PrimePi[x], LogIntegral[x]}, {x, 1, 10000},
PlotStyle -> {{RGBColor[0, 0, 1]}, {RGBColor[1, 0, 0]}}];
```

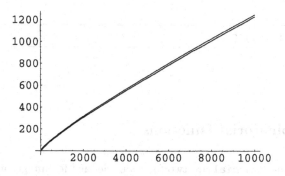

Fig. 2.1. *Graph of $\pi(x)$ and $\mathrm{li}(x)$ for $x \in [1, 10000]$.*

PrimeQ[n] yields True if n is a prime number, and False otherwise.

```
{PrimeQ[56509], PrimeQ[4745443]}
```

{True, False}

We can also use n ∈ Primes, where ∈ is entered as \[Element].

```
56509 ∈ Primes
4745443 ∈ Primes
```

True

False

PrimeQ[n, GaussianIntegers → True] yields True if n is a Gaussian prime, and False otherwise

```
PrimeQ[2, GaussianIntegers → True]
```

False

which is correct because

```
FactorInteger[2, GaussianIntegers → True]
```

{-I, 1}, {1 + I, 2}}

that is,

```
2 == - I (1 + I)^2
```

True

2.5 Combinatorial Functions

Factorial and Binomial are two functions defined for integer arguments.

2.5.1 Factorial

n! or Factorial[n] gives the factorial of n equal to $n(n-1)(n-2)\ldots 1$. The notation $n! = n(n-1)(n-2)\ldots.3.2.1$ is due to Kamp [27].

```
Table[n!, {n, 1, 10}]
```

$\{1, 2, 6, 24, 120, 720, 5040, 40320, 362880, 3628800\}$

For a noninteger or complex argument z, the numerical value of $z!$ is given by Gamma[1 + z].

```
3.4! == Gamma[4.4]
```

True

```
(2.1 + 4.5 I)! == Gamma[3.1 + 4.5 I]
```

True

n!! or Factorial2[n] gives the double factorial of n equal to $n(n-2)(n-4)\ldots$.

```
5!! == 5 3 1 == Factorial2[5]
```

True

```
6!! == 6 4 2 == Factorial2[6]
```

True

```
5!! 6!! == 6!
```

True

2.5.2 Binomial Coefficients

Binomial[n, m] gives the binomial coefficient $\binom{n}{m}$. Here are the first rows of Pascal's triangle.

```
Table[Binomial[n,k], {n, 0, 10}, {k, 0, n}] // TableForm
```

```
1
1    1
1    2    1
1    3    3    1
1    4    6    4    1
1    5    10   10   5    1
1    6    15   20   15   6    1
1    7    21   35   35   21   7    1
1    8    28   56   70   56   28   8    1
1    9    36   84   126  126  84   36   9    1
1    10   45   120  210  252  210  120  45   10   1
```

Mathematica can evaluate exactly a few sums involving binomial coefficients.

```
Sum[Binomial[n, k], {k, 0, n}]
```

2^n

```
Sum[(Binomial[n, k])^2, {k, 0, n}]
```

$$\frac{4^n \ \text{Gamma}[\frac{1}{2} + n]}{\text{Sqrt}[\text{Pi}]\text{Gamma}[1 + n]}$$

Can we simplify this result?

```
FullSimplify[Sum[(Binomial[n, k])^2, {k, 0, n}]]
```

$$\frac{4^n \ \text{Gamma}[\frac{1}{2} + n]}{\text{Sqrt}[\text{Pi}]\text{Gamma}[1 + n]}$$

Apparently not. But, knowing some properties of the Γ function, we can show that the result can take a much simpler form.

First

```
FullSimplify[Gamma[1+n] == n!]
```

True

then

```
FullSimplify[2^n Gamma[1/2 + n] == (2*n - 1) !! Sqrt[Pi], n ∈
Integers]
```

True

Hence

$$\frac{4^n \Gamma\left(\frac{1}{2} + n\right)}{\sqrt{\pi}\gamma(1+n)} = \frac{2^n(2n-1)!!}{n!} = \frac{(2n)!}{(n!)^2}.$$

That is, the result above is just `Binomial[2n, n]`, and *Mathematica* can verify it.

```
FullSimplify[Sum[(Binomial[n, k])^2, {k, 0, n}] ==
Binomial[2 n, n]]
```

True

Mathematica can also evaluate exactly the binomial sums:

```
Table[Sum[k^r Binomial[n, k], {k, 0, n}],
{r, 1, 5}]//TableForm
```

$2^{-1+n}\, n$

$2^{-2+n}\, n(1 + n)$

$2^{-3+n}\, n^2(3 + n)$

$2^{-4+n}\, n(1 + n)(-2 + 5\, n + n^2)$

$2^{-5+n}\, n^2(-10 + 15\, n + 10\, n^2 + n^3)$

Here are sums involving the inverse of central binomial coefficients that *Mathematica* can, surprisingly, evaluate exactly.

```
Sum[1 / Binomial[2n, n], {n,1, Infinity}]
```

$\dfrac{9 + 2\ \text{Sqrt}[3]\ \text{Pi}}{27}$

```
Sum[1/(n Binomial[2n,n]), {n, 1, Infinity}]
```

$$\frac{Pi}{3\,Sqrt[3]}$$

2.6 Rational Numbers

Is it possible to convert real numbers into rational numbers? Rationalize[x] converts x into a fraction a/b such that $|a/b - x| < 10^{-4}/b^2$. Rationalize[x, Δx] converts x into a fraction with the smallest denominator that lies within Δx.

```
Rationalize[3.14]
Rationalize[Pi, 10^(-6)]
```

$$\frac{157}{50}$$

$$\frac{355}{113}$$

```
N[Pi - 355/113]
```

$$-\,2.66764 \times 10^{-7}$$

The symbol $\{n_1, n_2, n_3, \ldots\}$ associated with a real number x is its continued fraction representation, which means that $x = n_1/(n_2 + 1/(n_3 + \cdots$.

The command ContinuedFraction[x, n] generates a list of the first n terms in the continued fraction representation of x.

```
ContinuedFraction[Pi, 7]
```

$$\{3, 7, 15, 1, 292, 1, 1\}$$

This list of n terms represents the fraction:

```
3 + 1 / (7 + 1 / (15 + 1 / (292 + 1 / 2)))
```

$$\frac{194849}{62024}$$

approximately equal to

```
N[3 + 1/(7 + 1/(15 + 1/(292 + 1/2)))]
```

3.14151

`FromContinuedFraction[list]` reconstructs a number from the list of its continued fraction terms.

```
FromContinuedFraction[{3, 7, 15, 1, 292, 1, 1}]//N
```

3.14159

2.7 Complex Numbers

To enter the complex number we can type I equal to Sqrt[-1].

```
Sqrt[-1]
```

I

Here are a few basic functions.

```
z = 2 - 5 I;
{Re[z], Im[z], Conjugate[z], Abs[z], Arg[z]}
```

$$\{2, - 5, 2 + 5 \; I, \; Sqrt[29], \; - \; ArcTan[\frac{5}{2}]\}$$

Evaluating expressions containing complex numbers is easy.

```
(2 + I) + (5 - 3 I)
```

7 - 2 I

```
(7 - I) (2 + 3 I)
```

```
17 + 19 I
```

```
(3 + 2 I) / (5 - 3 I)
```

$$\frac{9}{34} + \frac{19I}{34}$$

```
5.42 Exp[2.6 I] + 3.27 Exp[3.1 I]
```

```
-7.91151 + 2.92999 I
```

```
ComplexExpand[(2 - 3 I)^5]
```

```
122 + 597 I
```

```
Sqrt[(3 - 5 I)^2]
```

```
3 - 5 I
```

2.8 Different Bases

Mathematica can deal with numbers in different bases. BaseForm[number, b] gives number in base b. number can be an integer, a rational, a real, or a complex number.

```
{BaseForm[1457, 2], BaseForm[1457,18]}
```

$\{10110110001_2, \ 48h_{18}\}$

The largest base is 36. It uses the 10 digits 0, 1, 2, ..., 9 and the 26 letters a, b, c, ..., z

```
BaseForm[2/3, 2]
```

$$\frac{10_2}{11_2}$$

```
BaseForm[3.27, 2]
```

11.0100010100011111011_2

```
BaseForm[21 + 57 I, 3]
```

$210_3 + 2010_3$ I

We can also enter a number n in any base b (less than 36) as b^^n. Each digit of n must be less than b.

```
{5^^234101, 5^^34.21}
```

{8651, 19.44}

```
{3^^1201 + 2^^101, 3^^1201 2^^101, 3^^1201 / 2^^101}
```

$\{51, \ 230, \ \frac{46}{5}\}$

```
{3^^1201, 2^^101}
```

{46, 5}

Numerical calculations in any base can be done. The result is given in base 10.

```
{3^^1022 + 3^^2011, 3^^1022 3^^2011, 3^^1022 / 3^^2011}
```

$\{93, \ 2030, \ \frac{35}{58}\}$

We can even mix bases. The result is again given in base 10.

```
{3^^212, 7^^432, 3^^212 + 7^^432}
```

{23, 219, 242}

2.9 Calendars

A calendar may be viewed as some kind of positional number system (see next section). For example, the date {1789, 7, 14} is the 14th day of the 7th month (July) of the 1789th year. The package Miscellaneous'Calendar' provides the following functions to help perform basic calendar operations.

```
<<Miscellaneous'Calendar'
```

```
?Miscellaneous'Calendar'*
```

Miscellaneous'Calendar'

Calendar	Islamic
CalendarChange	JewishNewYear
DayOfWeek	Julian
DaysBetween	Monday
DaysPlus	Saturday
EasterSunday	Sunday
EasterSundayGreekOrthodox	Thursday
Friday	Tuesday
Gregorian	Wednesday

Calendar is an option for calendar functions indicating which calendar system to use: Gregorian, Julian, or Islamic. If set to Automatic, the Julian calendar is used before 2 September 1752, and the Gregorian calendar afterwards.

CalendarChange[date, calendar1, calendar2] converts a date in calendar1 to a date in calendar2.

```
CalendarChange[{1789, 7, 14}, Gregorian, Islamic]
```

{1203, 10, 20}

DayOfWeek[date] gives the day of the week for the date.

```
DayOfWeek[{1789, 7, 14}]
```

Tuesday

DaysBetween[date1, date2] gives the number of days between date1 and date2.

```
DaysBetween[{1769, 8, 15}, {1821, 5, 5}]
```

18890

EasterSunday[year] gives the date of Easter Sunday in the Gregorian calendar.

```
EasterSunday[2006]
```

{2006, 4, 16}

JewishNewYear[year] gives the date of the Jewish New Year occurring in the Christian year where $1900 \leq year < 2100$. Add 3761 to the Christian year to get the corresponding new Jewish Year.

```
{JewishNewYear[2006], DayOfWeek[JewishNewYear[2006]]}
```

{{2006, 9, 23}, Saturday}

2.10 Positional Number Systems

Positional number systems give compact representations of numbers. In such systems, each number has a unique representation by an ordered sequence of digits, the value of the number being determined by the position of the digits and the base b of the system. For example, in base b the sequence of digits $d_n d_{n-1} \ldots d_2 d_1 d_0$ (more exactly, this is a word in the b-letter alphabet $\{0, 1, 2, \ldots\}$) represents the number

$$N_b = d_n \times b^n + d_{n-1} \times b^{n-1} + \cdots + d_2 \times b^2 + d_1 \times b + d_0.$$

This representation is very economical because we need only b different symbols to represent a number in base b.

In the previous section we mentioned that a calendar is some kind of positional number system. The difference with standard positional number systems is that the representation of a date does not use a unique base b but a sequence of bases $(b_1, b_2, b_3, \ldots)$. In a multibase system the sequence of digits $d_n d_{n-1} \ldots d_2 d_1 d_0$ represents the number

$$
\begin{aligned}
N_{b_1 b_2 b_3 \ldots} = {} & d_n \times (b_n \times b_{n-1} \times \cdots \times b_2 \times b_1) \\
& + d_{n-1} \times (b_{n-1} \times b_{n-2} \times \cdots \times b_2 \times b_1) + \cdots \\
& + d_2 \times (b_2 \times b_1) + d_1 \times b_1 + d_0.
\end{aligned}
$$

Time intervals are usually represented in a multibase system. For example, the time interval of 1,668,214 seconds is more conveniently represented in the multibase (week, day, hour, minute, second) by the sequence $(2, 5, 7, 23, 34)$ meaning that 1668214 seconds = 2 weeks + 5 days + 7 hours + 23 minutes + 34 seconds because $34 + 23 \times 60 + 7 \times 60 \times 60 + 5 \times 24 \times 60 \times 60 + 2 \times 7 \times 24 \times 60 \times 60 = 1{,}668{,}214$.

Note that we did not define specific symbols but used decimal numbers to represent the numbers of weeks, days, hours, minutes, and seconds. In a multibase system, for all indices k, the "digit" d_k is in the range 0 to b_{k-1}. When the multibase is finite, the first digit of the sequence—the number of weeks in the example above—is in the range 0 to ∞.

Let us build up a function of two variables: a number n and a finite multibase b which gives the representation of the decimal number n in the multibase b.

We denote this function toMultibase[n_Integer, b_List]. The image of a pair (n, b) will be a finite sequence of length equal to Length[b] + 1.

We first note that if either the list b is empty or the number n is less than the last digit of the list b the sequence of digits representing n is just the one-element list n.

If the list b is not empty and the number n is larger than the last element of the list b, then the rightmost element of the sequence of digits is given by Mod[n, Last[b]].

To find the next element we have to replace n by Floor[n / Last[b]] and to prepend to the digit sequence the new digit obtained reapplying the function Mod[n, Last[b]] where the original b list has been replaced by the list Drop[b, -1] which is the command that deletes the last element of the original list b. To iterate this process and avoid confusion with the original variables n and b, we use the local variables tempNumber and tempBase. In the program below, we use the command Prepend[sequence,number] to add number at the beginning of sequence.

```
toMultibase[n_Integer, b_List] :=
Module[{tempNumber = n, tempBase = b, digitSequence = {}, d},
While[Length[tempBase] > 0,
d = Mod[tempNumber, Last[tempBase]];
digitSequence = Prepend[digitSequence, d];
tempNumber = Floor[tempNumber/Last[tempBase]];
tempBase = Drop[tempBase, -1]];
digitSequence = Prepend[digitSequence, tempNumber];
digitSequence]
```

For example,

```
toMultibase[1789, {12, 11, 10}]
```

$\{1, 4, 2, 9\}$

which means that

```
1789 == 9 + 2 × 10 + 4 × 11 × 10 + 1 × 12 × 11 × 10
```

True

We can use this toMultibase function to represent a time interval expressed in seconds in the multibase $\{7, 24, 60, 60\}$. For example, the time interval of 1,668,214 seconds is given by

```
toMultibase[1668214, {7, 24, 60, 60}]
```

$\{2, 5, 7, 23, 34\}$

On calendars, positional number systems, and how to write elegant *Mathematica* programs see Vardi [61].

2.11 Zeckendorf's Representation

The Belgian medical doctor Edouard Zeckendorf (1901–1923) published several mathematical papers. His most famous one deals with the representation of any integer as a sum of nonconsecutive Fibonacci numbers [69]. More precisely, Zeckendorf's theorem states that every integer n has a unique representation of the form

$$n = F_{k_1} + F_{k_2} + \cdots + F_{k_r},$$

where $k_j \geq k_{j+1} + 2$ for $j = 1, 2, \ldots, r - 1$ and $k_r \leq 2$. The *Mathematica* command Fibonacci[n] gives the Fibonacci number F_n. On how to write a program generating the Fibonacci sequence, refer to Part I, Chapter 8.

To find the Zeckendorf representation we use the so-called "greedy" algorithm [24], choosing F_{k_1} to be the largest Fibonacci number less than or equal to n, that is, such that $F_{k_1} \leq n < F_{k_1+1}$, then choosing F_{k_2} such that $F_{k_2} \leq n - F_{k_1} < F_{k_2+1}$, and so on. Another example of the greedy algorithm can be found in the chapter dedicated to Egyptian fractions. Note that the relation $0 \leq n - F_{k_1} < F_{k_1+1} - F_{k_1} = F_{k_1-1}$ implies the inequality $n - F_{k_1} < F_{k_1-1}$ so $F_{k_2} < F_{k_1-1}$. That is, the greedy algorithm leads to a representation of n by a sum of nonconsecutive Fibonacci numbers. Moreover, the algorithm shows that the representation is unique. The Zeckendorf theorem leads, therefore, to what can be called the *Fibonacci number system*. Any nonnegative integer n can, therefore, be represented by a unique sequence of 0s and 1s writing

$$n = (d_m d_{m-1} \ldots d_2)_{\text{Fibonacci}}.$$

where all the digits d_j are equal to 0 or 1, and this representation, which is similar to the binary representation, is equivalent to the equality

$$n = \sum_{j=2}^{m} d_j F_j.$$

Note that, according to the Zeckendorf theorem in the sequence of 0s and 1s $(d_m d_{m-1} \ldots d_2)$ we never find two adjacent 1s.

In the *Mathematica* Help Menu, looking for Fibonacci, in the Further Examples section we find the definitions of a few functions that can generate the Zeckendorf representation. Here are these functions.

The function LeadingIndex[n] gives the largest integer k such that F_k does not exceed n.

```
LeadingIndex[n_Integer?( #  0 &)]  :=
Module[{k}, If[n == 0, k = 2,
For[k=2, Fibonacci[k]  n, k++]; k--;]; k]
```

```
LeadingIndex[45]
```

9

```
Fibonacci[9]
```

34

The function ZeckendorfRepresentation[n] gives the coefficients of the expansion descending from the leading index.

```
ZeckendorfRepresentation[n_Integer?(#  0 &)] :=
Module[{i, k, 1, m, addon, r},
k = 0; If[n == 0, r = {0}, If[n == 1, r = {1},
1 = LeadingIndex[n]; m = n - Fibonacci[1];
k = LeadingIndex[m]; addon = Flatten[{1,Table[0, {i, k + 2,
1}]}];
r = Flatten[{addon, ZeckendorfRepresentation[m]}]]]; r]
```

Here is the Zeckendorf representation of 45.

```
ZeckendorfRepresentation[45]
```

{1, 0, 0, 1, 0, 1, 0, 0}

The function ff[Zrepresentation] gives the Fibonacci numbers corresponding to the Zeckendorf representation Zrepresentation.

```
ff[Zrepresentation_] :=
Fibonacci[2 + Length[Zrepresentation] -
First / @Position[Zrepresentation, 1]]
```

```
ff[{1, 0, 0, 1, 0, 1, 0, 0}]
```

{34, 8, 3}

3

Algebra

3.1 Algebraic Expressions

Expand[expression] expands products and positive integer powers in **expression**.

```
Expand[3^4]
```

81

```
Expand[(a + b)^3]
```

$a^3 + 3\,a^2\,b + 3\,a\,b^2 + b^3$

```
Expand[(a - b) (a + b)]
```

$a^2 - b^2$

```
Expand[(t^2 - 5)^3 /. t -> (a - b)^2]
```

$-\,125 + 75\,a^4 - 15\,a^8 + a^{12} - 300\,a^3\,b + 120\,a^7\,b - 12\,a^{11}\,b +$
$450\,a^2\,b^2 - 420\,a^6\,b^2 + 66\,a^{10}\,b^2 - 300\,a\,b^3 + 840\,a^5\,b^3 -$
$220\,a^9\,b^3 + 75\,b^4 - 1050\,a^4\,b^4 + 495\,a^8\,b^4 + 840\,a^3\,b^5 -$
$792\,a^7\,b^5 - 420\,a^2\,b^6 + 924\,a^6\,b^6 + 120\,a\,b^7 - 792\,a^5\,b^7 -$

$15\ b^8 + 495\ a^4\ b^8 - 220\ a^3\ b^9 + 66\ a^2\ b^{10} -$
$12\ a\ b^{11} + b^{12}$

ExpandAll[expression] expands everything including denominators.

```
ExpandAll[(a + b)^3 / (a - b)^2]
```

$$\frac{a^3}{a^2 - 2ab + b^2} + \frac{3a^2b}{a^2 - 2ab + b^2} + \frac{3ab^2}{a^2 - 2ab + b^2} +$$
$$\frac{b^3}{a^2 - 2ab + b^2}$$

FunctionExpand[expression] expands special functions. It accepts assumptions.

```
{FunctionExpand[Log[a b]],
FunctionExpand[Log[a b], a > 0 && b > 0]}
```

{Log[a b], Log[a] + Log[b]}

In the first case, FunctionExpand cannot find a simpler result. In order to give the second result, it needs more information about a and b.

Factor[polynomial] factors a polynomial over the integers.

```
Factor[a^3 + 3 a^2 b + 3 a b^2 + b^3]
```

$(a + b)^3$

ComplexExpand[expression] expands expression assuming that all variables are real.

```
ComplexExpand[(a + I b)^3]
```

$a^3 - 3\ a\ b^2 + I\ (3\ a^2\ b - b^3)$

Cancel[expression] cancels out common factors in the numerator and denominator of expression. Simplify[expression] returns the simplest form it finds.

```
Cancel[ (a + b)^3 / (a^2 - b^2]
Simplify[ (a + b)^3 / (a^2 - b^2]
```

$$\frac{(a+b)^2}{a-b}$$

$$\frac{(a+b)^2}{a-b}$$

Simplify accepts assumptions.

```
{Simplify[Sqrt[a^2]], Simplify[Sqrt[a^2], a < 0]}
```

$\{Sqrt[a^2], -a\}$

We can also use Assuming

```
Assuming[a < 0, Simplify[Sqrt[a^2]]]
```

- a

FullSimplify[expression] tries a wider range of transformations on expression involving elementary and special functions.

```
Simplify[(n + 1) Factorial[n]]
```

(1 + n) n!

```
FullSimplify[(n + 1) Factorial[n]]
```

Gamma[2 + n]

Together[expression] puts terms in a sum over a common denominator, and cancels factors in the result.

```
Together[a^2 / (a + b) + b^2 / (a - b)]
```

$$\frac{a^3 - a^2b + ab^2 + b^3}{(a-b)(a+b)}$$

`Numerator[expression]` and `Denominator[expression]` give, respectively, the numerator and the denominator of `expression`.

$$\text{Numerator}\left[\frac{a^3 - a^2b + ab^2 + b^3}{(a-b)(a+b)}\right]$$

$$a^3 - a^2b + ab^2 + b^3$$

$$\text{Denominator}\left[\frac{a^3 - a^2b + ab^2 + b^3}{(a-b)(a+b)}\right]$$

$$(a-b)(a+b)$$

`Apart[expression]` rewrites a rational `expression` as a sum of terms with minimal denominators.

$$\text{Apart}\left[\frac{a^3 - a^2b + ab^2 + b^3}{(a-b)(a+b)}\right]$$

$$-a - b - \frac{a^2}{-a+b} + \frac{a^2}{a+b}$$

`Apart[expression, variable]` treats all variables other than `variable` as constants.

$$\text{Apart}\left[\frac{ax + b(x-1)}{x(x-1)}, x\right]$$

$$\frac{a}{-1+x} + \frac{b}{x}$$

We have seen that `Factor[polynomial]` factors `polynomial` over the integers; using the option `GaussianIntegers → True` we can also factor over Gaussian integers.

```
{Factor[a^2+b^2], Factor[a^2+b^2, GaussianIntegers -> True]}
```

$\{a^2 + b^2, (a - I\ b)\ (a + I\ b)\}$

Collect[expression, x] collects together terms involving the same powers of objects matching x.

```
Collect[(a + b x) (c + d x), x]
```

$a\ c + (b\ c + a\ d)\ x + b\ d\ x^2$

Collect[expression, x, f] applies f to the expression that forms the co-efficient of each term obtained.

```
Collect[(a + b x) (c + d x), x, f]
```

$f[a\ c] + x^2\ f[b\ d] + x\ f[b\ c + a\ d]$

If the function f is defined, Collect uses this definition.

```
f = Function[y, y^3];
Collect[(a + b x) (c + d x), x, f]
```

$a^3\ c^3 + (bc + ad)^3\ x + b^3\ d^3\ x^2$

Coefficient[expression, x] gives the coefficient of x in expression.

```
Coefficient[a c + (b c + a d) x + b d x^2, x]
```

$b\ c + a\ d$

Coefficient[expression, x, n] gives the coefficient of x^n in expression.

```
Coefficient[a c + (b c + a d) x + b d x^2, x, 0]
```

$a\ c$

```
Coefficient[a c + (b c + a d) x + b d x^2, x, 2]
```

b d

CoefficientList[polynomial, x] gives a list of coefficients of powers of x in polynomial starting with power 0.

```
CoefficientList[a c + (b c + a d) x + b d x^2, x]
```

{a c, b c + a d, b d}

Exponent[expression, variable] indicates the maximum power of variable in expression.

```
Exponent[3 x^3 + 5 x^2 - 2 x, x]
```

3

PowerExpand may give an incorrect result and does not take into account assumptions!

```
{PowerExpand[Sqrt[a^2]], PowerExpand[Sqrt[a^2], a < 0]
```

{a, Sqrt[a^2]}

3.2 Trigonometric Expressions

TrigFactor[expression] and TrigExpand[expression] factor trigonometric functions in expression. They both work on circular and hyperbolic functions.

```
TrigFactor[Sin[2x]]
```

2 Cos[x] Sin[x]

```
TrigFactor[Sinh[2x]]
```

2 Cosh[x] Sinh[x]

We obtain the same result with **Factor[expression]** if we use the option **Trig → True**.

```
Factor[Sin[2 x], Trig → True]
```

2 Cos[x] Sin[x]

Compare the results obtained with either **TrigFactor** or **TrigExpand**.

```
TrigFactor[Cos[2x]]
TrigExpand[Cos[2x]]
```

(Cos[x] − Sin[x]) (Cos[x] + Sin[x])
$Cos[x]^2 - Sin[x]^2$

```
TrigExpand[Sinh[2x]]
```

$Cosh[x]2 + Sinh[x]^2$

```
Factor[Sin[3 x], Trig → True]
```

(1 + 2 Cos[2 x]) Sin[x]

```
TrigExpand[Sin[3 x]]
```

$3 Cos[x]^2 Sin[x] - Sin[x]^3$

```
TrigExpand[Cosh[3 x]]
```

$Cosh[x]^3 + 3 Cosh[x] Sinh[x]^2$

TrigReduce[expression] rewrites products and powers of trigonometric functions in expression in terms of trigonometric functions with combined arguments. It works on circular and hyperbolic functions.

```
TrigReduce[Cosh[x]^2 - Sinh[x]^2]
```

1

```
TrigReduce[Cosh[x]^4 - Sinh[x]^4]
```

Cosh[2 x]

TrigToExp[expression] converts trigonometric functions in expression to exponentials and ExpToTrig[expression] converts exponentials in expression to trigonometric functions.

```
TrigToExp[Cos[2 x] + Sinh[2 x]]
```

$$\frac{1}{2} E^{-2\,x} + \frac{1}{2} E^{-2\,I\,x} + \frac{1}{2} E^{2\,I\,x} + \frac{1}{2} E^{2\,x}$$

```
ExpToTrig[(Exp[2 x] - 1) / (Exp[2 x] + 1)]
```

$$\frac{-1 + \text{Cosh}[2x] + \text{Sinh}[2x]}{1 + \text{Cosh}[2x] + \text{Sinh}[2x]}$$

Sometimes we have to use Simplify for a simpler result.

```
{ExpToTrig[(Exp[2 x] - 1) / (Exp[2 x] + 1)], ExpToTrig[(Exp[2
x] - 1) / (Exp[2 x] + 1)] // Simplify}
```

$$\left\{\frac{-1 + \text{Cosh}[2x] + \text{Sinh}[2x]}{1 + \text{Cosh}[2x] + \text{Sinh}[2x]}, \text{Tanh}[x]\right\}$$

TrigToExp[expression] also works in the following cases.

```
TrigToExp[ArcTan[x]]
```

$$\frac{1}{2} \ \text{I Log}[1 - \text{I x}] \ - \ \frac{1}{2} \ \text{I Log}[1 + \text{I x}]$$

```
TrigToExp[ArcTanh[x]]
```

$$- \ \frac{1}{2} \ \text{Log}[1 - x] \ + \ \frac{1}{2} \ \text{Log}[1 + x]$$

```
Simplify[Sin[n Pi]]
```

```
Sin[n Pi]
```

Here *Mathematica* does not assume any specific property for the symbol n. If we tell *Mathematica* that n is an integer, then we get the expected result.

```
Simplify[Sin[n Pi], n ∈ Integers]
```

0

We can also use complex arguments.

```
{Sin[I Pi], Exp[I Pi / 2]}
```

```
{I Sinh[Pi], I}
```

When using approximate numerical values, we can obtain spurious small values due to round-off errors. Chop[expression] replaces approximate real numbers less than 10^{-10} in expression by 0. The default tolerance of 10^{-10} can be changed to Δ using the command Chop[expression, Δ].

```
{Exp[I N[Pi] / 2], Chop[Exp[I N[Pi] / 2]]}
```

$\{6.12323 \ 10^{-17} + 1. \ \text{I}, \ 1. \ \text{I}\}$

Here is a simple relation that has been used to calculate π.

```
FullSimplify[ArcTan[1 / 2] + ArcTan[1 / 5] + ArcTan[1 / 8]]
```

$$\frac{Pi}{4}$$

3.3 Solving Equations

3.3.1 Solving Polynomial Equations Exactly

Linear, quadratic, cubic, and quartic polynomial equations can be solved in terms of radicals. There is no solution for general polynomial higher-degree equations in terms of radicals. Some particular polynomial equations of degree higher than four can however be solved in terms of radicals. In 1829 Évariste Galois (1811–1832) submitted articles to the French *Académie des Sciences* on the algebraic solution of equations, and a new article in 1830 entitled *Mémoire sur les conditions de résolubilité des équations par radicaux* (On the condition that an equation be soluble by radicals) which gave rise to the field of Galois theory.

Solve[equations, variables] attempts to solve an equation, or a system of equations, where variables stands for the list of unknowns. *Mathematica* gives exact solutions to linear, quadratic, cubic, and quartic equations. Equations are given in the form lhs == rhs. Solve gives solutions in terms of rules of the form {x → solution}.

```
Solve[a x^2 + b x + c == 0, x]
```

$$\left\{\left\{x \to \frac{-b - \sqrt{b^2 - 4ac}}{2a}\right\}, \left\{x \to \frac{-b + \sqrt{b^2 - 4ac}}{2a}\right\}\right\}$$

The general quintic equation is not solvable by radicals.

```
Solve[a x^5 + b x^4 + c x^3 + d x^2 + e x + f == 0, x]
```

$$\{\{ x \to \text{Root}[f + e \#1 + d \#2^2 + c \#1^3 + b \#4^4 + a \#5^5 \&, 1]\},$$
$$\{ x \to \text{Root}[f + e \#1 + d \#2^2 + c \#1^3 + b \#4^4 + a \#5^5 \&, 2]\},$$
$$\{ x \to \text{Root}[f + e \#1 + d \#2^2 + c \#1^3 + b \#4^4 + a \#5^5 \&, 3]\},$$
$$\{ x \to \text{Root}[f + e \#1 + d \#2^2 + c \#1^3 + b \#4^4 + a \#5^5 \&, 4]\},$$
$$\{ x \to \text{Root}[f + e \#1 + d \#2^2 + c \#1^3 + b \#4^4 + a \#5^5 \&, 5]\}\}$$

But some particular quintic equations are solvable by radicals. Here is an example.

```
Solve[x^5 - x^4 - x +1 == 0, x]
```

$\{\{x \rightarrow -1\}, \{x \rightarrow -I\}, \{x \rightarrow I\}, \{x \rightarrow 1\}, \{x \rightarrow 1\}\}$

If the equations are not polynomial, *Mathematica* has to use inverse functions and warns us that some solutions might be missing.

```
Solve[Exp[x] == 2, x]
```

Solve::ifun:

Inverse functions are being used by Solve,

so some solutions may not be found; use Reduce for

complete solution information. More

$\{\{x \rightarrow Log[2]\}\}$

Let us follow the advice and use Reduce.

```
Reduce[Exp[x] == 2, x]
```

C[1] ∈ Integers && x == (2 I) Pi C[1] + Log[2]

Because if n is an integer, $\exp(2ni\pi) = 1$, the command Reduce does give us all the solutions! Here is another example.

```
Solve[Sin[x] == Cos[x], x]
```

Solve::ifun: Inverse functions are being used by Solve,

so some solutions may not be found; use Reduce for complete

solution information. More . . .

$\{\{x \rightarrow \frac{-3 Pi}{4}\}, \{x \rightarrow \frac{Pi}{4}\}\}$

```
Reduce[Sin[x] == Cos[x], x]
```

```
C[1] ∈ Integers && (x == - 2 ArcTan[1 + Sqrt[2]] + 2 Pi C[1]
|| x == - 2 ArcTan[1 - Sqrt[2]] + 2 Pi C[1])
```

```
Reduce[Sin[x] == Cos[x], x] // FullSimplify
```

```
(Pi + 8 Pi C[1] == 4 x
|| Pi (- 3 + 8 C[1]) == 4 x) && C[1] ∈ Integers
```

Reduce can also be used to solve inequalities.

```
Reduce[- x^2 + 2 x + 3 > 0, x]
```

$$- 1 < x < 3$$

```
Reduce[{x^2 + 3 x - 1 > 0, x^2 + x - 5 < 0}, x]
```

$$\frac{-3 + \mathrm{Sqrt}[13]}{2} < x < \frac{-1 + \mathrm{Sqrt}[21]}{2}$$

Mathematica can also solve equations containing rational powers. For example,

```
eqn = (x - 1)^(1 / 2) == (x + 1)^(1 / 3);
Solve[eqn]
```

$$\{\{x \to \frac{4}{3} + \frac{(73 - 6 \, \mathrm{Sqrt}[87])^{1/3}}{3} + \frac{(73 + 6 \, \mathrm{Sqrt}[87])^{1/3}}{3}\}\}$$

Actually, during the solving process of this type of equation, as a result of algebraic manipulations such as raising expressions to various powers, spurious solutions may arise. The command Solve verifies all solutions and eliminates the wrong ones. We may exhibit all solutions including the wrong ones using the option VerifySolutions → False.

```
sol = Solve[eqn, VerifySolutions -> False]
```

$$\{\{x \to \frac{4}{3} + \frac{(73 - 6\ \mathrm{Sqrt}[87])^{1/3}}{3} + \frac{(73 + 6\ \mathrm{Sqrt}[87])^{1/3}}{3}\},$$

$$\{x \to \frac{4}{3} - \frac{(1 + I\ \mathrm{Sqrt}[3])\ (73 - 6\ \mathrm{Sqrt}[87])^{1/3}}{6} -$$

$$\frac{(1 - I\ \mathrm{Sqrt}[3])\ (73 + 6\ \mathrm{Sqrt}[87])^{1/3}}{6}\},$$

$$\{x \to \frac{4}{3} - \frac{(1 - I\ \mathrm{Sqrt}[3])\ (73 - 6\ \mathrm{Sqrt}[87])^{1/3}}{6} -$$

$$\frac{(1 + I\ \mathrm{Sqrt}[3])\ (73 + 6\ \mathrm{Sqrt}[87])^{1/3}}{6}\}\}$$

We can verify that the extra solutions were wrong using the command
expression /. rule that applies a rule (or a list of rules) to expression.
Note that solutions are given by rules.

```
{eqn /. sol[[2]], eqn /. sol[[3]]}
```

{False, False}

3.3.2 Numerical Solutions

NSolve[equations, variables] gives a list of numerical approximations to
the roots of a polynomial equation or a system of polynomial equations.

```
NSolve[x^7 - 5 x^5 + 2 x^4 - x^3 + 6 x - 12 == 0, x]
```

{{x -> − 2.3818}, {x -> − 1.29259}, {x -> − 0.16742 − 1.17335 I}
{x -> − 0.16742 + 1.17335 I }, {x -> 0.973189 − 0.630825 I},
{x -> 0.973189 + 0.630825 I }, {x -> 2.06286}}

```
NSolve[{x - 2 y == 2, x - y + 3 z == - 2, x + y - 2 z == 0},
{x, y, z}]
```

{{x -> − 0.545455, y -> − 1.27273, z -> − 0.909091}}

```
NSolve[{ x^2 + 3 y^3 == 2, 2 x^3 - y == 1}, {x, y}]
```

$\{\{x \to -0.511668 + 0.849668\ I,\ y \to 0.948437 + 0.107866\ I\},$

$\{x \to -0.511668 - 0.849668\ I,\ y \to 0.948437 - 0.107866\ I\ \},$

$\{x \to -0.152442 + 0.779731\ I,\ y \to -0.450997 - 0.839403\ I\},$

$\{x \to -0.152442 - 0.779731\ I,\ y \to -0.450997 + 0.839403\ I\},$

$\{x \to 0.755431 - 0.214877\ I,\ y \to -0.347066 - 0.715907\ I\},$

$\{x \to 0.755431 + 0.214877\ I\ ,\ y \to -0.347066 + 0.715907\ I\},$

$\{x \to 0.950077,\ y \to 0.715169\},$

$\{x \to -0.56636 - 0.501715\ I,\ y \to -0.507959 - 0.713011\ I\},$

$\{x \to -0.56636 + 0.501715\ I,\ y \to -0.507959 + 0.713011\ I\}\}$

If equations involve more complicated functions we use FindRoot[eqn, {x, x0}] that searches a solution to eqn starting with x = x0. Using Plot helps to find the starting approximate solution x0 to eqn.

```
Plot[{2 Cos[x], Tan[x]}, {x, - 1, 1}];
```

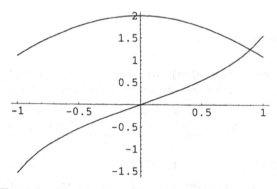

Fig. 3.1. *Graphs of* $2\cos(x)$ *and* $\tan(x)$ *for* $x \in [-1, 1]$.

The output (Figure 3.1) shows that the solution is close to 1. We therefore choose x0 = 1.

```
FindRoot[2 Cos[x] == Tan[x], {x, 1}]
```

$\{ x \to 0.895907\}$

Note that NSolve would have given correct solutions with, however, the usual warning.

```
NSolve[2 Cos[x] == Tan[x], x]
```

Solve::ifun: Inverse functions are being used by Solve, so some solutions may not be found; use Reduce for complete solution information. More . . .

$\{\{x \to -1.5708 - 0.732858\ I\}, \{x \to -1.5708 + 0.732858\ I\},$
$\{x \to 0.895907\}, \{x \to 2.24569\}\}$

If here again we follow the advice and use Reduce, we find

```
Reduce[2 Cos[x] == Tan[x], x]
```

C[1] ∈ Integers &&
(x == 2 ArcTan[Root[1 - #1 - 2 #1^2 - #1^3 + #1^4 & , 1]] +
2 Pi C[1]
|| x == 2 ArcTan[Root[1 - #1 - 2 #1^2 - #1^3 + #1^4 & , 2]] +
2 Pi C[1]
|| x == 2 ArcTan[Root[1 - #1 - 2 #1^2 - #1^3 + #1^4 & , 3]] +
2 Pi C[1]
|| x == 2 ArcTan[Root[1 - #1 - 2 #1^2 - #1^3 + #1^4 & , 3]] +
2 Pi C[1])

Reduce tells us that the equation has four solutions, but to find them we first have to solve the polynomial equation $1 - x - 2x^2 - x^3 + x^4 = 0$.

```
sol = NSolve[1 - x - 2 x^2 - x^3 + x^4, x]
```

$\{\{x \to -0.780776 - 0.624811\ I\}, \{x \to -0.780776 + 0.624811\ I\},$
$\{x \to 0.480534\}, \{x \to 2.08102\}\}$

Hence, the solutions are

```
{2 ArcTan[x] /. sol[[1]], 2 ArcTan[x] /. sol[[2]], 2
ArcTan[x] /. sol[[3]], 2 ArcTan[x] /. sol[[3]]
```

$\{-1.5708 - 0.732858\ I, -1.5708 + 0.732858\ I, 0.895907,$

$2.24569\}$

which are the solutions found using NSolve except that for the missing additive factor $2n\pi$ where n is an integer.

Consider another example.

```
Plot[{2 Cos[x], (x-1)^2}, {x, - 0.5, 2}];
```

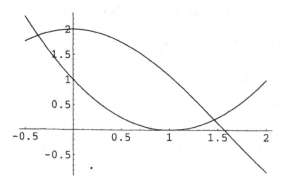

Fig. 3.2. *Graphs of* $2\cos(x)$ *and* $(x-1)^2$ *for* $x \in [-0.5, 2]$.

The graphs above show that the two solutions of the equation $2\cos(x) = (x-1)^2$ are close to 1.5 and −0.4. We can increase the precision by including the option WorkingPrecision $\rightarrow$ n to specify the number of digits that should be used to carry out computations.

```
FindRoot[2 Cos[x] == (x-1)^2, {x, 1.5},
WorkingPrecision → 40]
```

$\{x \rightarrow 1.4632766541815926451237275376283009964609\}$

```
FindRoot[2 Cos[x] == (x-1)^2, {x, - 0.4}, WorkingPrecision →
40]
```

$\{x \rightarrow - 0.3664571903448024777581452462984821489343\}$

In order to use ImplicitPlot to help solve simultaneous equations, we first have to load the corresponding package:

```
<< Graphics'ImplicitPlot'
```

```
pl1 = ImplicitPlot[2 x^2 + y^3 == 3, {x, - 3, 3}];
```

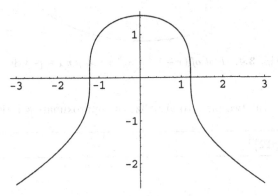

Fig. 3.3. *Plot of* $2x^2 + y^3 == 3$ *for* $x \in [-3, 3]$.

```
p12 = ImplicitPlot[(x - 1)^2 + 3 y^2 == 4, x, - 3, 3];
```

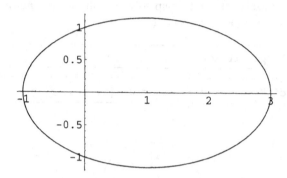

Fig. 3.4. *Plot of* $(x - 1)^2 + 3y^2 == 4$ *for* $x \in [-3, 3]$.

Let us combine the two plots to visualize the approximate solutions.

```
Show[{pl1, pl2}];
```

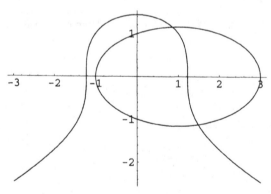

Fig. 3.5. *Plots of* $2x^2 + y^3 == 3$ *and* $(x - 1)^2 + 3y^2 == 4$ *for* $x \in [-3, 3]$.

```
eqns = {2 x^2 + y^3 == 3, (x - 1)^2 + 3 y^2 == 4};
```

The plot above shows that there exist two solutions close to $(x = 1, y = 1)$ and $(x = 1, y = -1)$.

```
sol1 = FindRoot[eqns, {x, 1}, {y, 1}]
```

$\{x \rightarrow 0.857916, y \rightarrow 1.15178\}$

```
sol2 = FindRoot[eqns, {x, 1}, {y, - 1}]
```

$\{x \rightarrow 1.48428, y \rightarrow - 1.12034\}$

We can check the solutions as above using the command **expression /. rule**.

```
eqns /. sol1
```

$\{$True, True$\}$

```
eqns /. sol2
```

$\{$True, True$\}$

If we know only that the value of a coefficient lies in a given interval, when solving the equation, we can use **Interval**; the solutions will then be expressed in terms of intervals.

```
Solve[ x^2 - 3 == Interval[{1, 6}], x]
```

$\{\{x \rightarrow Interval[\{- 3, - 2\}]\}, \{x \rightarrow Interval[\{2, 3\}]\}\}$

3.4 Vectors and Matrices

Matrix algebra was first developed in the mid-nineteenth century by the English mathematician Arthur Cayley (1821–1895).

Vectors and matrices are, respectively, represented by lists and lists of lists.

Here are a two-dimensional vector **v** and a 2×2 matrix **m**.

```
v = {x, y}
m = {{a, b}, {c, d}}
```

```
{x, y}
{{a, b}, {c, d}}
```

```
TableForm[m]
```

```
a    b
c    d
```

Matrix elements are parts of the list of lists

```
m[[1,1]]
m[[1,2]]
m[[2,1]]
m[[2,2]]
```

a

b

c

d

We can also use the function **Array** to define vectors, matrices, and, more generally, tensors.

```
Array[V, 5]
```

```
{V[1], V[2], V[3], V[4], V[5]}
```

The third element of **Array** gives the index origin if different from one.

```
Array[V, 5, 0]
```

```
{V[0], V[1], V[2], V[3], V[4]}
```

```
Array[M, {3, 5}]
```

```
{{M[1, 1], M[1, 2], M[1, 3], M[1, 4], M[1, 5]},
{M[2, 1], M[2, 2], M[2, 3], M[2, 4], M[2, 5]},
{ M[3, 1], M[3, 2], M[3, 3], M[3, 4], M[3, 5]}}
```

```
Array[T, {4, 3, 5}]
```

```
{{{T[1, 1, 1], T[1, 1, 2], T[1, 1, 3], T[1, 1, 4], T[1, 1, 5]},
{T[1, 2, 1], T[1, 2, 2], T[1, 2, 3], T[1, 2, 4], T[1, 2, 5]},
{T[1, 3, 1], T[1, 3, 2], T[1, 3, 3], T[1, 3, 4], T[1, 3, 5]}},
{{T[2, 1, 1], T[2, 1, 2], T[2, 1, 3], T[2, 1, 4], T[2, 1, 5]},
{T[2, 2, 1], T[2, 2, 2], T[2, 2, 3], T[2, 2, 4], T[2, 2, 5]},
{T[2, 3, 1], T[2, 3, 2], T[2, 3, 3], T[2, 3, 4], T[2, 3, 5]}},
{{T[3, 1, 1], T[3, 1, 2], T[3, 1, 3], T[3, 1, 4], T[3, 1, 5]},
{T[3, 2, 1], T[3, 2, 2], T[3, 2, 3], T[3, 2, 4], T[3, 2, 5]},
{T[3, 3, 1], T[3, 3, 2], T[3, 3, 3], T[3, 3, 4], T[3, 3, 5]}},
{{T[4, 1, 1], T[4, 1, 2], T[4, 1, 3], T[4, 1, 4], T[4, 1, 5]},
{T[4, 2, 1], T[4, 2, 2], T[4, 2, 3], T[4, 2, 4], T[4, 2, 5]},
{T[4, 3, 1], T[4, 3, 2], T[4, 3, 3], T[4, 3, 4], T[4, 3, 5]}}}}
```

Dimensions[expression] gives a list of the dimensions of expression.

```
Dimensions[%]
```

```
{4, 3, 5}
```

We can enter true subscripts using Subscript.

```
Table[Subscript[v, i], {i, 1, 5}]
Table[Subscript[t, i, j], {i, 1, 3}, {j, 1, 5}]
```

$\{v_1, v_2, v_3, v_4, v_5\}$

$\{\{t_{1,1}, t_{1,2}, t_{1,3}, t_{1,4}, t_{1,5}\},$
$\{t_{2,1}, t_{2,2}, t_{2,3}, t_{2,4}, t_{2,5}\},$
$\{t_{3,1}, t_{3,2}, t_{1,3}, t_{3,4}, t_{3,5}\}\}$

Using BasicTypesetting of the Palettes submenu we can enter indices and Greek letters. Greek letters can also be written entering \[LetterName], where LetterName is Alpha, Beta, and so on.

All the following command names have pretty clear meaning.

```
DiagonalMatrix[ α, β, γ] // MatrixForm
```

$$\begin{pmatrix} \alpha & 0 & 0 \\ 0 & \beta & 0 \\ 0 & 0 & \gamma \end{pmatrix}$$

```
IdentityMatrix[3]
```

{{1, 0, 0}, {0, 1, 0}, {0, 0, 1}}

```
v1 = {x1, y1}
v2 = {x2, y2}
v1 + v2
```

{x1, y1}

{x2, y2}

{x1 + x2, y1 + y2}

```
m.v
```

{a x + b y, c x + d y}

```
Det[m]
```

- b c + a d

```
Transpose[m]
```

{{a, c}, {b, d}}

```
Inverse[m]
```

$$\left\{\left\{\frac{d}{-b\ c+a\ d},\ -\frac{b}{-b\ c+a\ d}\right\},\ \left\{-\frac{c}{-b\ c+a\ d},\ \frac{a}{-b\ c+a\ d}\right\}\right\}$$

```
MatrixPower[m, 2]
```

$$\{\{a^2 + b\ c,\ a\ b + b\ d\},\ \{a\ c + c\ d,\ b\ c + d^2\}\}$$

```
Eigenvalues[m]
```

$$\left\{\frac{1}{2}\left(a+d-\sqrt{a^2+4\ b\ c-2\ a\ d+d^2}\right),\right.$$
$$\left.\frac{1}{2}\left(a+d+\sqrt{a^2+4\ b\ c-2\ a\ d+d^2}\right)\right\}$$

```
Eigenvectors[m]
```

$$\left\{\left\{-\frac{-a+d+\sqrt{a^2+4\ b\ c-2\ a\ d+d^2}}{2\ c},1\right\},\right.$$
$$\left.\left\{-\frac{-a+d-\sqrt{a^2+4\ b\ c-2\ a\ d+d^2}}{2\ c},1\right\}\right\}$$

The product of an (n_1, n_2)-matrix by an (n_2, n_3)-matrix is an (n_1, n_3)-matrix. An $(n, 1)$-matrix is a column vector, and a $(1, n)$-matrix is a row vector.

```
mA = Array[A, {3, 2}]
mB = Array[B, {2, 1}]
mA.mB
```

$\{\{A[1, 1], A[1, 2]\}, \{A[2, 1], A[2, 2]\}, \{A[3, 1], A[3, 2]\}\}$

$\{\{B[1, 1]\}, \{B[2, 1]\}\}$

$\{\{A[1, 1]\ B[1, 1] + A[1, 2]\ B[2, 1]\},$
$\{A[2, 1]\ B[1, 1] + A[2, 2]\ B[2, 1]\},$
$\{A[3, 1]\ B[1, 1] + A[3, 2]\ B[2, 1]\}\}$

```
MatrixForm[mA]
MatrixForm[mB
```

$$\begin{pmatrix} A[1,1], A[1,2] \\ A[2,1], A[2,2] \\ A[3,1], A[3,2] \end{pmatrix}$$

$$\begin{pmatrix} B[1,1], B[1,2] \\ B[2,1], B[2,2] \end{pmatrix}$$

```
Array[X,3,1]
Array[Y,1,3]
MatrixForm[Array[X,3,1]]
MatrixForm[Array[Y,1,3]]
```

{{X[1, 1]}, {X[2, 1]}, {X[3, 1]}}

{{Y[1, 1], Y[1, 2], Y[1, 3]}}

$$\begin{pmatrix} X[1,1] \\ X[2,1] \\ X[3,1] \end{pmatrix}$$

(Y[1, 1] Y[1, 2] Y[1, 3])

For more evolved matrix manipulations, load the package:

```
<< LinearAlgebra'MatrixManipulation'
```

Here are a few examples.

```
A = {{a11, a12}, {a21, a22}};
B = {{b11, b12}, {b21, b22}};
```

```
AppendColumns[A, B] // MatrixForm
```

```
a11   a12
a21   a22
b11   b12
b21   b22
```

AppendRows[A, B] //[1mm] MatrixForm

```
a11   a12   b11   b12
a21   a22   b21   b22
```

BlockMatrix[{{A, B}, {B, A}}] // MatrixForm

```
a11   a12   b11   b12
a21   a22   b21   b22
b11   b12   a11   a12
b21   b22   a21   a22
```

UpperDiagonalMatrix[M, 3] // MatrixForm

```
M[1, 1]   M[1, 2]   M[1, 3]
0         M[2, 2]   M[2, 3]
0         0         M[3, 3]
```

LowerDiagonalMatrix[M, 3] // MatrixForm

```
M[1, 1]   0         0
M[2, 1]   M[2, 2]   0
M[3, 1]   M[3, 2]   M[3, 3]
```

LinearEquationsToMatrices[
a11 x + a12 y == b1,
a21 x + a22 y == b2, x, y]

{{{a11, a12}, {a21, a22}}, {b1, b2}}

We don't need to load the package to execute the inverse command, we just
use the command Thread.

```
A = {{a11, a12}, {a21, a22}};
v = {x, y};
b = {b1, b2};
Thread[A. {x, y} == b]
```

{a11 x + a12 y == b1, a21 x + a22 y == b2}

4

Analysis

4.1 Differentiation

4.1.1 Partial Derivative

The function D[f[a,x,y],x] does partial differentiation with respect to a variable (here x).

```
D[x^(3 / 2) y^2, x]
```

$$\frac{3 \sqrt{x} \ y^2}{2}$$

If we define a function f of one variable, we can use, for the derivative, the notation f' but, as shown below, we have to be careful.

```
f[x_] := x^(2 / 3)
f'[x]
f'
```

$$\frac{2}{3 \ x^{1/3}}$$

$$\frac{2}{3 \ \#^{1/3}} \ \&$$

The last representation, associated with f', is a pure function.

Using the submenu Palettes $\rightarrow$ BasicInput, $D[f[x], x]$ can also be entered as $\partial_x f[x]$.

$$\partial_x\ x^{\wedge}(3/2)\ y^{\wedge}2$$

$$\frac{3\ \text{Sqrt}[x]\ y^2}{2}$$

D also does multiple differentiation with respect to different variables. The command below evaluates the second derivative with respect to x and the first one with respect to y.

$$\{D[x^{\wedge}(3/2)\ y^{\wedge}2,\ \{x,\ 2\},\ \{y,\ 1\}],\ D[x^{\wedge}(3/2)\ y^{\wedge}]2,\ x,\ x,\ y]\}$$

$$\left\{\ \frac{3\ y}{2\ \sqrt{x}},\ \frac{3\ y}{2\ \sqrt{x}}\ \right\}$$

Here are a few more examples illustrating how, using *Mathematica*, we can avoid doing rather tedious calculations.

$$D[\text{Cos}[\text{Log}[\text{Sqrt}[x]]],\ x,\ x]$$

$$\frac{-\text{Cos}[\text{Log}[\text{Sqrt}[x]]]}{4\ x^2}\ +\ \frac{\text{Sin}[\text{Log}[\text{Sqrt}[x]]]}{2\ x^2}$$

$$D[\text{Sin}[2\ x]^{\wedge}3\ \text{Cos}[x^{\wedge}2],\ x]$$

$$6\ \text{Cos}[2\ x]\ \text{Cos}[x^2]\ \text{Sin}[2\ x]^2\ -\ 2\ x\ \text{Sin}[2\ x]^3\ \text{Sin}[x^2]$$

$$D[\text{Sin}[2\ x]^{\wedge}3\ \text{Cos}[x^{\wedge}2],\ x,\ 10]$$

Long output suppressed.

$$D[f[x]^{\wedge}2,\ \{x,\ 5\}]$$

$$20\ f''[x]\ f^{(3)}[x]\ +\ 10\ f'[x]\ f^{(4)}[x]\ +\ 2\ f[x]\ f^{(5)}[x]$$

```
D[f[x^3], x]
```

$3 \ x^2 \ f'[x^3]$

Mathematica knows most special functions of mathematical physics.

```
D[BesselJ[3, x], x]
```

$\dfrac{1}{2} \ (BesselJ[2, \ x] - Bessel[4, \ x])$

```
D[LegendreP[5, x], {x, 2}
```

$-\dfrac{105 \ x}{2} + \dfrac{315 \ x^3}{2}$

The command Derivative[k, l][f][a, b] takes the derivative of $f(x,y)$ k times with respect to x and l times with respect to y and evaluates this derivative for $x = a$ and $y = b$.

```
f[x_, y_] := Tan[x - y] / Cos[x + y];
Derivative[2, 3][f][Pi / 6, Pi / 4] // Simplify
```

$19 \ Sqrt[2] \ (5009 - 5447 \ Sqrt[3])$

4.2 Total Derivative

Dt[f, x] gives the total derivative df/dx.

```
Dt[x^2 y^3]
```

$2 \ x \ y^3 \ Dt[x] + 3 \ x^2 \ y^2 \ Dt[y]$

We can set a variable (here a) to be a constant

```
SetAttributes[a, Constant]
```

```
Dt[a x^2 y^3]
```

$$2 \text{ a x } y^3 \text{ Dt[x]} + 3 \text{ a } x^2 \text{ } y^2 \text{ Dt[y]}$$

We can replace Dt[x] and Dt[y] by dx and dy, respectively, in order to obtain the following more traditional form.

```
% /. {Dt[x] → dx, Dt[y] → dy}
```

$$3 \text{ a dy } x^2 \text{ } y^2 + 2 \text{ a dx x } y^3$$

4.3 Integration

4.3.1 Indefinite Integrals

Integrate[f[x], x] gives the indefinite integral $\int f(x)\, dx$. Here are a few examples.

```
Integrate[1 / (1 + x^2)^2, x]
```

$$\frac{1}{2} \left(\frac{x}{1+x^2} + \text{ArcTan[x]} \right)$$

Using the submenu Palettes → BasicInput, Integrate[f[x], x] can also be entered as $\int$ f[x] dx.

Here are a few examples.

```
Integrate[Log[x^2], x]
```

$$- 2 \text{ x } + \text{ x Log}[x^2]$$

```
Integrate[x^2 Log[x], x]
```

$$\frac{1}{9} x^3 (- 1 + 3 \text{ Log[x]})$$

```
Integrate[(1 - Cos[x]) / (1 + Cos[x]), x]
```

$$- x + 2 \, \text{Tan}\left[\frac{x}{2}\right]$$

```
Integrate[Sin[Log[x]], x]
```

$$\frac{1}{2} \, x \, (\text{Cos}[\text{Log}[x]] - \text{Sin}[\text{Log}[x]])$$

```
Integrate[Cos[Log[x]], x]
```

$$\frac{x \, (\text{Cos}[\text{Log}[x]] + \text{Sin}[\text{Log}[x]])}{2}$$

```
Integrate[BesselJ[1, x], x]
```

$$- \text{BesselJ}[0, x]$$

4.3.2 Definite Integrals

Integrate[f[x], x, x1, x2] gives the definite integral $\int_{x1}^{x2} f(x) \, dx$.

```
Integrate[Exp[x] Sin[x], {x, 0, Pi}]
```

$$\frac{1 + E^{Pi}}{2}$$

Using the submenu Palettes → BasicInput, Integrate[f[x], x, x1, x2] can be entered as $\int_{x1}^{x2} f[x] \, dx$.

We can use *Mathematica* to display the output in TraditionalForm. Here are a few examples.

```
Integrate[( x - 1) / Log[x], {x, 0, 1}] // TraditionalForm
```

log(2)

Integrate[Log[x] / (x - 1), {x, 0, 1}] // TraditionalForm

$$\frac{\pi^2}{6}$$

(Log[x])^2 / (1 + x + x^2), {x, 0, Infinity}] //
TraditionalForm

$$\frac{16 \, \pi^3}{81 \, \sqrt{3}}$$

Mathematica can also deal with nonelementary functions.

Integrate[Cos[Exp[x]], {x, 0, 1}]

— CosIntegral[1] + CosIntegral[E]

CosIntegral[x] is defined by — Integrate[Cos[u] / u, {u, x, Infintextttity}] and SinIntegral[x] by Integrate[Sin[u] / u, {u, 0, x}]. These functions are, respectively, denoted $\mathrm{Ci}(x)$ and $\mathrm{Si}(x)$.

Plot[CosIntegral[x], {x, 1, E}];

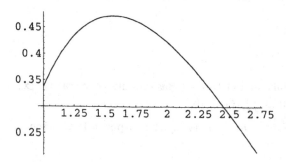

Fig. 4.1. *Graph of* $\mathrm{Ci}(x)$ *for* $x \in [1, \mathrm{e}]$.

Here are other nontrivial examples.

```
Integrate[Log[Cot[x]] , {x, 0, Pi/4}]
```

Catalan

Catalan is Catalan's constant C. Its numerical value is

```
Catalan // N
```

0.915966

This constant is defined by the series:

$$C = \sum_0^\infty \frac{(-1)^k}{(2k+1)^2}.$$

```
Integrate[Exp[- x^3] Sqrt[1 - x^3], {x, 0, 1}]
```

$$\frac{\text{Sqrt}[\text{Pi}] \ \text{Gamma}[\frac{1}{3}] \ \text{Hypergeometric1F1}[\frac{1}{3}, \frac{11}{6}, -1]}{6 \ \text{Gamma}[\frac{11}{6}]}$$

where Gamma[z] represents the Euler function Γ defined for $\text{Re}(z) > 0$, by

$$\Gamma(z) = \int_0^\infty t^{z-1} e^{-t} \, dt.$$

and Hypergeometric1F1 is the Kummer confluent hypergeometric function $_1F_1$ defined by

$$_1F_1(a; b; z) = \sum_{k=0-}^\infty \frac{(a)_k}{(b)_k} \frac{z^k}{k!}.$$

4.3.3 Numerical Integration

NIntegrate[f[x] , {x, x1, x2}] gives a numerical approximate value of the definite integral $\int_{x1}^{x2} f(x) \, dx$.

Here is a definite integral that cannot be evaluated exactly.

```
Integrate[1 / Cos[Tan[x]], {x, 0, Pi/4}]
```

$$\text{Integrate}[\text{Sec}[\text{Tan}[x]], \{x, 0, \frac{Pi}{4}\}]$$

Mathematica can, however, find a numerical approximate value of this definite integral.

```
NIntegrate[1 / Cos[Tan[x]], {x, 0, Pi/4}]
```

0.927485

WorkingPrecision is an option for various numerical operations that specifies how many digits of precision should be maintained in internal computations.

```
NIntegrate[Exp[-x^2], {x, - 5, 5}, WorkingPrecision → 50]
```

1.7724538509027909507649211099378135487892

Sometimes we may encounter problems of convergence as in the following example.

```
NIntegrate[Exp[-x^2], {x, - 500, 500}]
```

Numerical integration converging too slowly;
suspect one of the following: singularity,
value of the integration being 0, oscillatory
integrand, or insufficient WorkingPrecision.
If your integrand is oscillatory try using the
option Method -> Oscillatory in NIntegrate. More ...
NIntegrate::ncvb :
NIntegrate failed to converge to
prescribed accuracy after 7 recursive
bisections in x near x = -3.90625. More ...
0.88631

```
??NIntegrate
```

NIntegrate[f, {x, xmin, xmax}] gives a numerical
approximation to the integral of f with
respect to x from xmin to xmax. More ...
Attributes[NIntegrate] = {HoldAll, Protected}
Options[NIntegrate] = {AccuracyGoal → ∞, Compiled → True
EvaluationMonitor → None, GaussianPoints → Automatic,
MaxPoints → Automatic, MaxRecursion → 6,
Method → Automatic, MinRecursion → 0,
PrecisionGoal → Automatic, SingularityDepth → 4,
WorkingPrecision → MachinePrecision}

Increasing MinRecursion, MaxRecursion, and WorkingPrecision yields a much better result.

```
NIntegrate[Exp[- x^2], {x, - 500, 500},
MinRecursion → 5, MaxRecursion → 20,
WorkingPrecision → 30]
```

1.7724538509055160273

Breaking the integration interval into pieces explicitly covering the peak may help obtain an accurate answer.

```
NIntegrate[Exp[- x^2], {x, - 500, - 5, 5, 500},
WorkingPrecision → 30]
```

1.77245385090551602730

We can also evaluate the area between two curves as illustrated below.

```
<<Graphics`
```

```
f1[x_] := x^4
f2[x_] := 10 - x^3
NSolve[f1[x] == f2[x], x]
```

{{x -> - 2.09209}, {x -> - 0.240141 - 1.72692 I},
{x -> - 0.240141 + 1.72692 I}, {x -> 1.57237}}

FilledPlot[{f1[x], f2[x]}, {x, - 2.09209, 1.57237}];

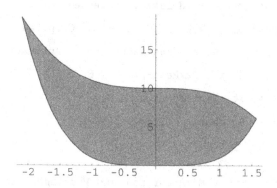

Fig. 4.2. *Area between two curves.*

The area between the two curves represented above is given by

---

NIntegrate[Abs[f2[x] - f1[x]], {x, - 2.09209, 1.57237}]

29.9679

or, because we have to integrate polynomial functions,

Integrate[f2[x], {x, - 2.09209, 1.57237}] - Integrate[f1[x],
{x, - 2.09209, 1.57237}]

29.9679

4.3.4 Multiple Integrals

Integrate[f[x, y], {x, x1, x2}, {y, y1, y2}] gives the multiple integral $\int_{x1}^{x2} dx \int_{y1}^{y2} dy f(x, y)$.

```
Integrate[x^2 y^2, {x, 0, a}, {y, 0, b}]
```

$$\frac{a^3 b^3}{9}$$

```
NIntegrate[Cos[x y], {x, 0, Pi / 2}, {y, 0, Pi / 2}]
```

1.77049

4.4 Differential Equations

Mathematica gives exact solutions to linear differential equations with constant coefficients.

```
DSolve[y''[x] - 3 y'[x] + 2 y[x] == Exp[x], y[x], x]
```

$$\{\{y[x] \rightarrow - E^x (1 + x) + E^x C[1] + E^{2x} C[2]\}\}$$

Mathematica can solve equations with nonconstant coefficients (Bernoulli).

```
DSolve[(1 - x^2) y'[x] - 2 x y[x] == x^2, y[x], x]
```

$$\left\{\left\{ y[x] \rightarrow \frac{-x^3}{3 (-1 + x^2)} + \frac{C[1]}{-1 + x^2}\right\}\right\}$$

```
DSolve[y'[x] == x y[x] + x^2 y[x]^2, y[x], x]
```

$$\left\{\left\{ y[x] \rightarrow \frac{-2 E^{x^2/2}}{2 E^{x^2/2} x - 2 C[1] - \text{Sqrt}[2Pi] \text{Erfi}[\frac{x}{\text{Sqrt}[2]}]}\right\}\right\}$$

Erfi[z] gives the imaginary error function erf(iz)/i. The error function erf is defined by

$$\mathrm{erf}(z) = \frac{2}{\sqrt{\pi}} \int_0^z e^{-t^2}\, dt.$$

The above equation can be written $y'/y^2 = x/y + x^2$ and if $z = 1/y$ we obtain the linear equation $z' + xz + x^2 = 0$, easily solved using *Mathematica*.

4.4.1 Solving nonelementary ODE

Mathematica recognizes a Bessel-type differential equation.

```
DSolve[x y''[x] + y'[x] + x y[x] == 0, y[x], x]
```

$\{\{y[x] \rightarrow \mathtt{BesselJ}[0, x]\ C[1] + \mathtt{BesselY}[0, x]\ C[2]\}\}$

We can plot the solutions.

```
Plot[{BesselJ[0, x], BesselY[0, x]}, {x, 0.1, 10},
PlotStyle → {RGBColor[1, 0, 0], RGBColor[0, 0, 1]}];
```

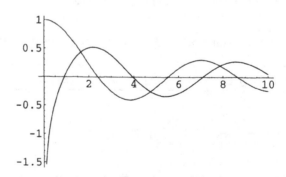

Fig. 4.3. *Graph of the Bessel functions $J_0(x)$ and $Y_0(x)$ for $x \in [0.1, 10]$.*

We took the initial point of the x-interval equal to 0.1 because—as we can infer from its graph—$Y_0(x)$ is singular at the origin.

4.4.2 Numerical Solutions

The following system of differential equations can only be solved numerically.

```
solution = NDSolve[{x'[t] == - 2 y[t] + (x[t])^2,
y'[t] == x[t] - y[t], x[0] == y[0] == 1}, {x, y}, {t, 0, 10}]
```

```
{{x -> InterpolatingFunction[{{0., 10.}}, <>],
y -> InterpolatingFunction[{{0., 10.}}, <>]}}
```

We can plot the solution.

```
ParametricPlot[Evaluate[{x[t], y[t]} /. solution], {t, 0,
10}]
```

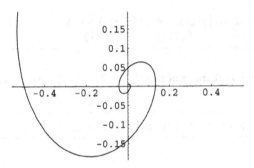

Fig. 4.4. *Parametric plot of the solution of the system* $x' = -2y + x^2, y' = x - y$
for the initial conditions $x(0) = y(0) = 1$ *in the interval* $t \in [0, 10]$.

Here is an example of multiple solutions.

```
Clear[solution]
solution = NDSolve[y'[x]^2 == Sin[2x], y[0] == 0, y[x], {x, 0,
1}]
```

```
{{y[x] -> InterpolatingFunction[{{0., 1.}}, <> ][x]},
{y[x] -> InterpolatingFunction[{{0., 1.}}, <> ][x]}}
```

```
Plot[Evaluate[y[x] /. solution], {x,0,1}];
```

We can solve a differential equation with a discontinuous derivative.

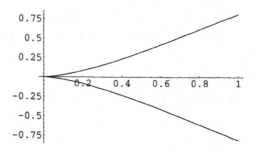

Fig. 4.5. *Plot of the solution of the ODE* $(y')^2 = \sin(x)$ *for the initial condition* $y(0) = 0$ *in the interval* $x \in [0, 1]$.

```
Clear[solution]
solution = NDSolve[{y'[x] == If[x < 0, - 1 / (x - 2)^2, 1 / (x
- 2)^2], y[0] == 0}, y[x], {x, - 2, 1}]
```

```
{{y[x] -> InterpolatingFunction[{{- 2., 1.}}, <> ][x]}}
```

```
Plot[Evaluate[y[x] /. solution], {x, - 2, 1}];
```

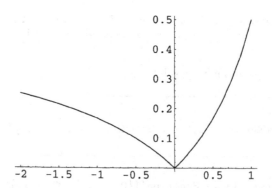

Fig. 4.6. *Plot of the solution of the ODE* $y' = -1/(x-2)^2$, *if* $x < 0$ *and* $1/(x-2)^2$, *if* $x > 0$, *for the initial condition* $y(0) = 0$ *in the interval* $x \in [-2, 1]$.

4.4.3 Series Solutions

Series solutions assume that y[x] is given by a power series with unknown coefficients in x. This series is built up with SeriesData[x, x0, {a0, a1, a2, ...}]

```
y[x_] := SeriesData[x, 0, Table[a[i], {i, 0, 7}]]
seriesDE = (y'[x])^2 - y[x] == x
```

$(- 1 + a[1]^2) + (- a[1] + 4\ a[1]\ a[2])\ x\ +$

$(- a[2] + 4\ a[2]^2 + 6\ a[1]\ a[3])\ x^2\ +$

$(- a[3] + 12\ a[2]\ a[3] + 8\ a[1]\ a[4])\ x^3\ +$

$(9\ a[3]^2 - a[4] + 16\ a[2]\ a[4] + 10\ a[1]\ a[5])\ x^4\ +$

$(24\ a[3]\ a[4] - a[5] + 20\ a[2]\ a[5] + 12\ a[1]\ a[6])\ x^5\ +$

$(16\ a[4]^2 + 30\ a[3]\ a[5] - a[6] + 24\ a[2]\ a[6]\ +$

$14\ a[1]\ a[7])\ x^6 + O[x]^7$

LogicalExpand[expression] expands expression containing logical operators such as && and || standing, respectively, for AND and OR.

```
coeffEqn = LogicalExpand[seriesDE]
```

$- 1 + a[1]^2 == 0\ \&\&\ - 1 - a[1] + 4\ a[1]\ a[2] == 0\ \&\&$

$- a[2] + 4\ a[2]^2 + 6\ a[1]\ a[3] == 0\ \&\&$

$- a[3] + 12\ a[2]\ a[3] + 8\ a[1]\ a[4] == 0\ \&\&$

$9\ a[3]^2 - a[4] + 16\ a[2]\ a[4] + 10\ a[1]\ a[5] == 0\ \&\&$

$24\ a[3]\ a[4] - a[5] + 20\ a[2]\ a[5] + 12\ a[1]\ a[6] == 0\ \&\&$

$16\ a[4]^2 + 30\ a[3]\ a[5] - a[6] + 24\ a[2]\ a[6] + 14\ a[1]\ a[7]$

$== 0$

We solve using the initial condition: a[0] == 1.

```
?Solve
```

Solve[eqns, vars] attempts to solve an equation

or set of equations for the variables vars. Solve[eqns, vars, elims] attempts to solve the equations for vars, eliminating the variables elims. More ...

```
coeffSol = Solve[{coeffEqn, a[0] == 1},
Table[a[i], {i, 1, 7}]]
```

$$\{\{a[7] \rightarrow -\frac{6889}{161280}, \ a[6] \rightarrow \frac{469}{11520}, \ a[5] \rightarrow -\frac{41}{960}, \ a[4] \rightarrow \frac{5}{96},$$
$$a[3] \rightarrow -\frac{1}{12}, \ a[1] \rightarrow 1, \ a[2] \rightarrow \frac{1}{2}\}, \ \{a[7] \rightarrow 0 \ , \ a[6] \rightarrow 0,$$
$$a[5] \rightarrow 0, \ a[4] \rightarrow 0, \ a[3] \rightarrow 0, \ a[1] \rightarrow -1, \ a[2] \rightarrow 0 \ \}\}$$

We find two solutions.

```
coeffSol[[1]]
```

$$\{a[7] \rightarrow -\frac{6889}{161280}, \ a[6] \rightarrow \frac{469}{11520}, \ a[5] \rightarrow -\frac{41}{960}, \ a[4] \rightarrow \frac{5}{96},$$
$$a[3] \rightarrow -\frac{1}{12}, \ a[1] \rightarrow 1, \ a[2] \rightarrow \frac{1}{2}\}$$

Substituting we obtain

```
seriesSol1 = y[x] /. Join[coeffSol[[1]], {a[0] → 1}]
```

$$1 + x + \frac{x^2}{2} - \frac{x^3}{12} + \frac{5\ x^4}{96} - \frac{41\ x^5}{960} + \frac{469\ x^6}{11520} - \frac{6889\ x^7}{161280} + O[x]^8$$

```
seriesSol1 = y[x] /. Join[coeffSol[[2]], {a[0] → 1}]
```

$$1 - x + O[x]^8$$

Note in the input cell above the use of the command Join to concatenate two lists.

Checking:

```
seriesDE /. coeffSol
```

$$\{x + O[x]^7 == x, \; x + O[x]^7 == x\}$$

The function **Normal** applied to an expression converts this expression to a normal form, that is, a polynomial form without the term $O[x]^8$.

```
sol1 = Normal[seriesSol1]
```

$$1 + x + \frac{x^2}{2} - \frac{x^3}{12} + \frac{5\,x^4}{96} - \frac{41\,x^5}{960} + \frac{469\,x^6}{11520} - \frac{6889\,x^7}{161280}$$

```
Plot[sol1, {x, 0, 3}]
```

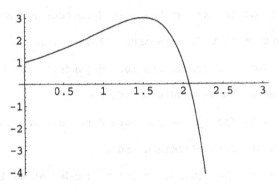

Fig. 4.7. *Plot of the first series solution of the ODE* $(y')^2 - y = x$, *for the initial condition* $y(0) = 1$ *in the interval* $x \in [0,3]$.

4.4.4 Differential Vector Equations

DSolve[equation,y,x] does not work when the unknown function **y** is a vector, that is a list, as shown below.

We first load the package **Calculus'VectorAnalysis** in order to be able to use the command **CrossProduct**.

```
<< Calculus'VectorAnalysis'
```

We then define the vector position of the electric charge, the electric and magnetic fields, and the vector force.

```
r[t_] := {x[t], y[t], z[t]}
electricField = 0, Efield, 0;
magneticField = {0, 0, Bfield};
force = q (electricField + CrossProduct[r'[t],
magneticField]);
```

Finally we enter the equation that determines the motion of the electric point charge in the electromagnetic field.

```
DSolve[{m r''[t] == force, r[0] == {0, 0, 0},
r'[0] == {v1, v2, v3}}, r[t], t] //.{Bfield → m ω / q,
Efield → V Bfield} // ExpandAll // Simplify
```

DSolve::underdet: The system has more dependent variables than

equations, and so is underdetermined. More . . .

DSolve::underdet: The system has more dependent variables than

equations, and so is underdetermined. More . . .

DSolve::underdet: The system has more dependent variables than

equations, so is underdetermined. More . . .

General::stop: Further output of DSolve::underdet will be

suppressed during this calculation. More . . .

DSolve[{m r''[t] == force, r[0] == {0, 0, 0},

r'[0] == {v1, v2, v3}}, r[t], t] //.{Bfield → m ω / q,

Efield → V Bfield} // ExpandAll // Simplify

One has to transform the vector equation into a list of scalar equations using Thread.

```
Clear[eqns]
eqns = Map[Thread, {m r''[t] == force, r[0] == {0, 0, 0},
r'[0] == {v1, v2, v3}}] //. {Bfield → m ω / q,
Efield → V Bfield} // ExpandAll
// Simplify
```

$\{\{m\ x''[t] == m\ \omega\ y't], m\ y''[t] == m\ \omega\ (V - x'[t]),$
$m\ z''[t] == 0\}, \{x[0] == 0, y[0] == 0, z[0] == 0\},$
$\{v1 == x'[0], v2 == y'[0], v3 == z'[0]\}\}$

We can then enter the equation under the form above and DSolve will work.

```
solution1 = DSolve[eqns , {x[t], y[t], z[t]}, t] // ExpandAll
// Simplify
```

$$\{\{x[t] \rightarrow \frac{v2 + t\ V\ \omega - v2\ Cos[t\ \omega] + (-V + v1)\ Sin[t\ \omega]}{\omega},$$

$$y[t] \rightarrow \frac{V - v1 + (-V + v1)\ Cos[t\ \omega] + v2\ Sin[t\ \omega]}{\omega},$$

$$z[t] \rightarrow t\ v3\}\}$$

We can check that this method gives the correct result.

```
solution2 = DSolve[{m x''[t] == q y'[t] Bfield,
m y''[t] == q Efield - q x'[t] Bfield,
m z''[t] == 0, x[0] == y[0] == z[0] == 0,
x'[0] == v1, y'[0] == v2, z'[0] == v3},
{x[t], y[t], z[t]}, t] //. {Bfield → m ω / q,
Efield → V Bfield} // ExpandAll // Simplify
```

$$\{\{x[t] \rightarrow \frac{v2 + t\ V\ \omega - v2\ Cos[t\ \omega] + (-V + v1)\ Sin[t\ \omega]}{\omega},$$

$$y[t] \rightarrow \frac{V - v1 + (-V + v1)\ Cos[t\ \omega] + v2\ Sin[t\ \omega]}{\omega},$$

$$z[t] \rightarrow t\ v3\}\}$$

```
solution1 == solution2
```

True

There exists a comprehensive *Mathematica* package called `VisualDSolve` that provides a wide variety of tools for the visualization of solutions to ordinary differential equations. The interested reader should refer to Schwalbe and Wagon [50].

4.5 Sum and Products

4.5.1 Exact Results

`Sum[f[k], {k, n}]` evaluates the sum $\sum_{k=1}^{n} f(k)$.

```
Sum[1 / n^2, {n, Infinity}]
```

$$\frac{Pi^2}{6}$$

```
Sum[1/ n^4, {n, Infinity}]
```

$$\frac{Pi^4}{90}$$

We mentioned above that Catalan's constant is defined by the series $C = \sum_{k=0}^{\infty} (-1)^k/(2k+1)^2$. We can use *Mathematica* to check this definition.

```
Sum[(-1)^k /(2 k +1)^2, {k, 0, Infinity}] == Catalan
```

True

`Sum[f[k], {k, n1, n2}]` evaluates the sum $\sum_{k=n_1}^{n_2} f(k)$.

```
Sum[1 / (1 + n^2), {n, - Infinity, Infinity}]
```

```
Pi Coth[Pi]
```

```
Sum[(- 1)^n / (2 n + 1), {n, 0, Infinity}]
```

$$\frac{Pi}{4}$$

Product[f[k], {k, n}] evaluates the product $\prod_{k=1}^{n} f(k)$.

```
Product[1 + 1 / n^2, {n, 1, Infinity}]
```

$$\frac{Sinh[Pi]}{Pi}$$

Product[f[k], {k, n1, n2}] evaluates the product $\prod_{k=n_1}^{n_2} f(k)$.

```
Product[ 1 - 1 / n^3, {n, 2, Infinity}]
```

$$\frac{Cosh[\frac{Sqrt[3]\ Pi}{2}]}{3\ Pi}$$

4.5.2 Numerical Results

```
Sum[1/ n^3, {n, 1, Infinity}]
```

Zeta[3]

```
?Zeta
```

Zeta[s] gives the Riemann zeta function. Zeta[s, a]
gives the generalized Riemann zeta function. More ...

The zeta and generalized zeta functions are, respectively, defined by

$$\zeta(s) = \sum_{k=1}^{\infty} k^{-s}, \text{and } \zeta(s,a) = \sum_{k=0}^{\infty} (k+a)^{-s}, (k+a \neq 0).$$

To obtain the numerical value of sum we have to use NSum.

```
NSum[1 / n^3, {n, 1, Infinity}]
```

1.20206

Because *Mathematica* recognizes that $\sum_{k=1}^{\infty} k^{-s}$ is $\zeta(s)$, to obtain a numerical value, we could also have entered

```
NSum[1 / n^3, {n, 1, Infinity}] // N
```

1.20206

```
Sum[1 / (1 + n^2), {n, - Infinity, Infinity}]
NSum[1 / (1 + n^2), {n, - Infinity, Infinity}]
```

Pi Coth[Pi]

3.15335

```
Sum[1 / (1 + n^3), {n, 0, Infinity}]
```

$$- \frac{1}{3} \ \text{RootSum}[1 + \#1^3 \ \&, \ \frac{\text{PolyGamma}[0, -\#1]}{\#1^2} \ \&]$$

```
?PolyGamma
```

`PolyGamma[z]` gives the digamma function psi(z).

`PolyGamma[n, z]` gives the nth derivative of the digamma function.

More ...

The `PolyGamma` function ψ is defined by $\psi(z) = \Gamma'(z)/\Gamma(z)$, where Γ is the Euler function.

`RootSum[f, form]` represents the sum of `form[x]` for all x that satisfy the polynomial equation `f[x] == 0`. In the present case, we obtain

```
- Apply[Plus, Map[PolyGamma[0, -#] / #^2 &,
{- 1, (-1)^(1/3), - (-1)^(2/3)}] // N] / 3
```

$1.6865 - 2.59052 \ 10^{-16} \ I$

and using Chop

```
Chop[%]
```

1.6865

This result can be obtained directly using NSum

```
NSum[1 / (1 + n^3), {n, 0, Infinity}]
```

1.6865

4.6 Power Series

Series[f[x], {x, x0, n}] gives a power series expansion for f[x] in a neighborhood of the point x0 to order $(x - x0)^n$.

```
Series[Sin[x], {x, 0, 10}]
```

$$x - \frac{x^3}{6} + \frac{x^5}{120} - \frac{x^7}{5040} + \frac{x^9}{362880} + O[x]^{11}$$

```
Series[Log[x], {x, 1, 5}]
```

$$(x - 1) - \frac{1}{2} (x - 1)^2 + \frac{1}{3} (x - 1)^3 - \frac{1}{4} (x - 1)^4 +$$

$$\frac{1}{5} (x - 1)^5 + O[x - 1]^6$$

```
Series[Exp[- x] / (1+x), {x, 0, 5}]
```

$$1 - 2x + \frac{5x^2}{2} - \frac{8x^3}{3} + \frac{65x^4}{24} - \frac{163x^5}{60} + O[x]^6$$

```
Series[BesselJ[0, x], {x, 0, 10}]
```

$$1 - \frac{x^2}{4} + \frac{x^4}{64} - \frac{x^6}{2304} + \frac{x^8}{147456} - \frac{x^{10}}{14745600} + O[x]^{11}$$

As mentioned above, Normal transforms the series expansion into an ordinary expression (polynomial).

```
Normal[Series[Exp[- x] / (1+x), {x, 0, 5}]]
```

$$1 - 2x + \frac{5x^2}{2} - \frac{8x^3}{3} + \frac{65x^4}{24} - \frac{163x^5}{60}$$

4.7 Limits

Limit[expression, x → x0] finds the limit of expression when x tends to x0.

```
Limit[Sin[x] / x, x → 0]
```

1

```
Limit[(1 - Cos[x]) / x^2, x → 0]
```

$$\frac{1}{2}$$

```
Limit[(Cosh[x] - 1) / (Sinh[x] Tanh[x]), x → 0]
```

$$\frac{1}{2}$$

```
Limit[BesselJ[0, x] / Cos[x], x → 0]
```

1

If Limit cannot handle a limit, it is worthwhile to load the package Calculus `Limit` that greatly expands the capabilities of Limit and try using Limit again. The built-in function Limit computes limits using symbolic and analytic methods. The command NLimit[f[x], x → x0] contained in the package NumericalMath`NLimit` calculates a sequence of values for the function $f(x)$ with successively smaller step sizes and then extrapolates to the limit to find an approximate numerical value. Here are two illustrative examples.

The golden ratio, that is, the ratio $r = a/b < 1$ such that

$$\frac{a}{b} = \frac{b}{a+b} \text{ or } r = \frac{1}{1+r},$$

which, according to the Italian mathematician Luca Pacioli (1445–1517), is the *Divine Proportion*.[1] Its numerical value is obtained solving the equation.

```
Solve[r == 1 / (1+r), r]
```

$$\{\{r \to \frac{1}{2} (-1 - \text{Sqrt}[5])\}, \{r \to \frac{1}{2} (-1 + \text{Sqrt}[5])\}\}$$

r is the positive root, $r = \sqrt{5} - 1/2 \approx 0.618034$. From the relation

$$r = \frac{1}{1+r},$$

we get the following sequence of equalities related to the continued fraction representation of r, that is,

---

[1]*De Divina Proportione* is the title of a book Pacioli wrote in 1496 but published in Venice in 1509. The book is illustrated with drawings by Leonardo da Vinci. Pacioli, a Franciscan monk, who taught mathematics at the University of Perugia, worked with all major artists of the Quattrocento such as Leonardo da Vinci (1452–1519), Piero Della Francesca (1420–1492), and Leon Battista Alberti (1404–1472). Pacioli is also considered the father of accounting.

$$r = \frac{1}{1+r} = \cfrac{1}{1+\cfrac{1}{1+r}}$$

$$= \cfrac{1}{1+\cfrac{1}{1+\cfrac{1}{1+r}}}$$

$$= \cfrac{1}{1+\cfrac{1}{1+\cfrac{1}{1+\cfrac{1}{r}}}}$$

$$= \cdots.$$

Note. The golden ratio is, sometimes, defined by the inverse of r (i.e., $1+r$). The result above can also be obtained using `NestList`.

```
golden[r_] := 1 / (1+r)
NestList[golden, r, 3]
```

$$\left\{r,\ \frac{1}{1+r},\ \cfrac{1}{1+\cfrac{1}{1+r}},\ \cfrac{1}{1+\cfrac{1}{1+\cfrac{1}{1+r}}}\right\}$$

The representation of r as a continued fraction is therefore $(0,1,1,1,\ldots)$ which can be also found using the *Mathematica* function `FromContinuedFraction`.

```
FromContinuedFraction[Join[0, Table[1, {100}]]] // N
```

0.618034

Let us see if the command `Limit` gives us the exact result.

```
Limit[Nest[1 / (1 + #) &, 1, n],
n → Infinity]
```

Nest::intnm:

Nonnegative machine-size integer expected at position 3

in Nest[$\frac{1}{1+\#1}$] & , 1, n]. More . . .

$$\texttt{Limit[Nest[}\frac{1}{1+\#1}\texttt{] \& , 1, n], n} \rightarrow \texttt{Infinity]}$$

Let us try again loading first the package `Calculus'Limit'`

```
<<Calculus'Limit'
```

```
Limit[Nest[1 / (1 + #) &, 1, n], n -> Infinity]
```

$$\frac{-1 + \text{Sqrt}[5]}{2}$$

In the limit $n \rightarrow \infty$, it is clear that the continued fraction $(0, a, a, a, \ldots)$ does not depend upon the value of a. This can be verified using `Limit`.

```
Limit[Nest[1 / (1 + #) &, Random[Integer, {1, 100}], n], n ->
Infinity]
```

$$\frac{-1 + \text{Sqrt}[5]}{2}$$

The Euler constant γ—called `EulerGamma` in *Mathematica*—is defined by

$$\gamma = \lim_{m \to \infty} \left(\sum_{k=1}^{m} \frac{1}{k} - \log(m) \right)$$

Because the finite sum represents the harmonic number H_m

```
Sum[1 / k, {k, 1, m}]
```

HarmonicNumber[m]

we can write

```
EulerConstant = Limit[HarmonicNumber[m] - Log[m],
m → Infinity]
```

EulerGamma

Mathematica gives the correct result. To obtain an approximate numerical value, we first load the package

```
<<NumericalMath'NLimit'
```

and then use `NLimit`

```
NLimit[Sum[1 / k, {k, 1, m}] - Log[m], m -> Infinity]
```

0.577216

which is indeed the correct numerical value of the constant `EulerGamma`.

```
N[EulerGamma]
```

0.577216

4.8 Complex Functions

A function f defined on an open domain $\mathcal{D}$ in the complex plane is said to be *analytic*, if for any $z_0 \in \mathcal{D}$ we can write $f(z)$ as the sum of a convergent power series in a neighborhood of z_0; that is,

$$f(z) = \sum_{n=0}^{\infty} c_n(z - z_0)^n,$$

where the coefficients c_n ($n = 0, 1, 2, \ldots$) are complex numbers.

A point such as z_0 in the neighborhood of which a function f can be written as the sum of a convergent power series is said to be *regular*. Any point that is not regular is *singular*.

A power series converges in an open disk, that is, a domain whose boundary is a circle of radius r centered at z_0. The largest value of r is the *radius of convergence* of the power series. It is denoted by R, hence, a power series is convergent if its radius of convergence is positive. For example, the power series representing the exponential e^z in a neighborhood of the origin, is $\sum_{n=0}^{\infty} z^n/n!$. Its radius of convergence is infinite. In other words, the exponential function is analytic in the whole complex plane. This is, in general, not the case. Given a complex function f analytic in a neighborhood of a point z_0 the power series $\sum_{n=0}^{\infty} c_n(z - z_0)^n$ has a finite radius of convergence whose value is limited by the existence of the singularities of the function

f. The value of the radius of convergence can be determined by the *Cauchy rule* which says that the radius of convergence of a power series of the form $\sum_{n=0}^{\infty} c_n (z - z_0)^n$ is given by the relation:

$$\frac{1}{R} = \text{limitsup}_{n \to \infty} |c_n|^{1/n}.$$

Consider for example, the function $1/(1 - z)$. Its power series expansion in a neighborhood of the origin is given by $\sum_{n=0}^{\infty} z^n = 1 + z + z^2 + \cdots$. This function has only one singular point at finite distance, namely $z = 1$, and we can easily verify that its radius of convergence is equal to 1, precisely the distance from the origin to this singular point. In fact,

$$\text{limitsup}_{n \to \infty} 1^{1/n} = \lim_{n \to \infty} 1^{1n} = 1.$$

Analytic functions are *differentiable*; that is, if $f(z) = \sum_{n=0}^{\infty} c_n (z - z_0)^n$ is a power series converging in an open disk $D(0, R)$ of radius $R > 0$ centered at the origin, then its derivative defined in the complex plane by

$$f'(z) = \lim_{\zeta \to 0} \frac{f(z + \zeta) - f(z)}{\zeta},$$

exists for all z in $D(0, R)$. If a complex function f is differentiable at a point z_0, it is infinitely differentiable at that point. Differentiable complex functions are also called *holomorphic*. In the case of functions of a complex variable, the adjectives analytic and holomorphic are synonymous.

The singular point of the function $1/(1 - z)$ is *isolated*. That is, there exists, centered at that point, an open disk with a nonzero radius in which $z = 1$ is the only singular point. If, starting from any point z_0, where the function is equal to $1/(1 - z_0)$, we follow a circular path around the point $z = 1$, when we come back to the point z_0, the function is again equal to $1/(1 - z_0)$. We can therefore say that the function is well defined in the whole complex plane except at $z = 1$. Again, this property is not shared by all complex functions. Consider the function $\sqrt{z}$. The origin is a singular point because $\sqrt{z}$ cannot be represented by the sum of a convergent power series in any neighborhood of the origin. This singular point is, however, not isolated. The squares of the two complex numbers $z_1 = \rho e^{i\theta/2}$ and $z_2 = \rho e^{i(\theta/2+\pi)}$ are equal, therefore the function $\sqrt{z}$ cannot be defined without precaution. If, for instance, we choose $\sqrt{z}$ to be the function equal to 1 when $z = 1$, then if we write $z = \rho e^{i\theta}$, and consider that the value $z = 1$ corresponds to the choice $\rho = 1$ and $\theta = 0$, following a circular path of radius 1 centered at the origin, the argument θ increases continuously from 0 to 2π and, consequently $\sqrt{z} = \sqrt{\rho} e^{i\theta/2}$ is found to be equal to -1. Such a behavior is not acceptable for a function. Thus, to correctly define the function $\sqrt{z}$ we should not be able to follow a continuous path around the singular point $z = 0$. The domain in which the function $\sqrt{z}$ is well-defined could be, for instance, the set

$$\{z \mid z = \rho e^{i\theta}, \rho > 0, -\pi < \theta < \pi\}$$

that is, the complement in the complex plane of the negative real semi-axis. Along this so-called *branch cut* the function $\sqrt{z}$ is discontinuous. All the points of the branch cut are singular. For the function $\sqrt{z}$ the origin is not an isolated singular point. Such a singular point is called a *branch point*. The branch cut could have been chosen in many different ways. For instance, any semi-axis whose equation is $\rho e^{i\alpha}$ where α is a given argument between 0 and 2π, and ρ is a parameter varying from 0 to ∞, is an acceptable branch cut. Our choice, which corresponds to $\alpha = \pi$ is the traditional one, also adopted by *Mathematica*.

The tridimensional plots of the real and imaginary parts of $\sqrt{z}$ clearly show the discontinuity along the negative real semi-axis.

```
Plot3D[Re[Sqrt[x + I y]], {x, - 3, 3}, {y, - 3, 3}];
```

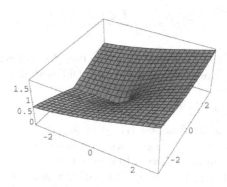

Fig. 4.8. *Plot of the real part of $\sqrt{x + iy}$ in the domain $\{x, y\} \in [-3, 3] \times [-3, 3]$.*

```
Plot3D[Im[Sqrt[x + I y]], {x, - 3, 3}, {y, - 3, 3}];
```

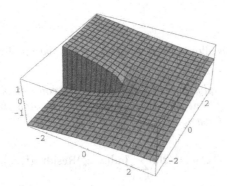

Fig. 4.9. *Plot of the imaginary part of $\sqrt{x + iy}$ in the domain $\{x, y\} \in [-3, 3] \times [-3, 3]$.*

More generally, if a function g defined in a domain $\mathcal{D}$ of the complex plane is not injective, the equation $g(z) = u$ may have more than one solution and g has, in this case, more than one inverse function: $f_1, f_2, f_3, \ldots$ such that, for $k = 1, 2, 3, \ldots$, $g(f_k(u)) = u$, and these functions cannot be continuous in $g(\mathcal{D})$. This is exactly what happened for the noninjective function $g(z) = z^2$, defined in the whole complex plane and having two noncontinuous inverse functions in the whole complex plane.

Here is another interesting result concerning the representation of functions in the vicinity of an isolated singularity. Let S be the open annulus $\{z \mid r_1 < |z - z_0|, r_2\}$; if f is analytic in S then, for all $z \in S$, we have

$$f(z) = \sum_{n=-\infty}^{\infty} c_n (z - z_0)^n,$$

where the power series $\sum_{n=0}^{\infty} c_n (z - z_0)^n$ is convergent for $|z - z_0| < r_2$ and the series $\sum_{n=1}^{\infty} c_{-n} (z - z_0)^{-n}$ is convergent for $|z - z_0| > r_1$. This particular expansion is called the *Laurent series* of f in S.

If the Laurent series is defined in $\{z \mid 0 < |z - z_0| < r\}$, called the *punctured neighborhood* of the singular point z_0, the point z_0 is a *pole* if the series with negative exponents has a finite number of terms, and the greatest value of n is the *order* of the pole; if the series with negative exponents has an infinite number of terms, the point z_0 is an *essential singularity*.

If z_0 is an isolated singularity of $f(z)$, the coefficient c_{-1} of the Laurent series is called the *residue* of f at $z = z_0$. The *Mathematica* command **Residue[f[z], {z, z0}]** finds the residue of f at the point **z0**.

```
Residue[Exp[z] / Sin[z]^2, {z, Pi}]
```

E^{Pi}

The *residue theorem* allows, in particular, the evaluation of contour integrals. It states:

Let $\mathcal{D}$ be an open simply connected domain in the complex plane and $\{z_1, \ldots, z_n\}$ a set of n different isolated points. For any analytic function f in $\mathcal{D}\backslash\{z_1, \ldots, z_n\}$, we have

$$\int_\gamma f(z)\, dz = 2\mathrm{i}\pi \sum_{k=1}^{n} \mathrm{Ind}(\gamma, z_k)\mathrm{Residue}(f, z_k),$$

where γ is a closed path in $\mathcal{D}\backslash\{z_1, z_2, \ldots, z_n\}$.

The symbol $\mathrm{Ind}(\gamma, z_k)$ is the index of the point path z_k with respect to the path γ. It is defined by

$$\mathrm{Ind}(\gamma, z_k) = \frac{1}{2\mathrm{i}\pi} \int_\gamma \frac{dz}{z - z_k}.$$

This formula is a consequence of the relation

$$\int_\gamma \frac{dz}{z} = 2\mathrm{i}\pi,$$

where γ is the circular path $t \mapsto e^{2\mathrm{i}\pi t}$ with $t \in [0, 1]$.

If we integrate along the circular path γ, the result above is obvious because replacing z by $e^{2\mathrm{i}\pi t}$ yields $2\mathrm{i}\pi \int_0^1 dt = 2\mathrm{i}\pi$. But, as a consequence of Cauchy's theorem, we can change the circle into the square with vertices $1, \mathrm{i}, -1, -\mathrm{i}$ without changing the result. Using *Mathematica* we can evaluate the contour integral directly:

```
Integrate[1 / z, {z, 1, I, - 1, - I, 1}]
```

(2 I) Pi

For instance, the real integral $\int_{-\infty}^{\infty} f(x)\, dx$ of the function f whose only singularities, as a function of the complex variable z are poles, none of them being real, can be easily evaluated using the residue theorem. Consider the contour integral $\int_\gamma f(z)\, dz$ where γ is the closed path equal to the union of the paths $\gamma_1 : t \mapsto R(2t - 1)$ and $\gamma_2 : t \mapsto Re^{\mathrm{i}\pi t}$ for t varying from 0 to 1.

This integral is equal to

$$\int_{-R}^{R} f(x)\,dx + \int_{\gamma_2} f(z)\,dz = \frac{1}{2i\pi} \sum_k \text{Residue}(f, z_k),$$

where the sum is over all the residues of the poles of f in the upper half plane such that $|z_k| < R$. When $R \to \infty$, the first integral becomes the real integral we have to evaluate and the second one tends to zero if $\lim_{x\to\infty} xf(x) = 0$. When this condition is satisfied, the real integral $\int_{-\infty}^{\infty} f(x)\,dx$ is simply equal to the sum of the residues of f at all poles of the upper half-plane. For instance,

$$\int_{-\infty}^{\infty} \frac{dx}{1+x^2} = 2i\pi\text{Residue}\left(\frac{dx}{1+x^2}, i\right),$$

because the index of i with respect to the path γ is 1. The residue is given by

```
Residue[1 / (1 + x^2), {x, I}]
```

$$\frac{I}{2}$$

We thus obtain the classical result: $\int_{-\infty}^{\infty} dx/(1+x^2) = \pi$.

We can also evaluate very simply integrals of the form $\int_0^{2\pi} f(\cos(x), \sin(x))\,dx$, where f is a rational fraction in $\cos(x)$ and $\sin(x)$. If we put $z = e^{ix}$, the integral becomes $1/iz \int_\gamma f((z + z^{-1})/2, (z - z^{-1})/2i)\,dz$, where γ is the circular path of radius 1 $\theta \mapsto e^{i\theta}$ for θ varying from 0 to 2π. Therefore,

$$\int_0^{2\pi} f(\cos(x0), \sin(x))\,dx = 2\pi \sum_k \text{Residue}\left(\frac{1}{z} f\left(\frac{z + z^{-1}}{2}, \frac{z - z^{-1}}{2i}\right), z_k\right),$$

where the sum is over the residues of all the poles inside the unit circle. For instance, if $a > |b|$,

$$\int_0^{2\pi} \frac{d\theta}{a + b\sin\theta} = \int_\gamma \frac{2\,dz}{bz^2 + 2aiz - b}.$$

The complex function has only one pole inside the unit circle, namely $i(-a + \sqrt{a^2 - b^2})/b$. Because

```
Residue[2 / (b z^2 + 2 a I z - b),
  {z, I (- a + Sqrt[a^2 - b^2]) / b}]
```

$$\frac{-I}{\text{Sqrt}[a^2 - b^2]}$$

we find

$$\int_0^{2\pi} \frac{d\theta}{a + b\sin\theta} = \frac{2\pi}{\sqrt{a^2 - b^2}}.$$

For a detailed study with historical references to the properties of functions of a complex variable see [7].

4.9 Fourier Transforms

4.9.1 Discrete Fourier Transform

Fourier[list] gives the discrete Fourier transform (or frequency spectrum) v_s of a list of complex numbers $\{u_1, u_2, \ldots, u_n\}$. By default it is defined by

$$v_s = \frac{1}{\sqrt{n}} \sum_{r=1}^{n} u_r e^{2i\pi(r-1)(s-1)/n},$$

where n is the list length.

```
data = Table[Random[], {10}]
FT1 = Fourier[data]
```

{0.818232, 0.582609, 0.51494, 0.146173, 0.693925,

0.867643, 0.72016, 0.11342, 0.663217,0.822117}

{1.87916 + 0. I, 0.0717421 − 0.0841385 I, 0.44075 − 0.097791 I,

− 0.24971 − 0.0584497 I, − 0.0475295 + 0.0148031 I, 0.27781+ 0. I,

− 0.0475295 − 0.0148031 I, − 0.24971 + 0.0584497 I,

0.44075 + 0.097791 I, 0.0717421 + 0.0841385 I}

Chop may be used to discard the tiny imaginary parts that appear due to numerical error.

```
FT2 = Chop[FT1]
```

{1.87916, 0.0717421 − 0.0841385 I, 0.44075 − 0.097791 I,

− 0.24971 − 0.0584497 I, − 0.0475295 + 0.0148031 I,

$- 0.0475295 - 0.0148031$ I, $- 0.24971 + 0.0584497$ I,

$0.44075 + 0.097791$ I, $0.0717421 + 0.0841385$ I}

The inverse discrete Fourier transform is, by default, given by

$$u_r = \frac{1}{\sqrt{n}} \sum_{r=1}^{n} v_s e^{-2i\pi(r-1)(s-1)/n}.$$

InverseFourier[FT2]

{0.818232, 0.582609, 0.51494, 0.146173, 0.693925,

0.867643, 0.72016, 0.11342, 0.663217,0.822117}

4.9.2 Fourier Transform

FourierTransform[f[x], x, t] gives the Fourier transform F of the function f of the variable x as a function of t . By default it is defined by

$$F(t) = \frac{1}{\sqrt{2\pi}} \int_{-\infty}^{\infty} f(x) e^{itx} \, dx.$$

Different choices of definitions can be specified using the option FourierParameters.

??FourierParameters

FourierParameters is an option to Fourier transform

functions that specifies the convention to follow for

the overall constant and the frequency constant. For

FourierParameters -> {a, b}, FourierTransform[expr, t, w]

is equivalent to Sqrt[Abs[b] / ((2 Pi)^(1 - a))]

Integrate[expr Exp[I b w t], {t, - Infinity, Infinity}]. More ...

Attributes[FourierParameters] = {Protected, ReadProtected}

ReadProtected is an attribute that prevents values associated with a symbol from being seen, and Protected is an attribute that prevents any values associated with a symbol from being modified.

```
FourierTransform[x^2 Exp[- x^2], x, t]
```

$$- \frac{2 \, E^{-t^2/4} + t^2 \, E^{-t^2/4}}{4 \, \text{Sqrt}[2]}$$

InverseFourierTransform[F[t], t, x] gives the inverse Fourier transform f of the function F of the variable t as a function of x. It is given by

$$f(x) = \frac{1}{\sqrt{2\pi}} \int_{-\infty}^{\infty} F(t)e^{-itx} \, dx.$$

```
InverseFourierTransform[%, t, x]
```

$$E^{-x^2} x^2$$

If the function f is either odd or even, we can use either the FourierSinTransform or the FourierCosTransform, respectively, defined by

$$\sqrt{\frac{2}{\pi}} \int_0^{\infty} f(x) \sin(tx) \, dx \text{ and } \sqrt{\frac{2}{\pi}} \int_0^{\infty} f(x) \cos(tx) \, dx.$$

```
{sinFT, cosFT} = {FourierSinTransform[x / (1 + x^4), x, t],
1 / (1 + x^2), x, t]}
```

$$\{E^{t/\text{Sqrt}[2]} \, \text{Sqrt}[Pi/2] \, \text{Sin}[t/\text{Sqrt}[2]], E^{-t} \, \text{Sqrt}[Pi/2]\}$$

The InverseFourierSinTransform and the InverseFourierCosTransform of a function F are given, respectively, by

$$\sqrt{\frac{2}{\pi}} \int_0^{\infty} F(t) \sin(tx) \, dt \text{ and } \sqrt{\frac{2}{\pi}} \int_0^{\infty} F(t) \cos(tx) \, dt.$$

```
InverseFourierSinTransform[sinFT, t, x]
InverseFourierCosTransform[cosFT, t, x]
```

$$\frac{x}{1+x^4}$$

$$\frac{1}{1+x^2}$$

4.10 Fourier Series

We first have to load the package Calculus`FourierTransform`.

```
<<Calculus`FourierTransform`
```

- FourierCoefficient[f[x], x, n] gives the nth coefficient of the exponential Fourier series of function f of period 1 on the interval $[-\frac{1}{2}, \frac{1}{2}]$.

- FourierSinCoefficient[f[x], x, n] gives the nth coefficient of the sine Fourier series of the odd function f of the variable x.

- FourierCosCoefficient[f[x], x, n] gives the nth coefficient of the cosine Fourier series of the even function f of the variable x.

- FourierSeries[f[x], x, n] gives the exponential Fourier series of the function f of the variable x on the interval $[-\frac{1}{2}, \frac{1}{2}]$ to order n.

- FourierTrigSeries[f[x], x, n] gives the trigonometric Fourier series of the function f of the variable x on the interval $[-\frac{1}{2}, \frac{1}{2}]$ to order n.

For different definitions use different settings in the option FourierParameters → {a,b}.

$$\int_{-1/2}^{1/2} f(x)e^{2inx}\,dx \qquad \text{with setting } \{0, 1\},$$

$$|b|^{(1-a)/2} \int_{-1/2b}^{1/2b} f(x)e^{2ibnx}\,dx \qquad \text{with setting } \{a, b\}.$$

```
{FourierSinCoefficient[x, x, 1, FourierParameters → {0, 1}],
FourierSinCoefficient[x, x, 1, FourierParameters -> {0, 2}]}
```

$$\{\frac{1}{\text{Pi}}, \frac{1}{2 \text{ Sqrt[2] Pi}}\}$$

```
Table[FourierSinCoefficient[x, x, k], {k, 1, 4}]
```

$$\{\frac{1}{Pi}, \ -\frac{1}{2\ Pi}, \ +\frac{1}{3\ Pi}, \ -\frac{1}{4\ Pi}\}$$

```
FourierTrigSeries[x, x, 4]
```

$$\frac{Sin[2\ Pi\ x]}{Pi} - \frac{Sin[4\ Pi\ x]}{2\ Pi} + \frac{Sin[6\ Pi\ x]}{3\ Pi} - \frac{Sin[8\ Pi\ x]}{4\ Pi}$$

```
sawFunctionPlot = Plot[x - Round[x], {x, - 0.5, 1.5},
PlotStyle → {RGBColor[1,0,0]}];
```

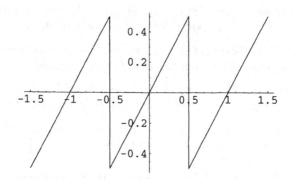

Fig. 4.10. *Plot of the sawtooth function for $x \in [-1.5, 1.5]$.*

Round[x] gives the integer closest to x.

In the following graphs note the options PlotStyle used here to specify the curve color, and DisplayFunction used with the setting Identity causing the objects to be returned, but no display to be generated.

```
pl1 = Plot[Sin[2 Pi x] / Pi, {x, - 1.5, 1.5},
PlotStyle → {RGBColor[0, 0, 1]},
DisplayFunction → Identity];
pl2 = Plot[Sin[2 Pi x] / Pi - Sin[4 Pi x] / (2 Pi),
{x, - 1.5, 1.5}, PlotStyle → {RGBColor[0, 0, 1]},
DisplayFunction → Identity];
pl3 = Plot[Sin[2 Pi x] / Pi - Sin[4 Pi x] / (2 Pi) + Sin[6 Pi
x] / (3 Pi), {x, - 1.5, 1.5},
PlotStyle → {RGBColor[0, 0, 1]},
DisplayFunction → Identity];

pl4 = Plot[Sin[2 Pi x] / Pi - Sin[4 Pi x] / (2 Pi) +
Sin[6 Pi x] / (3 Pi) - Sin[8 Pi x] / (4 Pi), {x, - 1.5, 1.5},
PlotStyle → {RGBColor[0, 0, 1]},
DisplayFunction → Identity];
Show[GraphicsArray[{{pl1, pl2}, {pl3, pl4},
DisplayFunction → $DisplayFunction]];
```

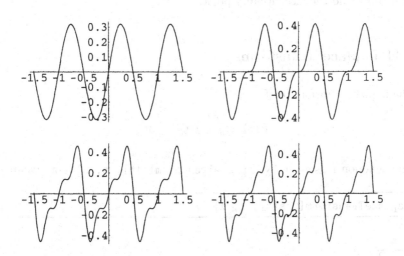

Fig. 4.11. *Plot of the first Fourier series approximating the sawtooth function for* $x \in [-0.5, 1.5]$.

```
Show[{pl4, sawFunctionPlot}, DisplayFunction ->
$DisplayFunction];
```

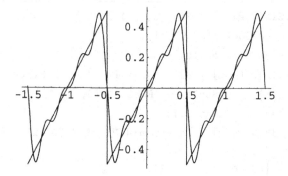

Fig. 4.12. *Plot of sawtooth function and its four-term Fourier sine series for* $x \in [-1.5, 1.5]$.

Figure 4.12 above exhibits the Gibbs phenomenon, that is, the overshoot of partial Fourier series that occurs at first-order discontinuities moving towards the jump as the number of terms of the series increases. It was first described by H. Wilbraham in 1848 [67] and then analyzed in detail 50 years later by Josiah Willard Gibbs (1839–1903) [20, 11]. Note that the Fourier series of a discontinuous function does not converge uniformly in an arbitrary small interval around the discontinuity point.

4.11 Laplace Transforms

The Laplace transform of $f(t)$ is

$$F(z) = \int_0^\infty f(t)e^{-zt}\,dt.$$

This function F is given by `LaplaceTransform[f[t], t, s]`. For example:

```
LaplaceTransform[Cos[t], t, z]
```

$$\frac{z}{1 + z^2}$$

The command `InverseLaplaceTransform[F[z], z, t]` gives the inverse Laplace transform of $F(z)$] as a function of t.

```
InverseLaplaceTransform[z / (1 + z^2), z, t]
```

```
Cos[t]
```

The Heaviside function (also called unit step) is represented by the function `UnitStep[t]`, equal to 0 for $t < 0$ and to 1 for $t \geq 0$. Its Laplace transform is

```
LaplaceTransform[UnitStep[t], t, z]
```

$$\frac{1}{z}$$

Mathematica can also give the Laplace transform of the so-called "Dirac δ function" defined usually as a function equal to zero everywhere except at the origin where it is infinite and such that

$$\int_{-\infty}^{\infty} \delta(t)\varphi(t)\,dt = \varphi(0),$$

where φ is any well-behaved function defined at the origin. From a purely mathematical point of view, this definition makes no sense because δ being null almost everywhere the (Lebesgue) integral above should be zero. Actually δ is a singular *distribution*, that is, a continuous linear form defined on a space $\mathcal{D}$ of test functions such that, for all $\varphi \in \mathcal{D}$, $\delta(\varphi) = \varphi(0)$. On distribution theory and its history see [8].

The Laplace transform of the Dirac δ is given by

```
LaplaceTransform[DiracDelta[t], t, z]
```

1

and

```
InverseLaplaceTransform[1, z, t]
```

```
DiracDelta[t]
```

If F is the Laplace transform of a function f defined on $[0, \infty[$, the Laplace transform of its derivative f' is equal to $-f(0) + zF(z)$.

```
LaplaceTransform[f'[t], t, s] // Simplify
```

$-$ f[0] + s LaplaceTransform[f[t], t, z]

On distribution theory see Chapter 4 of Boccara [4]. For a more detailed study of distribution theory with historical references see Boccara [8].

4.12 Recurrence Equations

RSolve[{rEquation, initValues},f[n], n] solves the recurrence equation rEquation for the initial values initValues.

The following recurrence equation with the initial values $f(1) = b$ and $f(2) = a+b$ defines the generalized Fibonacci series $G(a, b)$. The traditional Fibonacci series corresponds to $G(0, 1)$.

```
sol = RSolve[{f[n] == f[n-1] + f[n-2],
f[1] == b, f[2] == a + b}, f[n], n] // FullSimplify
```

{{f[n] -> $(2^{-1-n}$ $((1 + $ Sqrt[5]$)^n$

$(- ((- 5 + $ Sqrt[5]$)$ a) $+ 2$ Sqrt[5] b) $+$

$(- 1 + $ Sqrt[5] $)^n$ $((5 + $ Sqrt[5]$)$ a $- 2$ Sqrt[5] b)

$E^{I\ n\ Pi}$)) / 5}}

```
Flatten[Table[f[n] /. sol, {n, 1, 10}]] // Simplify
```

{b, a + b, a + 2 b, 2 a + 3 b, 3 a + 5 b, 5 a + 8 b,

8 a + 13 b, 13 a + 21 b, 21 a + 34 b, 34 a + 55 b}

Without specifying initial values we obtain the general solution containing arbitrary constants.

```
Clear[f, sol]
sol = RSolve[f[n] == f[n - 1] + f[n - 2], f[n], n] //
FullSimplify
```

$$\{\{f[n \to \left(\frac{(1 - \text{Sqrt}[5])}{2}\right)^n C[1] + \left(\frac{(1 + \text{Sqrt}[5])}{2}\right)^n C[2] \}\}$$

4.13 Z Transforms

The Z transform $F(z)$ of $f(n)$ is

$$F(z) = \sum_{n=0}^{\infty} f(n)z^{-n}.$$

$F(z)$ is therefore the generating function of the sequence $(f(0), f(1), f(2), \ldots)$. It is given by the *Mathematica* command ZTransform[f[n], n, z]. For example:

```
ZTransform[a^n / n!, n, z]
```

$E^{a/z}$

The command InverseZTransform[F[z], z, n] gives the inverse Z transform of $F(z)$ as a function of n.

```
InverseZTransform[Exp[a / z], z, n]
```

$$\frac{a^n}{n!}$$

Z transforms are used to solve recurrence equations. Consider, for example, the recurrence equation defining the generalized Fibonacci sequence:

```
Clear[equation]
equation = f[n+2] == f[n+1] + f[n];
```

with the initial values

```
Clear[initialValues]
initialValues = {f[0] → a, f[1] -> b};
```

Taking the Z transform of the equation gives

```
transfEquation = ZTransform[equation, n, z]
```

$- (z^2 \, f[0]) - z \, f[1] + z^2 \, \text{ZTransform}[f[n] , n, z] ==$

$- (z \, f[0]) + \text{ZTransform}[f[n], n, z] + z \, \text{ZTransform}[f[n], n, z]$

Taking into account the initial values, we obtain

```
zEquation = transfEquation /. initialValues
```

$- (b\ z) - a\ z^2 + \text{ZTransform}[f[n],\ n,\ z] ==$

$- (a\ z) + \text{ZTransform}[f[n],\ n,\ z] + z\ \text{ZTransform}[f[n],\ n,\ z]$

Solving for the ZTransform yields

```
Solve[zEquation, ZTransform[f[n], n, z]]
```

$$\{\{\text{ZTransform}[f[n],\ n,\ z]\ ->\ \frac{-(a\ z) + b\ z + a\ z^2}{-1 - z + z^2}\}\}$$

and taking the inverse transform, we finally obtain the expression of the generalized Fibonacci sequence:

```
Clear[solution]
solution = InverseZTransform[(a z^2 + (b - a) z) /
(z^2 - z - 1), z, n] // Simplify
```

$(2^{-1-n}\ ((-((-5 + \text{Sqrt}[5])\ (1 + \text{Sqrt}[5])^n) +$

$(1 - \text{Sqrt}[5])^n\ (5 + \text{Sqrt}[5]))\ a - 2\ \text{Sqrt}[5]$

$((1 - \text{Sqrt}[5])^n - - (1 + \text{Sqrt}[5]\ )^n\ )\ b))\ /\ 5$

```
Flatten[Table[solution, {n, 1, 10}]] //Simplify
```

$\{b,\ a + b,\ a + 2\ b,\ 2\ a + 3\ b,\ 3\ a + 5\ b,\ 5\ a + 8\ b,$

$8\ a + 13\ b,\ 13\ a + 21\ b,\ 21\ a + 34\ b,\ 34\ a + 55\ b\}$

which is the sequence we obtained in the previous section.

4.14 Partial Differential Equations

DSolve[equation, y, {x1, x2, ...}] and NDSolve[equation, y, {x1, x2, ...}] can also be used to solve partial differential equations.

The one-dimensional wave equation

$$\frac{\partial^2 u}{\partial t^2} = c^2 \frac{\partial^2 u}{\partial x^2}$$

can be solved, its solution being expressed as a pure function.

```
Clear[eqn, sol, f]
eqn = D[u[x, t], {t, 2}] == c^2 D[u[x, t], {x, 2}];
sol = Simplify[DSolve[eqn, u[x, t], {x, t}], c > 0]
```

$$\{\{u[x, t] \rightarrow C[1][t - \frac{x}{c}] + C[2][t + \frac{x}{c}]\}\}$$

Here C[1] and C[2] are arbitrary functions of $t - x/c$ and $t + x/c$, respectively. To solve a specific boundary problem, DSolve is not useful. We can use the Laplace transform.

First, let us take the Laplace transform of the equation to get rid of the time derivative.

```
LTeqn1 = LaplaceTransform[eqn, t, s]
```

s^2 LaplaceTransform[u[x, t], t, s] $- s\, u[x, 0] - u^{(0, 1)}[x, 0] ==$

c^2 LaplaceTransform[$u^{(2, 0)}$[x, t], t, s]

If x belongs to the interval $[0, 1]$ with the boundary conditions $u(0, t) = u(1, t) = 0$, and the initial conditions $u(x, 0) = f(x)$, $u_t(x, 0) = 0$, we replace in the equation above LaplaceTransform[u[x, t], t, s] by U(x), LaplaceTransform[D[u[x, t], t, 2, t, s] by U''[x], [u[x, 0]] by f[x], and Derivative[0,1][u][x, 0] by 0. It yields

```
Simplify[DSolve[{c^2 U''[x] - s^2 U[x] ==
s an Sin[n Pi x], U[0] == 0, U[1] == 0}, U[x], x],
n ∈ Integers]
```

$$\{\{U[x] \rightarrow -(\frac{an\, s\, Sin[n\, Pi\, x]}{c^2\, n^2\, Pi^2 + s^2})\}\}$$

The inverse Laplace transform of this solution is

```
InverseLaplaceTransform[s Sin[n Pi x] /
(c^2 n^2 Pi^2 + s^2), s, t]
```

Cos[c n Pi t]   Sin[n Pi x]

Finally, the solution is

$$u(x,t) = \sum_{n=1}^{\infty} a_n \sin(n\pi x) \cos(cn\pi t),$$

where

$$a_n = 2 \int_0^1 f(x) \sin(n\pi x)\, dx.$$

If, for example, we consider the function f defined by

$$f(x) = \begin{cases} mx/a, & \text{if } 0 \le x < a, \\ m(x-1)/(a-1), & \text{if } a \le x \le 1, \end{cases}$$

which corresponds to a plucked string instrument, we find

```
an = Simplify[2 Integrate[m (x / a) Sin[n Pi x], {x, 0, a}] +
2 Integrate[ m (x - 1) / (a - 1) Sin[n Pi x],
{x, a, 1}], n ∈ Integers]
```

$$\frac{-\,2\ m\ \text{Sin}[a\ n\ \text{Pi}]}{(-\,1+a)\ a\ n^2\ \text{Pi}^2}$$

That is,

$$\frac{2m}{\pi^2(a-1)} \sum_{n=1}^{\infty} \frac{1}{n^2} \sin(n\pi a) \sin(n\pi x) \cos(cn\pi t).$$

Using **TrigReduce** we can write the product of trigonometric functions as a sum of cosines.

```
TrigReduce[Sin[n Pi a] Sin[n Pi x] Cos[n c Pi t] ]
```

(Cos[a n Pi − c n Pi t − n Pi x] +
Cos[a n Pi + c n Pi t − n Pi x] −
Cos[a n Pi − c n Pi t + n Pi x] −
Cos[a n Pi + c n Pi t + n Pi x]) / 4

which shows that the solution could be expressed in terms of the polylogarithm function dilog Li_2 defined by

$$\text{Li}_2(x) = \sum_{n=1}^{\infty} \frac{z^n}{n^2}, \quad \text{where } |z| < 1,$$

whose *Mathematica* notation is `PolyLog[2, z]`.

5

Lists

Mathematica lists are powerful objects. They provide an efficient way of manipulating groups of expressions as a whole.

5.1 Creating Lists

Here is a list of elements.

```
{a, b, c, d, e, f, g}
```

{a, b, c, d, e, f, g}

```
FullForm[{a, b, c, d, e, f, g}]
```

List[a, b, c, d, e, f, g]

Table[expression, {n}] generates a list of n copies of expression.

Table[expression, {k, n}] generates a list of the values of expression when k runs from 1 to n.

Table[expression, {k, n1, n2}] generates a list of the values of expression when k runs from n1 to n2.

Table[expression, {k, n1, n2, Δk}] is the same as above using steps Δk.

Table[expression, {j, m1, m2}, {k, n1, n2}] generates a nested list.

Here are a few examples.

```
Table[k (k + 1) / 2, {k, 10}]
```

$\{1, 3, 6, 10, 15, 21, 28, 36, 45, 55\}$

```
Table[k, {k, 0, 1, 1 / 10}]
```

$\{0, \dfrac{1}{10}, \dfrac{1}{5}, \dfrac{3}{10}, \dfrac{2}{5}, \dfrac{1}{2}, \dfrac{3}{5}, \dfrac{7}{10}, \dfrac{4}{5}, \dfrac{9}{10}, 1\}$

```
Table[i^2 + j^2, {i, 3}, {j, 4}]
```

$\{\{2, 5, 10, 17\}, \{5, 8, 13, 20\}, \{10, 13, 18, 25\}\}$

Range[n] generates the list $\{1, 2, \ldots, n\}$.

Range[m, n] generates the list $\{m, m{+}1, \ldots, n\}$,

Range[m, n, Δk] is the same as above but with steps Δk.

```
Range[12]
```

$\{1, 2, 3, 4, 5, 6, 7, 8, 9, 10, 11, 12\}$

```
Range[3, 17]
```

$\{3, 4, 5, 6, 7, 8, 9, 10, 11, 12, 13, 14, 15, 16, 17\}$

```
Range[3, 27, 3]
```

$\{3, 6, 9, 12, 15, 18, 21, 24, 27\}$

```
Range[0, 1, 0.1]
```

$\{0, 0.1, 0.2, 0.3, 0.4, 0.5, 0.6, 0.7, 0.8, 0.9, 1.\}$

`Array[f, n]` generates the list $\{f[1], f[2], \ldots, f[n]\}$. `Array[f, {m, n}]` generates an $m \times n$ list with elements `f[i, j]`.

```
Array[f, 5]
```

$\{f[1], f[2], f[3], f[4], f[5]\}$

```
Array[f, {3, 4}]
```

$\{\{f[1,1], f[1,2], f[1,3], f[1,4]\},$
$\{f[2,1], f[2,2], f[2,3], f[2,4]\},$
$\{f[3,1], f[3,2], f[3,3], f[3,4]\}\}$

We can also change the origin.

```
Array[f, 5,- 1]
```

$\{f[-1], f[0], f[1], f[2], f[3]\}$

Here is a 4×6 array in which the indices start from 2 and 4, respectively.

```
Array[f, {4, 6}, {2, 4}]
```

$\{\{f[2, 4], f[2, 5], f[2, 6], f[2, 7], f[2, 8], f[2, 9]\},$
$\{f[3, 4], f[3, 5], f[3, 6], f[3, 7], f[3, 8], f[3, 9]\},$
$\{f[4, 4], f[4, 5], f[4, 6], f[4, 7], f[4, 8], f[4, 9]\},$
$\{f[5, 4], f[5, 5], f[5, 6], f[5, 7], f[5, 8], f[5, 9]\}\}$

There exist various commands giving better visual presentations of lists.

`TableForm[list]` presents the list elements as a rectangular array. This command accepts a variety of `Options`.

```
A = {{12, 3456, 752}, {3, 586, 87}, {17645, 98, 3}};
TableForm[A]
```

12	3456	752
3	586	87
17645	98	3

```
TableForm[A, TableHeadings → Automatic]
```

	1	2	3
1	12	3456	752
2	3	586	87
3	17645	98	3

```
TableForm[A, TableHeadings → {{"row 1", "row 2", "row 3"},
None}]
```

row 1	12	3456	752
row 2	3	586	87
row 3	17645	98	3

ColumnForm[list] presents the list elements as a vertical array. There also exist various Options.

```
ColumnForm[A]
```

{12, 3456, 752}
{3, 586, 87}
{17645, 98, 3}

```
B = {13, 8145, 37};
ColumnForm[B]
```

13

8145

37

```
ColumnForm[B, Center]
```

```
 13
8145
 37
```

```
ColumnForm[B, Right]
```

```
 13
8145
 37
```

5.2 Extracting Elements

```
lis = Table[Random[], {12}]
```

{0.0622006, 0.672395, 0.296732, 0.0864989, 0.506992, 0.247055,
0.266085, 0.0842416, 0.360267, 0.336073, 0.709386, 0.388325}

Part[expression, n] or expression[[n]] gives the nth part of expression.

Part[expression, - n] or expression[[- n]] gives the nth part of expression starting from the end.

```
{Part[lis, 3], lis[[3]]}
```

{0.296732, 0.296732}

```
{Part[lis, - 2], lis[[- 2]]}
```

{0.709386, 0.709386}

```
Part[lis, {3, 6, 8}]
```

{0.296732, 0.247055, 0.0842416}

gives the third, sixth, and eighth parts of lis.

Note that lis[[3, 6, 8]] is not equivalent. It will give the (3, 6, 8) element of a nested list which, in this specific case, does not exist.

```
lis[[3, 6, 8]]
```

Part : : partd ;

Part specification {0.0622006', 0.672395', 0.296732', << 7 >>,

0.709386', 0.388325'} [[<< 1 >.]] is longer than depth of

object. More ...

The symbol << 7 >> indicates the number of nondisplayed elements of lis.

{0.0622006, 0.672395, 0.296732, 0.0864989, 0.506992,

0.247055, 0.266085,0.0842416,0.360267,0.336073,0.709386,

0.388325}[[3, 6, 8]]

Extract[expression, list] extracts the part of expression at the positions specified by list. Note in the following commands the number and position of curly braces.

```
Extract[lis, {2}]
Extract[lis, {{2}}]
```

0.672395

{0.672395}

```
Extract[lis, {{3}, {6}, {8}}]
(* same as Part[lis, {3, 6, 8}] *)
```

{0.296732, 0.247055, 0.0842416}

Here is a nested list.

```
l = {{a, b, c}, {d, e, f}}
```

{{a, b, c}, {d, e, f}}

```
{1[[1]], 1[[1, 2]]}
```

{{a, b, c}, b}

The same result is obtained with **Extract**.

```
Extract[1, {{1}, {1, 2}}]
```

{{a, b, c}, b}

Extract[expression,list, f] extracts the part of **expression** at the positions specified by **list** and finds the value of **f** at each of them.

```
f[x_] := x^2
Extract[lis, {3}, f]
```

0.0880499

which is the square of 0.296732, the third element of **lis**.

First[expression] and **Last[expression]** give, respectively, the first and last elements of **expression**.

```
{First[lis], Last[lis]}
```

{0.0622006, 0.388325}

Take[list, n] gives the first n elements of **list**.

Take[list, - n] gives the last n elements of **list**.

Take[list, {m, n}] gives elements m through n of **list**.

```
Take[lis, 2]
```

{0.0622006, 0.672395}

```
Take[lis, - 2]
```

{0.599309, 0.265092}

```
Take[lis, {3, 7}]
```

{0.0726532, 0.680859, 0.180326, 0.974102, 0.0301189}

Rest[expression] gives expression with the first element removed.

```
Rest[x^2 + y^4 + z^6]
```

$\{y^4 + z^6\}$

```
Rest[lis]
```

{0.672395, 0.296732, 0.0864989, 0.506992, 0.247055,0.266085,
0.0842416, 0.360267, 0.336073,0.709386,0.388325}

Rest[expression] is equivalent to Drop[expression,1]. Drop[expression,
n] gives expression with its first n elements dropped.

```
Drop[lis,1]
```

{0.672395, 0.296732, 0.0864989, 0.506992, 0.247055, 0.266085,
0.0842416, 0.360267, 0.336073,0.709386,0.388325}

```
Drop[x^2 + y^4 + z^6, 1]
```

$\{y^4 + z^6\}$

```
Drop[lis, 5]
```

{0.247055, 0.266085, 0.0842416, 0.360267, 0.336073,
0.709386, 0.388325}

```
Drop[a + b + c + d + e, 3]
```

d + e

Drop[expression, - n] gives expression with its last n elements dropped.

```
Drop[lis, - 3]
```

{0.0622006, 0.672395, 0.296732, 0.0864989,
0.506992, 0.247055, 0.266085, 0.0842416, 0.360267}

Drop[expression, {m, n}] gives expression with elements m through n dropped.

```
Drop[lis, {2, 10}]
```

{0.0622006, 0.709386, 0.388325}

5.3 Adding Elements

```
Clear[lis]
lis = Range[2, 12]
```

{2, 3, 4, 5, 6, 7, 8, 9, 10, 11,12}

Prepend[expression, element] adds element at the beginning of expression.

```
lis = Prepend[lis, 1]
```

{1, 2, 3, 4, 5, 6, 7, 8, 9, 10, 11, 12}

```
Prepend[y^2 + z^2, x^2]
```

$$x^2 + y^2 + z^2$$

Append[expression, element] adds element at the end of expression.

```
lis = Append[lis, 13]
```

$\{1, 2, 3, 4, 5, 6, 7, 8, 9, 10, 11, 12, 13\}$

```
Append[x^2 + y^2, z^2]
```

$$x^2 + y^2 + z^2$$

Insert[expression,element, {i, j, ...}] inserts element at position {i, j, ...} in expression.

```
lis = Insert[lis, A, {{2}, {4}, {6}, {8}, {10}, {12}}]
```

$\{1, A, 2, 3, A, 4, 5, A, 6, 7, A, 8, 9, A, 10, 11, A, 12, 13\}$

Cases[e1, e2, ..., pattern] gives a list of the elements ej matching pattern.

```
Cases[lis, _Integer]
```

$\{1, 2, 3, 4, 5, 6, 7, 8, 9, 10, 11, 12, 13\}$

The same result may be obtained with Delete.

```
Delete[lis, {{2}, {5}, {8}, {11}, {14}, {17}}]
```

$\{1, 2, 3, 4, 5, 6, 7, 8, 9, 10, 11, 12, 13\}$

DeleteCases[expression, pattern] removes all elements of expression that match pattern.

```
lis1 = DeleteCases[lis, A]
```

{1, 2, 3, 4, 5, 6, 7, 8, 9, 10, 11, 12, 13}

```
lis2 = DeleteCases[lis, _Integer]
```

{A, A, A, A, A, A}

We can concatenate lists using Join.

```
Clear[lis1, lis2, lis3]
lis1 = {a, b, c, d, e}
lis2 = {f, g, h, i}
lis3 = {j, k, l, m, n, o, p}
Join[lis1, lis2, lis3]
```

{a, b, c, d, e}

{f, g, h, i}

{j, k, l, m, n, o, p}

{a, b, c, d, e, f, g, h, i, j, k, l, m, n, o, p}

Lists can be viewed as sets

```
Clear[lis1, lis2, lis3]
lis1 = {a, c, b, e};
lis2 = {d, f, c, g};
lis3 = {c, b, h};
Union[lis1, lis2, lis3]] (* all different elements either in
lis1, or lis2, or lis3 *)
Intersection[lis1, lis2, lis3] (* all common elements of
lis1, lis2, and lis3 *)
Complement[lis1, lis2] (* elements of lis1 that are not in
lis2 *)
```

{a, b, c, d, e, f, g, h}

{c}

{a, b, e}

```
Clear[lis]
lis = {a, b, a, e, c, a, d, e, b};
Union[lis] (* eliminate repeated elements *)
```

{a, b, c, d, e}

5.4 Finding, Grouping, and Counting Elements

```
Clear[lis]
lis = Table[Random[Integer, 3], {20}]
```

{3, 1, 0, 0, 3, 3, 1, 1, 1, 0, 3, 2, 0, 0, 3, 1, 3, 1, 2, 3}

Position[list, pattern] gives the list of the positions at which elements of list match pattern.

```
{Position[lis, 0], Position[lis, 1], Position[lis, 2],
Position[lis, 3]}
```

{{{3}, {4}, {10}, {13}, {14}},
{{2}, {7}, {8}, {9}, {16}, {18}},
{{12}, {19}},
{{1}, {5}, {6}, {11}, {15}, {17}, {20}}}

```
Cases[lis, n_ /; n > 2]
```

{3, 3, 3, 3, 3, 3, 3}

pattern /; test tells *Mathematica* to match pattern only if the evaluation of test yields True.

Partition[list, n] generates nonoverlapping sublists of length n.

```
Partition[lis, 4]
```

{{3, 1, 0, 0}, {3, 3, 1, 1}, {1, 0, 3, 2},
{0, 0, 3, 1}, {3, 1, 2, 3}}

Partition[list, n, d] generates sublists of length n with offset d.

```
Partition[lis, 4, 1]
```

{{3, 1, 0, 0}, {1, 0, 0, 3}, {0, 0, 3, 3}, {0, 3, 3, 1},
{3, 3, 1, 1}, {3, 1, 1, 1}, {1, 1, 1, 0}, {1, 1, 0, 3},
{1, 0, 3, 2}, {0, 3, 2, 0}, {3, 2, 0, 0}, {2, 0, 0, 3},
{0, 0, 3, 1}, {0, 3, 1, 3}, {3, 1, 3, 1}, {1, 3, 1, 2},
{3, 1, 2, 3}}

```
Partition[lis, 4, 2]
```

{{3, 1, 0, 0}, {0, 0, 3, 3}, {3, 3, 1, 1}, {1, 1, 1, 0},
{1, 0, 3, 2}, {3, 2, 0, 0}, {0, 0, 3, 1}, {3, 1, 3, 1},
{3, 1, 2, 3}}

```
Partition[lis, 4, 3]
```

{{3, 1, 0, 0}, {0, 3, 3,1}, {1, 1, 1, 0}, {0, 3, 2, 0},
{0, 0, 3, 1}, {1, 3, 1, 2}}

```
Partition[lis, 4, 4]
```

{{3, 1, 0, 0}, {3, 3, 1, 1}, {1, 0, 3, 2}, {0, 0, 3, 1},
{3, 1, 2, 3}}

Count[list, pattern] counts the number of elements in list matching pattern.

```
{Count[lis, 0], Count[lis, 1], Count[lis, 2], Count[lis, 3]}
```

{5, 6, 2, 7}

5.5 Mathematical Operations on Lists

```
Clear[lis]
lis = Range[10]
```

{1, 2, 3, 4, 5, 6, 7, 8, 9, 10}

Apply[f,expression] or f @@ expression replaces the head of expression by f.

```
{Head[lis], Head[Apply[f, lis]]}
```

{List, f}

```
{Apply[Plus, lis], Plus @@ lis}
{Apply[Times, lis], Times @@ lis}
```

{55, 55}

{3628800, 3628800}

Total[list] is equivalent to Apply[Plus, list].

```
Total[lis]
```

55

Map[f, expression] or f /@ applies f to each element on the first level in expression. The following example shows how to apply a function f to different levels of a nested list.

```
Clear[lis, f]
lis = {{{u, v, w}, {x, y}, {z}}, {{a, b}, {c, d, e}}};
Map[f, lis] (* applies f to first level *)
Map[f,lis, {1}] (* applies f to first level *)
Map[f,lis, {2}] (* applies f to second level *)
Map[f,lis, {3}] (* applies f to third level *)
MapAll[f, lis] (* applies f to all levels *)
```

{f[{{u, v, w}, {x, y}, {z}}], f[{{a, b}, {c, d, e}}]}
{f[{{u, v, w}, {x, y}, {z}}], f[{{a, b}, {c, d, e}}]}
{{f[{u, v, w}], f[{x, y}], f[{z}]}, {f[{a, b}], f[{c, d, e}]}}
{{{f[u], f[v], f[w]}, {f[x], f[y]}, {f[z]}},
{{f[a], f[b]}, {f[c], f[d], f[e]}}}
f[{f[{f[{f[u], f[v], f[w]}], f[{f[x], f[y]}], f[{f[z]}]}],
f[{f[{f[a], f[b]}], f[{f[c], f[d], f[e]}]}]}]

```
Clear[lis, f]
lis = Range[1, 10]
f[x_] := x^2
{Map[f, lis], f / @ lis}
```

{1, 2, 3, 4, 5, 6, 7, 8, 9, 10}
{{1, 4, 9, 16, 25, 36, 49, 64, 81, 100},
{1, 4, 9, 1 6, 25, 36, 49, 64, 81, 100}}

We can also use pure functions.

```
Map[Function[x, x^2], lis]
Map[#^2 &, lis]
```

{1, 4, 9, 16, 25, 36, 49, 64, 81, 100}
{1, 4, 9, 16, 25, 36, 49, 64, 81, 100}

MapAt[g, expression, {part1, part2, ...}] applies g to specified parts
of expression.

```
MapAt[g, {a, {b, c}, d, e}, {{2, 1}, {4}}]
```

{a, {g[b], c}, d, g[e]}

To apply g to element b, we can first determine its Position.

```
pos = Position[{a, {b, c}, d, e}, b]
```

{{2,1}}

```
MapAt[g, {a, {b, c}, d, e}, pos]
```

{a, {g[b], c}, d, e}

MapThread[g, {{a1, a2, ...}, {b1, b2, ...}}] gives g[a1, b1],
g[a2, b2], ...

```
MapThread[g, {{1, 2, 3}, {1, 2, 3}}]
```

{g[1, 1], g[2, 2], g[3, 3]}

```
MapThread[Plus, {{1, 2, 3}, {1, 2, 3}}]
```

{2, 4, 6}

```
g[x_, y_] := x^2 + y^2
MapThread[g, {{1, 2, 3}, {1, 2, 3}}]
```

{2, 8, 18}

Here also we can use pure functions.

```
MapThread[Function[{x, y}, x + y], {{1, 2, 3}, {1, 2, 3}}]
```

{2, 4, 6}

```
MapThread[Function[{x ,y}, x^2 + y^2], {{1, 2, 3}, {1, 2,
3}}]
```

{2, 8, 18}

```
MapThread[#1 + #2 &, {{1, 2, 3}, {1, 2, 3}}]
```

{2, 4, 6}

```
MapThread[#1^2 + #2^2 &, {{1, 2, 3}, {1, 2, 3}}]
```

{2, 8, 18}

MapIndexed[g, expression] applies g to each element on the first level in expression giving the index of the element.

```
MapIndexed[g, {a, {b, c}, d, e}]
```

{g[a, {1}], g[{b, c}, {2}], g[d, {3}], g[e, {4}]}

5.6 Rearranging Lists

```
Clear[lis]
lis = {{{1, 2, 3}, {a, b, c}, {A, B, C}},
{{7, 8, 9}, {u, v, w}, {U, V, W}}}
```

{{{1, 2, 3}, {a, b, c}, {A, B, C}},
{{7, 8, 9}, {u, v, w}, {U, V, W}}}

Transpose[list] transposes the first two levels in list.

```
Transpose[lis]
```

{{{1, 2, 3}, {7, 8, 9}}, {{a, b, c}, {u, v, w}},
{{A, B, C}, {U, V, W}}}

Flatten[nestedLists] concatenates all sublists in one unique list.

```
Flatten[lis]
```

{1, 2, 3, a, b, c, A, B, C, 7, 8, 9, u, v, w, U, V, W}

This operation can be performed at different levels.

```
Flatten[lis,1]
```

{{1, 2, 3}, {a, b, c}, {A, B, C}, {7, 8, 9}, {u, v, w}, {U, V, W}}

```
Flatten[lis, 2]
```

{1, 2, 3, a, b, c, A, B, C, 7, 8, 9, u, v, w, U, V, W}

Permutations[list] generates the list of all permutations of the elements in list.

```
Permutations[{1, 2, 3}]
```

{{1, 2, 3}, {1, 3, 2}, {2, 1, 3}, {2, 3, 1}, {3, 1, 2}, {3, 2, 1}}

RotateLeft[expression, n] cycles the elements in expression n positions to the left and RotateRight[expression, n] n positions to the right.

```
RotateLeft[{a, b, c, d, e}]
RotateLeft[{a, b, c, d, e}, 1]
```

{b, c, d, e, a}
{b, c, d, e, a}

```
RotateLeft[{a, b, c, d, e}, 2]
```

{c, d, e, a, b}

```
RotateRight[{a, b, c, d, e}, 2]
```

{d, e, a, b, c}

Sort[list] sorts the elements of list into canonical order.

```
Sort[{d, b, o, j, p, q}]
```

{b, d, j, o, p, q}

```
Sort[{6, 8, 2, 1, 0, 5, 3}]
```

{0, 1, 2, 3, 5, 6, 8}

```
Sort[{9, 3, t, c, 7, n, 1, 4}]
```

{1, 3, 4, 7, 9, c, n, t}

The expression to be sorted need not have head List.

```
Sort[f[4, 6, a, 1, 7, j, 2, 9, 8]]
```

f[1, 2, 4, 6, 7, 8, 9, a, j]

5.7 Listability

Most *Mathematica* functions are listable. Here are a few examples.

```
Clear[lis]
lis = Table[0.5 + 0.1 k, {k, 0, 10}]
```

{0.5, 0.6, 0.7, 0.8, 0.9, 1., 1.1, 1.2, 1.3, 1.4, 1.5}

```
Sin[lis]
```

{0.479426, 0.564642, 0.644218, 0.717356, 0.783327, 0.841471,
0.891207, 0.932039, 0.963558, 0.98545, 0.997495}

```
Sinh[lis]
```

{0.521095, 0.636654, 0.758584, 0.888106, 1.02652, 1.1752,
1.33565, 1.50946, 1.69838, 1.9043, 2.12928}

```
Log[lis]
```

{−0.693147, −0.510826, −0.356675, −0.223144, −0.10536,
0.,0.0953102, 0.182322, 0.262364, 0.336472, 0.405465}

```
Exp[lis]
```

{1.64872, 1.82212, 2.01375, 2.22554, 2.4596, 2.71828,
3.00417, 3.32012, 3.6693, 4.0552, 4.48169}

Listability is also valid for nested lists. The listable function applies to the highest level.

```
Sin[{{a, b}, {{c, d}, {e}}}]
Sinh[{{a, b}, {{c, d}, {e}}}]
Log[{{a, b}, {{c, d}, {e}}}]
Exp[{{a, b}, {{c, d}, {e}}}]
```

{{Sin[a], Sin[b]}, {{Sin[c], Sin[d]}, {Sin[e]}}}
{{Sinh[a], Sinh[b]}, {{Sinh[c], Sinh[d]}, {Sinh[e]}}}
{{Log[a], Log[b]}, {{Log[c], Log[d]}, {Log[e]}}}
{{E^a, E^b}, {{E^c, E^d}, {E^e}}}

User-defined functions are, in general, not listable.

```
Clear[lis]
lis = Table[0.5 + 0.1 k, {k, 0, 10}];
f[lis]
```

f[{0.5, 0.6, 0.7, 0.8, 0.9, 1., 1.1, 1.2, 1.3, 1.4, 1.5}]

They can, however, be made Listable using the command:

```
Attributes[f] = Listable
```

Listable

```
f[lis]
```

{f[0.5], f[0.6], f[0.7], f[0.8], f[0.9], f[1.], f[1.1],
f[1.2], f[1.3], f[1.4], f[1.5]}

If a function is not Listable, it can applied to all the elements of a list using the command Map.

```
F[lis]
```

F[{0.5, 0.6, 0.7, 0.8, 0.9, 1., 1.1, 1.2, 1.3, 1.4, 1.5}]

```
Map[F, lis]
```

{F[0.5], F[0.6], F[0.7], F[0.8], F[0.9], F[1.], F[1.1], F[1.2],
F[1.3], F[1.4], F[1.5]}

Functions may be defined on lists. For example, the function ff first adds the elements of a list and then squares the result.

```
ff[lis_List] := (Apply[Plus, lis])^2
ff[{1, 2, 3, 4, 5}]
```

fff first adds the elements of the list, takes the closest integer of the sum, and returns 0 if the integer is even and 1 otherwise.

```
fff[lis_List] := Mod[Round[Apply[Plus, lis]], 2]
fff[{3.12, 7.53, 2.27, 5.11}]
fff[{3.12, 7.53, 2.27, 6.11}]
```

0

1

6

Graphics

Graphics are important components of many applications, and *Mathematica* provides powerful graphics capabilities. This chapter is rather detailed but a lot more can be found in [66] and [55].

6.1 2D Plots: Function Plotting

To graph $f(x)$ in the interval (x_1, x_2), we use the command Plot[f[x], { x, x1, x2}].

```
Plot[Cos[2x] + Sin[x], {x, - Pi, Pi}];
```

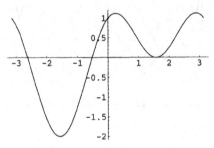

Fig. 6.1. *Graph of* $\cos(2x) + \sin(x)$ *for* $x \in [-\pi, \pi]$.

We can graph more than one function on the same interval using the command Plot[{f[x], g[x], h[x]}, {x, x1, x2}].

```
Plot[{Cos[x], Cos[3 x], Cos[5 x]}, {x, - Pi, Pi}];
```

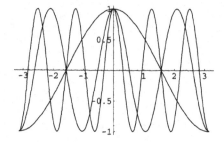

Fig. 6.2. *Graphs of* cos(x), cos(3x), *and* cos(5x) *for* $x \in [-\pi, \pi]$.

6.1.1 Parametric Plots

We can also plot curves represented by parametric coordinates using the command: `ParametricPlot[{x[t], y[t]}, {t, t1, t2}]` which generates a parametric plot where the x and y coordinates are functions of t varying between t_1 and t_2.

```
ParametricPlot[{Sin[3t], Sin[8 t]}, {t, 0, 2 Pi}];
```

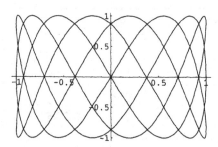

Fig. 6.3. *Parametric plot of* $(\sin(3t), \sin(8t))$ *for* $t \in [0, 2\pi]$.

6.1.2 Polar Plots

We can also plot a curve defined by its polar coordinate representation $r = f(\theta)$ using `ParametricPlot`, where the functions x and y are defined by

$$x(t) = f(t)\cos(t), y(t) = f(t)\sin(t).$$

Considering, for example, the curve defined by $r = \sin 4\theta$, we obtain

```
ParametricPlot[{Sin[4t] Cos[t], Sin[4t] Sin[t]}, {t, 0, 2Pi},
AspectRatio -> 1];
```

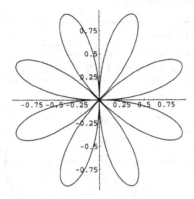

Fig. 6.4. *Parametric plot of the curve given in polar coordinates by* $r = \sin(4\theta)$ *for* $\theta \in [0, 2\pi]$.

Such a type of polar graph is called a *rose*. Note that we used the option AspectRatio → 1 to specify that the ratio of height to width of the plot should be equal to 1.

We do not need to use **ParametricPlot**. We can plot curves using the command **PolarPlot**.

```
PolarPlot[Sin[3 theta], {theta, 0, 2 Pi}];
```

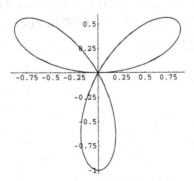

Fig. 6.5. *Polar plot of the curve defined by* $r = \sin(3\theta)$ *for* $\theta \in [0, 2\pi]$.

6.1.3 Implicit Plots

Loading the package Graphics'ImplicitPlot' we can also plot a curve defined by an implicit function $f(x, y) = 0$.

```
<< Graphics'ImplicitPlot
```

```
ImplicitPlot[(x^2 + y^2)^3 == (x^2 - y^2), {x, - 2, 2}];
```

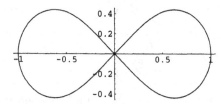

Fig. 6.6. *Implicit plot of the curve defined by* $(x^2 + y^2)^3 = (x^2 - y^2)$ *for* $x \in [-2, 2]$.

6.1.4 Color

To display color we have at our disposal various graphics directives: RGBColor[redLevel, greenLevel,blueLevel]. and CMYKColor[cyanLevel, magentaLevel, yellowLevel, blackLevel], where colorLevel is a real number in the range 0 to 1. PlotStyle → style specifies graphics directives for all lines or points.

```
Plot[{Cos[x], Cos[3 x], Cos[5 x]}, {x, - Pi, Pi},
PlotStyle → {RGBColor[0, 0, 1], RGBColor[0, 1, 0],
RGBColor[1, 0, 0]}];
```

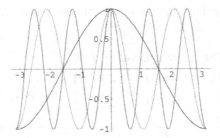

Fig. 6.7. *Graphs of* cos(x), cos(3x), *and* cos(5x) *for* $x \in [-\pi, \pi]$, *colored, respectively, in blue, green, and red.*

```
Plot[{Cos[x], Cos[3 x], Cos[5 x]}, {x, - Pi, Pi},
PlotStyle → {CMYKColor[1, 0, 0, 0], CMYKColor[0, 1, 0, 0],
CMYKColor[0, 0, 1, 0]}];
```

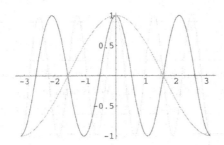

Fig. 6.8. *Same as above but colored, respectively, in cyan, magenta, and yellow.*

Hue[h], where h varies from 0 to 1, specifies color with a particular hue. As h increases, starting from h = 0, which corresponds to red, the color changes to yellow, green, cyan, blue, and red again for h = 1.

```
Show[Graphics[{Table[{Hue[j / 9],
Rectangle[{j, 0}, {j+1,5}]}, {j, 0, 9}],
Line[{{0, 0}, {10, 0}, {10, 5}, {0, 5}, {0,0}}]}],
AspectRatio → 0.2];
```

Fig. 6.9. *Rectangles of varying hue.*

To specify the gray level intensity, we can use `GrayLevel[Level]`, where Level varies from 0 (black) to 1 (white).

```
Show[Graphics[{Table[{GrayLevel[j / 9],
Rectangle[{j, 0}, {j+1,5}]}, {j, 0, 9}],
Line[{{0, 0}, {10, 0}, {10, 5}, {0, 5}, {0, 0}}]}],
AspectRatio → 0.2];
```

Fig. 6.10. *Rectangles of varying gray level.*

The package `Graphics'Colors'` contains a large number of predefined colors. Entering the command `?Graphics'Colors'*` gives the list of all predefined colors.

```
<<Graphics'Colors'
?Graphics'Colors'*
```

The output is suppressed but we can get the color code of each of these predefined colors by entering its name. For example:

```
Firebrick
```

`RGBColor[0.698004, 0.133305, 0.133305]`

6.1.5 Dashing

`Dashing[a, b,...]` draws dashed lines with segments of successive length a, b, ...

```
Plot[{Cos[x], Cos[3 x], Cos[5 x]}, {x, - Pi, Pi},
PlotStyle → {Dashing[{0.07, 0.07}], Dashing[{0.04, 0.04}],
Dashing[{0.01, 0.01}]}];
```

6.1.6 Text

`Text[expression,{x, y}]` is a graphics primitive that prints expression centered at the point {x, y}.

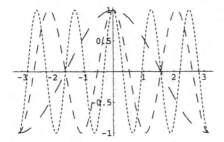

Fig. 6.11. *Graphs of* cos(*x*), cos(3*x*), *and* cos(5*x*) *for* $x \in [-\pi, \pi]$, *with different dashing plot styles.*

```
Plot[Exp[- 0.1 x] Cos[x], {x, 0, 6Pi},
PlotStyle → {RGBColor[0, 0, 1]},
Epilog → Text[''Damped Oscillations'', {3 Pi, 0.7}],
DefaultFont → {''Helvetica'', 12}];
```

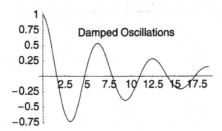

Fig. 6.12. *Graphs of* $e^{-0.1x}$ cos(3x) *for* $x \in [0, 6\pi]$, *with added text.*

Epilog is an option to be rendered after the main part of the graphics is rendered, DefaultFont controls the font used for text in graphics, and TextStyle specifies the default style and font options with which text should be rendered. The text has to be placed between double quotes (").

```
Plot[{Cos[x], Cos[3 x], Cos[5 x]}, {x, - Pi, Pi},
PlotStyle → {RGBColor[0, 0, 1], RGBColor[0, 1, 0],
RGBColor[1, 0, 0]},
TextStyle → {FontSlant → ''Italic'', FontSize → 12}];
```

FontSlant is an option that specifies how slanted characters should be.

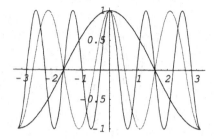

Fig. 6.13. *Same as Figure 6.7 with a different text style.*

6.1.7 Axes, Ticks and Labels

AspectRatio is an option which specifies the ratio of height to width for a plot. For two-dimensional plots, the default value of AspectRatio is 1 / GoldenRatio. AspectRatio → Automatic (equivalent to AspectRatio → 1) uses the same scale in x and y.

```
Plot[BesselJ[0, x], {x, 0, 20}];
```

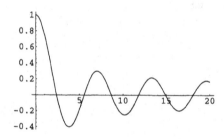

Fig. 6.14. *Graph of $J_0(x)$ for $x \in [0, 20]$.*

We can replace the traditional axes by frame axes, and label the plot.

```
Plot[BesselJ[0,x], {x, 0, 20}, Axes → False, Frame → True,
PlotStyle → {RGBColor[0,0,1]},
PlotLabel → ''Bessel Function of Order 0'',
DefaultFont → {''Courier'', 12}];
```

Here is another example using TraditionalForm to write a mathematical expression, and Ticks to specify tick mark positions.

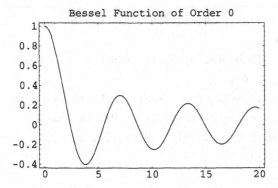

Fig. 6.15. *Same as Figure 6.14 with a plot label.*

```
lbl = Exp[- α x] Cos[x];
Plot[Exp[- 0.1 x] Cos[x], {x, 0, 6 Pi},
PlotStyle → {RGBColor[0, 0,1]},
Ticks → {{5, 10, 15}, {- 0.75, - 0.25, 0.25, 0.75}},
Epilog → Text[TraditionalForm[lbl], {3 Pi, 0.7}],
DefaultFont → {``Helvetica'', 12}];
```

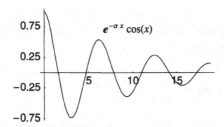

Fig. 6.16. *Same as Figure 6.12 with mathematical symbols in traditional form.*

In the following example, we use `Ticks` to specify the positions of tick marks only for one axis and let the tick marks for the other axis be placed automatically. We also use the graphics primitive `Point` to place a big red dot at the origin. We study in more detail graphics primitives in section 6.3 below.

```
plCosColor = Plot[{Cos[x], Cos[3 x], Cos[5 x]},
{x, - Pi, Pi},
PlotStyle → {RGBColor[0, 0, 1], RGBColor[0, 1, 0],
RGBColor[1, 0, 0]},
TextStyle → {FontSlant → "Italic", FontSize → 14},
Ticks → {{ - Pi, - Pi/2, 0, Pi/2, Pi} Automatic},
Epilog → {PointSize[0.05], RGBColor[1, 0, 0],
Point[{0, 0}]}];
```

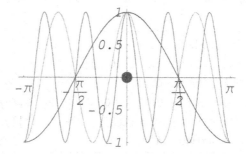

Fig. 6.17. *Same as Figures 6.7 and 6.13 with different options.*

Using the command Options[plCosColor] will give the explicit list of options of the above graphics object.

6.1.8 Graphics Array

To present a collection of plots we use GraphicsArray. The following command displays the traditional form of a mathematical symbol.

```
TraditionalForm[BesselJ[0, x]]
```

$J_0(x)$

We first generate four plots without displaying them, and use the option Ticks → None to eliminate ticks.

```
plJ0 = Plot[BesselJ[0, x], {x, 0, 10},
PlotStyle → {CMYKColor[1, 0, 0, 0]}, Ticks → None,
Epilog → Text[TraditionalForm[BesselJ[0, x]], {3, 0.5}],
DefaultFont → {''Helvetica'', 16},
DisplayFunction → Identity];
plJ1 = Plot[BesselJ[1, x], {x, 0, 10},
PlotStyle → {CMYKColor[0, 1, 0, 0]}, Ticks → None,
Epilog → Text[TraditionalForm[BesselJ[1, x]], {4.5, 0.35}],
DefaultFont → {''Helvetica'', 16},
DisplayFunction → Identity];
plJ2 = Plot[BesselJ[2, x], {x, 0, 10},
PlotStyle → {CMYKColor[0, 0, 1, 0.5]}, Ticks → None,
Epilog → Text[TraditionalForm[BesselJ[2, x]], {6, 0. 3}],
DefaultFont → {''Helvetica'', 16},
DisplayFunction → Identity];
plJ3 = Plot[BesselJ[3, x], {x, 0, 10},
PlotStyle → {CMYKColor[0, 0, 0, 1]}, Ticks → None,
Epilog → Text[TraditionalForm[BesselJ[3, x]], {7, 0. 3}],
DefaultFont → {''Helvetica'', 16},
DisplayFunction → Identity];
```

And we display the four plots in a 2×2 array.

```
plArray1 = Show[GraphicsArray[{{plJ0, plJ1}, {plJ2, plJ3}},
TextStyle → {FontSlant → ''Italic'', FontSize → 14},
PlotLabel → ''First Bessel Functions''],
DisplayFunction → $DisplayFunction];
```

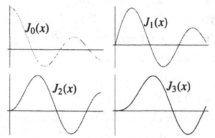

Fig. 6.18. *Graphics array of the Bessel functions J_0, J_1, J_2, and $J_3(x)$ in the interval $[0, 10]$.*

We can increase spacing with the option GraphicsSpacing,

```
plArray2 = Show[GraphicsArray[{{plJ0, plJ1}, {plJ2, plJ3}},
TextStyle → {FontSlant → ''Italic'', FontSize → 14},
GraphicsSpacing → {0.3, 0.3},
PlotLabel → ''First Bessel Functions'']];
```

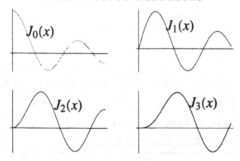

Fig. 6.19. *Graphics array of the Bessel functions J_0, J_1, J_2, and $J_3(x)$ in the interval $[0, 10]$ using the option* GraphicsSpacing.

and finally add a frame.

```
Show[plArray2, Frame → True];
```

First Bessel Functions

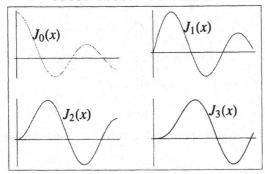

Fig. 6.20. *Graphics array above with a frame.*

6.1.9 Plot Range

Consider the following graphics.

```
plCosh = Plot[Cosh[2 x] Cos[10 x], {x, - 3, 3}];
```

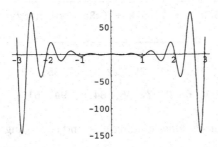

Fig. 6.21. *Graph of* $\cosh(2x)\cos(10x)$ *for* $x \in [-3,3]$.

We can restrict the plot range to analyze more closely a part of the plot using the option **PlotRange**.

```
Show[plCosh, PlotRange → {{-1, 1}, {- 2, 2}}];
```

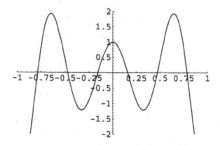

Fig. 6.22. *Graph of* cosh(2x) cos(10x) *in a reduced plot range.*

6.2 More 2D Plots

6.2.1 Plotting Lists

In this section various commands that help visualize numeric data are presented. Data are either of the form of a simple list $\{d_1, d_2, \ldots, d_n\}$, where each d_i is a single term or a list of lists of the form: $\{\{x_1, y_1\}, \{x_2, y_2\}, \ldots, \{x_n, y_n\}\}$, where each sublist $\{x_i, y_i\}$ represents the coordinates of point i. List[{d1, d2, ..., dn}] is equivalent to List[{{1, d1}, {2, d2}, ..., {n, dn}}].

```
data = Table[0.3 k^2 + 1.75 k + (2Random[Integer] - 1), {k,
1, 15}]
```

{3.05, 5.7, 6.95, 10.8, 17.25, 22.3, 27.95, 32.2,
39.05, 46.5, 56.55, 65.2, 72.45, 84.3, 94.75}

We first plot this list of values, adding text and specifying the point size (see Figure 6.23).

```
plData = ListPlot[data, PlotStyle → {PointSize[0.02],
RGBColor[1,0,0]}, Epilog → Text[''data points''  {4, 60}],
DefaultFont → {''Helvetica'', 14}];
```

We then find a quadratic function to fit the data and plot this function,

```
fitData = Fit[data, {1, x, x^2}, x]
```

$1.04396 + 1.43911\ x + 0.319877\ x^2$

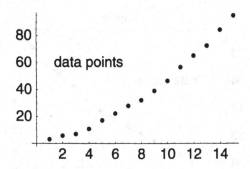

Fig. 6.23. *Plot of a list of data points.*

```
plFit = Plot[fitData, {x, 0, 15}];
```

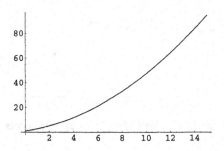

Fig. 6.24. *Graph of* $1.04396 + 1.43911x + 0.319877x^2$ *(* $x \in [0, 15]$ *) that fits the list of data points of Figure 6.23 above.*

and we finally represent on the same graph the fitting function and the data points (Figure 6.25).

```
Show[{plData, plFit}, PlotLabel → ''Data fit'',
DefaultFont → {''Helvetica'' , 14}, Frame → True];
```

Loading the package `Graphics'MultipleListPlot'` gives the possibility to plot several lists of data on the same graph.

```
<<Graphics'MultipleListPlot'
```

```
data1 = Table[Cos[Pi k / 20] + 0.1 Random[], {k, 1, 10}];
data2 = Table[Sin[Pi k / 20] + 0.1 Random[], {k, 1, 10}];
MultipleListPlot[data1, data2, PlotJoined → True];
```

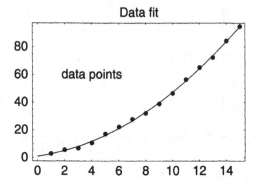

Fig. 6.25. *Plots of the list of data points and the quadratic fitting function.*

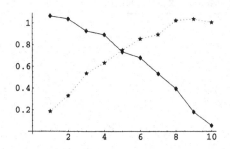

Fig. 6.26. *Plots of two lists of data points.*

The same plot with different options (Figure 6.27):

```
MultipleListPlot[data1, data2,
SymbolShape → {PlotSymbol[Triangle], PlotSymbol[Box]},
SymbolStyle → {RGBColor[1,0,0], RGBColor[0,0,1]},
PlotJoined → True];
```

6.2.2 Special Plots

Loading the package `Graphics`Graphics`` allows us to use a greater variety of graphics commands.

```
<<Graphics`Graphics`
```

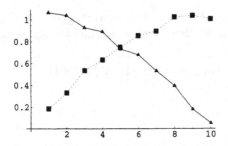

Fig. 6.27. *Same as Figure 6.26 above with different options.*

LogPlot[f[x], {x, x1, x2}] generates a log-linear plot in the interval [x1, x2] and LogLogPlot[f[x], {x, x1, x2}] generates a log-log plot in the same interval.

```
LogPlot[Exp[- 3 x], {x, 0, 6}];
```

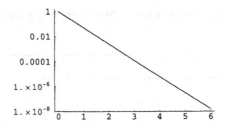

Fig. 6.28. *Logplot of* e^{-3x} *for* $x \in [0, 6]$.

```
LogLogPlot[x^(3/4), {x, 0, 1}];
```

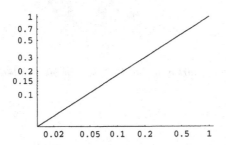

Fig. 6.29. *Loglogplot of* $x^{3/4}$ *for* $x \in [0, 1]$.

BarChart[list], PieChart[list], and Histogram[list] plot as their names explicitly suggest, a bar chart, a pie chart, and the histogram of a list of data.

```
rnd = Table[Random[Integer, {1, 5}], {20}]
```

{2, 4, 4, 4, 1, 4, 3, 2, 4, 1, 5, 5, 5, 2, 1, 5, 1, 4, 4, 2}

```
BarChart[rnd];
```

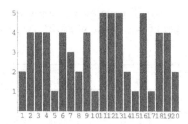

Fig. 6.30. *Bar chart of a list of 20 random integers between 1 and 5.*

```
PieChart[rnd];
```

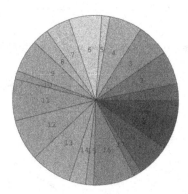

Fig. 6.31. *Pie chart of a list of 20 random integers between 1 and 5.*

```
Histogram[rnd];
```

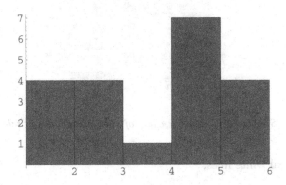

Fig. 6.32. *Histogram of a list of 20 random integers between 1 and 5.*

6.2.3 A Horizontal Bar Chart with Many Options

We represent the 2004 Car Sale Statistics in Maryland (data obtained from the Maryland Motor Vehicle Administration). We draw a horizontal bar chart increasing little by little the number of options.

```
Clear[data]
data = {{33361, ''January''}, {27780, ''February''},
{39340, ''March''}, {37478, ''April''}, {37819, ''May''},
{42758, ''June''}, {38329, ''July''}, {37175, ''August''},
{38712, ''September''}, {34839, ''October''},
{29859, ''November''}, {31058, ''December''}};
numbers = Map[First, data];
months = Map[Last, data];
months = Transpose[{Range[Length[months]], months}];
barch1 = BarChart[numbers, BarOrientation → Horizontal,
BarEdges → False, BarStyle → GrayLevel[0.5],
Ticks → {Automatic, months}];
```

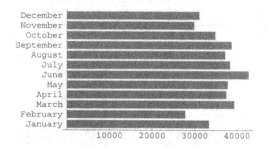

Fig. 6.33. *Simple horizontal bar chart of 2004 Maryland car sale statistics.*

We add vertical white lines,

```
barch2 = Show[barch1, Epilog → {GrayLevel[1],
AbsoluteThickness[0.25], Table[Line[{{i ,0.5}, {i, 12.5}}],
{i, 10000, 50000, 10000}]}];
```

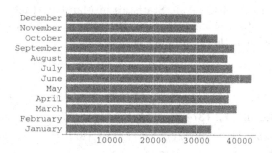

Fig. 6.34. *Horizontal bar chart of 2004 Maryland car sale statistics with vertical white lines.*

then add a title and change the default font,

```
barch3 = Show[barch2, AxesStyle → GrayLevel[1],
PlotLabel → FontForm[''Maryland 2004 Car Sales Statistics'',
{''Helvetica'', 16}], DefaultFont → {''Helvetica'', 11}];
```

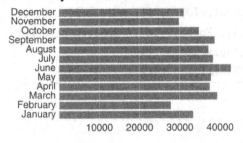

Fig. 6.35. *Adding a title to the figure above.*

and finally add a frame slightly thicker than the default thickness.

```
Show[GraphicsArray[{barch3}, Frame → True,
FrameStyle → Thickness[0.01]]];
```

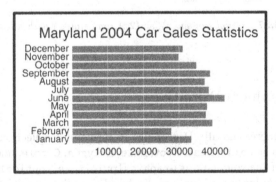

Fig. 6.36. *Horizontal bar chart of 2004 Maryland car sale statistics with vertical white lines, a title, and a frame.*

6.2.4 Labels

The option **PlotLegend** draws a legend beside the plot associating text with the plot style.

```
<< Graphics'Legend'
```

```
Plot[{Sin[x], Sin[2 x], Sin[3 x]}, {x, - 2 Pi, 2 Pi},
PlotStyle → {RGBColor[0, 0, 1], RGBColor[0, 1, 0],
RGBColor[1, 0, 0]}, FormatType → TraditionalForm,
DefaultFont → {"Helvetica" ,14},
PlotLegend → {"Sin[x]", "Sin[2x]", "Sin[3x]"}];
```

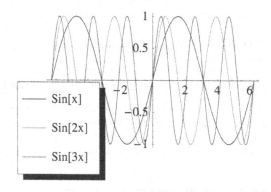

Fig. 6.37. *Graphs of* sin(*x*), sin(2*x*), *and* sin(3*x*) *with a legend.*

6.3 2D Graphical Primitives

Mathematica represents all 2D graphics in terms of a collection of graphics primitives such as Point, Line, Rectangle, Polygon, Circle, and so on. The general form of the command is Graphics[primitive, option1, option2, ...]. These graphics primitives are displayed with the command Show.

6.3.1 Point

Point[{x, y}] represents a single point located at {x, y}.

```
Show[Graphics[Table[{PointSize[0.04], RGBColor[0, 0, 1],
Point[{Cos[2 k Pi], Sin[2 k Pi]}]}, {k, 0, 0.9, 0.1}]],
AspectRatio → Automatic];
```

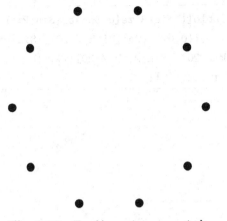

Fig. 6.38. *Ten blue points on a circle.*

6.3.2 Line

Line[{{x1, y1}, {x2, y2}, ..., {xn, yn}] represents a line through all points {{x1, y1}, {x2, y2}, ..., {xn, yn}.

```
Show[Graphics[{GrayLevel[0.3], Thickness[0.03],
Line[{{0, 0}, {0, 1}, {1, 1}, {1, 0}, {0, 0}}]}],
AspectRatio → Automatic];
```

Fig. 6.39. *A thick square drawn using the command* Line.

To combine colored lines and points,

```
Show[Graphics[Table[{PointSize[0.05 Abs[Sin[Pi k / 10]]],
Hue[k / 10], Line[{{0,0}, {Cos[Pi k / 20], Sin[Pi k / 20]}}],
Point[{Cos[Pi k / 20], Sin[Pi k / 20]}]}, {k, 1, 40}]],
AspectRatio → Automatic];
```

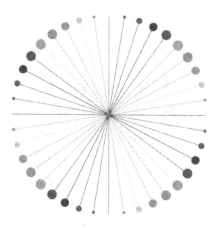

Fig. 6.40. *Colored lines and points.*

```
Show[Graphics[Table[{Hue[k / 60],
Line[{{k, 0}, {k, Sin[Pi k/30]}}]}, {k, 1, 60}]]];
```

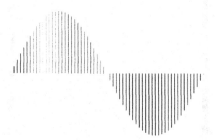

Fig. 6.41. *Colored lines of varying lengths.*

6.3.3 Rectangle

Rectangle[{{x1, y1}, {x2, y2}}] represents a filled rectangle where {x1, y1} are the coordinates of the bottom left corner, and {x2, y2} the coordinates of the top right corner.

```
Show[Graphics[{Rectangle[{0, 0}, {2, 1}],
Rectangle[{2.5, 0}, {4.5, 1}]}],
AspectRatio → Automatic];
```

Fig. 6.42. *Two filled rectangles.*

6.3.4 Polygon

Polygon[{x1,y1}, {x2, y2}, ..., {xn, yn}] represents a filled n-gon whose corners are the n points {xi, yi}.

Below we define a polygon by the position (x, y) of its center and the number n of equal sides.

```
centeredPoly[{x_, y_}, n_] :=
Polygon[Table[{Sin[2 k Pi / n] + x, Cos[2 k Pi / n] + y},
{k, 0, n}]]
```

```
Show[Graphics[{{Hue[0. 6], centeredPoly[{0.5, 0.5}, 6]},
{Hue[0.7], centeredPoly[{2.5, 0.5}, 7]},
{Hue[0.8], centeredPoly[{4.5, 0.5}, 8]},
{Hue[0.3], centeredPoly[{0.5, 3}, 3]},
{Hue[0.4], centeredPoly[{2.5, 3}, 4]},
{Hue[0.5], centeredPoly[{4.5, 3}, 5]}}],
AspectRatio → Automatic];
```

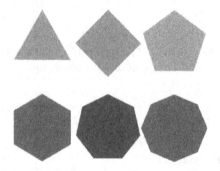

Fig. 6.43. *Six regular polygons whose positions are defined by their centers.*

6.3.5 Circle

Circle[{x, y}, r] represents a circle of radius r centered at {x, y}.

```
Show[Graphics[{
{GrayLevel[0.6], Thickness[0.01], Circle[{0, 0}, 1]},
{Hue[0.5], centeredPoly[{0, 0}, 5]}}],
AspectRatio → Automatic];
```

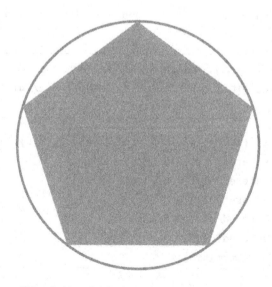

Fig. 6.44. *Circle with an inscribed pentagon.*

6.3.6 Text

`Text[expression, {x, y}]` is a graphics primitive that prints `expression` centered at the point {x, y}. We can use this command to label points as shown below.

```
pts = Table[{Random[], Random[]}, {12}];
Show[Graphics[{{PointSize[0.07], CMYKColor[0, 0, 1, 0],
Map[Point, pts]}, Table[Text[i, Part[pts, i]],
{i, 1, Length[pts]}]}], PlotRange → All];
```

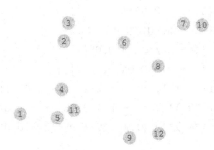

Fig. 6.45. *Labeled points.*

6.3.7 Golden Ratio

The shape of a picture is determined by the `AspectRatio` option. The default option is the golden ratio, that is, a ratio $r = a/b$ such that

$$\frac{a}{b} = \frac{b}{a+b} \text{ or } r = \frac{1}{1+r}.$$

The numerical value of r is the positive root of

```
Solve[r == 1 / (1+r), r]
```

$$\{\{r \rightarrow \frac{1}{2} (-1 - \text{Sqrt}[5])\}, \{r \rightarrow \frac{1}{2} (-1 + \text{Sqrt}[5])\}\}$$

that is,

$$r = \frac{1}{2}(-1 + \sqrt{5}) = 0.618034.$$

Starting from an initial golden rectangle, when we remove the largest square, the remaining rectangle is also a golden rectangle. The figure below represents a few iterations of this process.

```
r = 0.618034;
l1 = Line[{{0, 0}, {1 + r, 0}, {1 + r, 1}, {0, 1}, {0, 0}}];
l2 = Line[{{1, 0}, {1, 1}}];
l3 = Line[{{1, r}, {1 + r, r}}];
l4 = Line[{{1 + r^2, r}, {1 + r^2, 1}}];
l5 = Line[{{1 + r^2, 3 r - 1},{1 + r, 3 r - 1}}];
l6 = Line[{{1 + r^2 + r^4, r + r^3}, {1 + r^2 + r^4, 1}}];
t1 = Text["1",{0.5, - 0.05}];
t2 = Text["1",{- 0.05, 0.5}];
t3 = Text["r", {1 + r / 2, - 0.05}];
t4 = Text["r", {1 + r + 0.04, r / 2}];
t5 = Text["\!\(r\^2\)", {1 + r^2/2, r - 0.05}];
t6 = Text["\!\(r\^3\)", {1 + r + 0.05, r + r^3/2}];
t7 = Text["\!\(r\^4\)", {1 + r^2 + r^4/2 + 0.01,
r + r^3 - 0.05}];
Show[Graphics[{l1, l2, l3, l4, l5, l6,
t1, t2, t3, t4, t5, t6, t7}],
DefaultFont → {"Helvetica", 14}, AspectRatio →
Automatic];
```

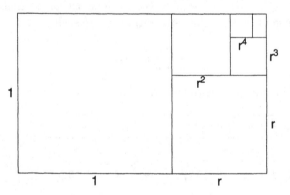

Fig. 6.46. *Sequence of golden rectangles.*

In the input above we used the form \!\(r\^2\) to display r^n as shown below. We could have obtained a similar result using TraditionalForm[Superscript-[r,n]] (see Figure 6.20) except that in this case the symbol would have been displayed using a different font.

```
\!\(r\^2\
```

r^n

Taking into account that $r = 1/(1 + r)$, that is, $r + r^2 = 1$, we can verify that

$$\sum_{n=0}^{\infty} r^{2n+1} = \frac{r}{1 - r^2} = 1 \text{ and } \sum_{n=0}^{\infty} r^{2n} = \frac{1}{1 - r^2} = 1 + r.$$

The golden ratio is closely related to Fibonacci numbers.

```
Table[Together[Nest[Function[x, 1 + 1 / x], a, k]],
{k, 1, 5}]
```

$$\left\{ \frac{1+a}{a}, \frac{1+2a}{1+a}, \frac{2+3a}{1+2a}, \frac{3+5a}{2+3a}, \frac{5+8a}{3+5a} \right\}$$

The sequence of fractions above is also given by

```
Table[(Fibonacci[k] + a Fibonacci[k+1]) /
(Fibonacci[k - 1] + a Fibonacci[k]), {k, 1, 5}]
```

$$\left\{ \frac{1+a}{a}, \frac{1+2a}{1+a}, \frac{2+3a}{1+2a}, \frac{3+5a}{2+3a}, \frac{5+8a}{3+5a} \right\}$$

We also have

```
NestList[Function[x, Together[1 / (1+x)]], a, 5]
```

$$\left\{ a, \frac{1}{1+a}, \frac{1+a}{2+a}, \frac{2+a}{3+2a}, \frac{3+2a}{5+3a}, \frac{5+3a}{8+5a} \right\}$$

which can be written

```
Table[(Fibonacci[k] + a Fibonacci[k - 1]) /
(Fibonacci[k + 1] + a Fibonacci[k]), {k, 0, 5}]
```

$$\left\{a, \frac{1}{1+a}, \frac{1+a}{2+a}, \frac{2+a}{3+2\,a}, \frac{3+2\,a}{5+3\,a}, \frac{5+3\,a}{8+5\,a}\right\}$$

6.4 Animation

6.4.1 Rolling Circle

The cycloid is the locus of a point of the circumference of a circle that rolls along a straight line. We first define a function that draws a dotted circle, and then animate the rolling motion of this circle along a straight line.

```
dottedCircle[{x_, y_}, angle_, rad_] := {Circle[{x, y}, rad],
Line[{{x, y}, rad {Sin[angle], Cos[angle]} + {x, y}}],
PointSize[0.03], RGBColor[1,0,0],
Point[rad {Sin[angle], Cos[angle]} + {x, y}]}
```

```
Show[Graphics[dottedCircle[{0, 0}, Pi / 2, 1]],
AspectRatio → Automatic];
```

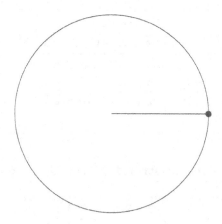

Fig. 6.47. *Dotted circle.*

To drawing the dotted circle and the straight line,

```
Show[Graphics[{dottedCircle[{0, 1}, Pi, 1],
Line[{{0, 0}, {4 Pi, 0}}]}],
AspectRatio → Automatic];
```

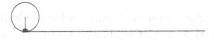

Fig. 6.48. *Dotted circle rolling on a straight line.*

To animate the rolling motion we generate the following sequence of images and then select an output cell and in the Cell menu go to Animate Selected Graphics.

```
Table[Show[Graphics[{dottedCircle[{t, 1}, Pi + t, 1],
Line[{{0, 0}, {4 Pi, 0}}]}], AspectRatio → Automatic],
{t, 0, 4 Pi, Pi / 10}];
```

In the next figure the dotted circle has reached the position corresponding to $t = 5$.

Fig. 6.49. *Position of the rolling dotted circle for $t = 5$.*

We can find the locus of the red point using Cases.

```
pt = Cases[dottedCircle[{t, 1}, Pi + t, 1], Point[_]]
```

{Point[{t - Sin[t], 1 - Cos[t]}]}

and represent the locus by a sequence of points.

```
ptList = Table[{PointSize[0.02], RGBColor[1, 0, 0], pt}, {t,
0, 4 Pi, Pi / 10}];
```

```
Show[Graphics[ptList], AspectRatio → Automatic];
```

Fig. 6.50. *Locus of the red dot.*

More details can be found on rolling circles in [64].

Another interesting application of Animate Selected Graphics is to visualize how a graph is traced. Consider again the polar coordinate representation $r = \sin 4\theta$ (Figure 6.4), and use the command

```
Do[ParametricPlot[{Sin[4t] Cos[t], Sin[4t] Sin[t]},
{t, 0, 2 k Pi / 20}, PlotRange → {{- 1, 1}, {- 1,1}},
AspectRatio → Automatic], {k, 1, 20}];
```

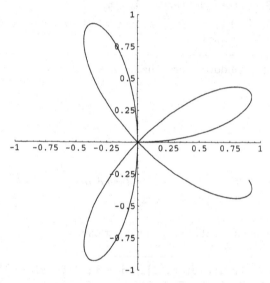

Fig. 6.51. *One image of the sequence generating the animated drawing of the rose* $r = 4\theta$.

6.5 2D Vector Fields

The package **Graphics'PlotField'** is used to plot two-dimensional vector fields.

```
<<Graphics'PlotField'
```

PlotVectorField[{f[x, y], g[x,y]}, {x, x1, x2}, {y, y1, y2}] plots the vector field defined by the two-dimensional vector function $(f(x,y), g(x,y))$ in the domain $[x_1, x_2] \times [y_1, y_2]$.

```
PlotVectorField[{Cos[2x], Sin[y]},
{x, - Pi, Pi}, {y,- Pi, Pi}];
```

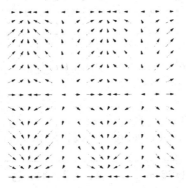

Fig. 6.52. *Vector field* $(\cos(2x), \sin(y))$ *in the domain* $[-\pi, \pi] \times [-\pi, \pi]$.

There exist many options. We can, for example, add colors and a frame. The color function (RGBColor[#, 1 - #, 0] &) makes long arrows red and short arrows green.

```
PlotVectorField[{Cos[2x], Sin[y]},
{x, - Pi, Pi}, {y,- Pi, Pi},
ColorFunction → (RGBColor[#, 1 - #, 0] &) , Frame → True];
```

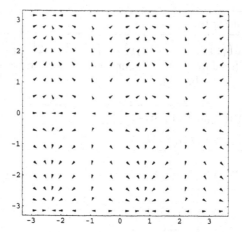

Fig. 6.53. *Vector field* $(\cos(2x), \sin(y))$ *in the domain* $[-\pi, \pi] \times [-\pi, \pi]$ *adding colors and a frame.*

PlotGradientField[V[x, y], {x, x1, x2}, {y, y1, y2}] plots the gradient field of the potential function $V(x, y)$ in the domain $[x_1, x_2] \times [y_1, y_2]$.

```
PlotGradientField[x^3 + y^3, {x, - 3, 3}, {y, - 3, 3},
ColorFunction → (RGBColor[#,1 - #, 0] &), Frame → True];
```

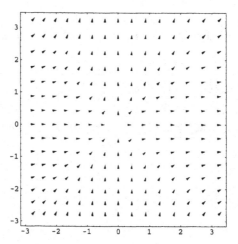

Fig. 6.54. *Gradient field of* $x^3 + y^3$ *in the domain* $[-3, 3] \times [-3, 3]$.

6.6 3D Plots

6.6.1 Plot3D

To plot the tridimensional surface represented by $z = f(x, y)$ in the domain $[x_1, x_2] \times [y_1, y_2]$, we enter the command `Plot3D[f[x,y], {x, x1, x2}, {y1, y2}]`.

```
Plot3D[x^2 + y^2, {x, - 3, 3}, {y, - 3, 3},
AspectRatio → 1]
```

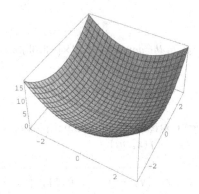

Fig. 6.55. *Surface $x^2 + y^2$ in the domain $[-3, 3] \times [-3, 3]$.*

There exist many options. A few are used in the following plot. The option `PlotPoints` specifies how many sample points to use, and `FaceGrids` specifies grid lines to draw on the faces of the bounding box. `Mesh` specifies whether an explicit (x, y) mesh should be drawn. `ViewPoint` gives the point in space from which the plotted objects are to be viewed.

```
Plot3D[x^2 + y^2, {x,- 3, 3}, {y, - 3, 3},
PlotPoints → 60, Mesh → False, FaceGrids → All,
AxesLabel → {"Length", "Width", "Height"},
AspectRatio → 1, ViewPoint → {1, 1, 0.3}];
```

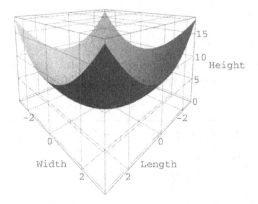

Fig. 6.56. *Surface $x^2 + y^2$ in the domain $[-3, 3] \times [-3, 3]$ from a different viewpoint.*

6.6.2 ListPlot3D

ListPlot3D[array] generates a tridimensional plot of a surface represented by an array of heights $z_{i,j} = f(x_i, y_j)$.

```
data = Table[x^2 + y^2, {y, - 1, 1, 0.1}, {x, - 1, 1, 0.1}];
ListPlot3D[data];
```

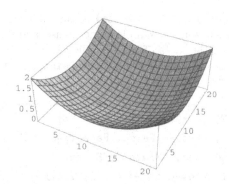

Fig. 6.57. *Tridimensional list plot of nested lists.*

Here is another example.

```
ListPlot3D[Table[Sin[x + y],
{x, 0, 3 Pi / 2, Pi / 10}, {y, 0, 3 Pi / 2, Pi / 10}]];
```

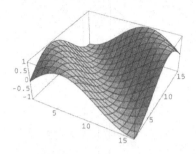

Fig. 6.58. *Tridimensional list plot.*

The same result could be obtained using the command Show[SurfaceGraphics-[array]].

```
Show[SurfaceGraphics[Table[Sin[x+ y],
{x, 0, 3 Pi / 2, Pi / 10}, {y, 0, 3 Pi / 2, Pi,/ 10}]]];
```

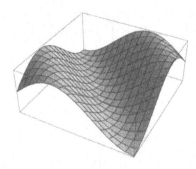

Fig. 6.59. *Same as above using* SurfaceGraphics.

Whereas ListPlot[points] takes a list of 2D points and plots them in a plane, ScatterPlot3D[points] takes a list of 3D points and plots them in a 3D space. First we have to load the package Graphics'Graphics3D'. As for ListPlot[points] we can use the option PlotJoined → True.

```
<<Graphics'Graphics3D'
```

```
ScatterPlot3D[Table[{Sin[2 t], Cos[2 t], t},
{t, 0, 4 Pi, Pi / 50}], Axes → False];
```

Fig. 6.60. ScatterPlot3D: *the 3D analogue of* ListPlot.

```
ScatterPlot3D[Table[{Sin[2 t], Cos[2 t], t},
{t, 0, 4Pi, Pi / 50}], PlotJoined → True, Axes → False];
```

Fig. 6.61. *Same as above with the option* PlotJoined → True.

ListContourPlot3D[array] generates a tridimensional contour plot from a tridimensional array $f(x_i, y_i, z_i)$.

We need first to load the package <<Graphics`ContourPlot3D`.

```
<<Graphics'ContourPlot3D'
```

```
data = Table[x^2 + y^2 - z, {z, - 1, 1, 0.1},
{y, - 1, 1, 0.1}, {x, - 1, 1, 0.1}];
ListContourPlot3D[data,Lighting → False,
Axes → True, ContourStyle → {Hue[0.15]}];
```

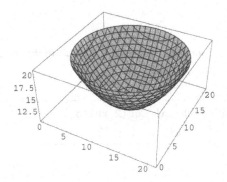

Fig. 6.62. *Tridimensional contour plot of nested lists.*

6.6.3 Different Coordinate Systems

We first have to load the package **Graphics'** to be able to use different systems of coordinates.

```
<<Graphics'
```

```
CylindricalPlot3D[r^2 Cos[2 phi], {r, 0, 1}, {phi, 0, 2 Pi}];
```

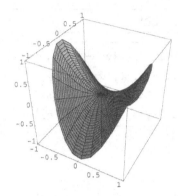

Fig. 6.63. *Cylindrical coordinates: surface* $r^2 \cos(2\varphi)$ *in the domain* $(r, \varphi) = [0, 1] \times [0, 2\pi]$.

```
SphericalPlot3D[Cos[theta] Cos[2 theta],
{theta, 0, Pi / 4}, {phi, 0, 2 Pi}];
```

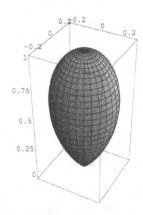

Fig. 6.64. *Spherical coordinates: surface* $\cos(\theta) \cos(2\theta)$ *in the domain* $(\theta, \varphi) = [0, \pi/4] \times [0, 2\pi]$.

6.6.4 ContourPlot

We can also generate a contour plot of the same surface, and eliminate either the contour shading or the contour lines.

```
ContourPlot[x^2 + y^2 , {x,- 3, 3}, {y,- 3, 3},
AspectRatio → 1];
```

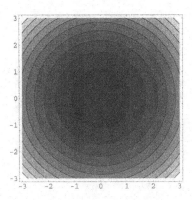

Fig. 6.65. *Contour plot of $x^2 + y^2$ in the domain $[-3,3] \times [-3,3]$.*

The option ContourShading specifies whether the regions between contour lines should be shaded.

```
ContourPlot[x^2 + y^2 , {x, - 3, 3}, {y, - 3, 3},
AspectRatio → 1, ContourShading → False];
```

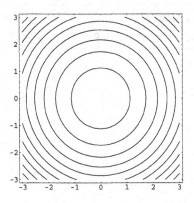

Fig. 6.66. *Contour plot of $x^2 + y^2$ in the domain $[-3,3] \times [-3,3]$ with* ContourShading → False.

```
ContourPlot[x^2 + y^2 , {x, - 3, 3}, {y, - 3, 3},
AspectRatio → 1, ContourLines → False];
```

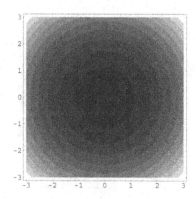

Fig. 6.67. *Contour plot of $x^2 + y^2$ in the domain $[-3, 3] \times [-3, 3]$ with* ContourLines → False.

6.6.5 DensityPlot

A density plot is a rectangular plot that consists of smaller rectangle each colored according to the value of a function.

```
DensityPlot[Sin[x] Cos[y],
{x, - Pi, Pi}, {y, - Pi / 2, 3 Pi / 2}, PlotPoints → 100];
```

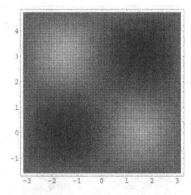

Fig. 6.68. *Density plot of* $\sin(x)\cos(y)$ *in the domain* $[-\pi, \pi] \times [-\pi/2, 3\pi/2]$.

We can improve the plot using options. Above we have already used above PlotPoints and Mesh. The option ColorFunction specifies a function to apply to z values to determine the color to use for a particular (x, y) region.

```
DensityPlot[Sin[x] Cos[y],
{x,- Pi, Pi}, {y,- Pi / 2, 3 Pi / 2}, PlotPoints → 100,
Mesh → False, ColorFunction → Hue];
```

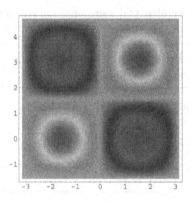

Fig. 6.69. *Same as above with different options.*

A density plot may be compared to a shaded contour plot. For rapidly changing functions, DensityPlot works slightly better than ContourPlot. For slowly changing functions, it is the opposite. Here is an example of a rapidly changing function:

```
DensityPlot[Sin[10 x] Cos[10 y], {x,- Pi, Pi},
{y,- Pi / 2, 3 Pi / 2},PlotPoints → 100,
Mesh → False, ColorFunction → Hue];
```

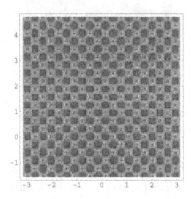

Fig. 6.70. *DensityPlot of* $\sin(10x)\cos(10y)$ *in the domain* $[-\pi, \pi] \times [-\pi/2, 3\pi/2]$.

6.6.6 ParametricPlot3D

ParametricPlot3D generates tridimensional curves and surfaces. They are defined by three functions of, respectively, one or two parameters.

```
ParametricPlot3D[Sin[3 omega], Cos[3 omega], omega,
{omega, 0, 2 Pi}, Ticks → None];
```

See output in Figure 6.71.

```
ParametricPlot3D[{Cos[x] Cos[y], Cos[x] Sin[y], Sin[x]},
{x, - Pi / 2, Pi / 2}, {y, 0, 2 Pi}, Axes → False, Boxed →
False];
```

See output in Figure 6.72.

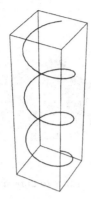

Fig. 6.71. *Tridimensional parametric plot of* $(\sin(3\omega), \cos(3\omega), \omega)$ *in the domain* $[0, 2\pi]$.

Fig. 6.72. *Parametric plot of* $(\cos(x)\cos(y), \cos(x)\sin(y), \sin(x))$ *in the domain* $[-\pi/2, \pi/2] \times [0, 2\pi]$.

6.7 3D Graphical Primitives

As for 2D graphics, there exists a collection of 3D graphics primitives such as Point, Line, Polygon, and so on. The general form of the command is Graphics[primitive, option1, option2,...]. These graphics primitives are displayed with the command Show.

Here is an example of a pyramid having an octagonal base and eight triangular faces.

```
Show[Graphics3D[Table[Polygon[{{Sin[n], Cos[n], 0},
{Sin[n + Pi/4], Cos[n + Pi/4], 0}, {0, 0, 1}}],
{n, Pi / 4, 2 Pi, Pi / 4}]]];
```

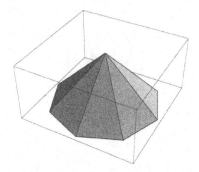

Fig. 6.73. *Using 3D graphics primitives to draw a pyramid with an octagonal base.*

We can modify ViewPoint.

```
Show[Graphics3D[Table[Polygon[{{Sin[n], Cos[n], 0},
{Sin[n + Pi/4], Cos[n + Pi/4], 0}, {0, 0, 1}}],
{n, Pi/4, 2 Pi, Pi/4}]], ViewPoint → {0.5, - 1, 0.5}];
```

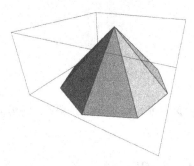

Fig. 6.74. *Same as above with a modified viewpoint.*

7

Statistics

7.1 Random Numbers

Random[] gives a pseudorandom number lying in the interval $[0, 1]$

```
Table[Random[], {10}]
```

{0.97168, 0.767369, 0.159879, 0.839617, 0.527752,
0.833297, 0.471409, 0.793496, 0.714053, 0.379495}

Random[type, range] gives a pseudorandom number of the specified type lying in the specified range. Possible types are: Integer, Real, and Complex.

```
Random[Real, {0, 1}, 50]
```

0.25001677073690337847614440066746917236028215705360

```
Table[Random[Real, {5.4, 7.3}], {10}]
```

{5.99996, 7.15631, 7.02153, 6.61911, 5.92165,
6.66404, 6.21742, 6.71256, 6.16074, 5.52855}

```
Table[Random[Integer], {10}]
```

{0, 0, 1, 1, 0, 1, 0, 0, 1, 0}

```
Table[Random[Integer, {6, 21}] , {10}]
```

{20, 7, 15, 19, 6, 16, 16, 13, 19, 8}

```
Table[Random[Complex], {10}]
```

{0.944269 + 0.71944 I, 0.916369 + 0.196858 I,
0.736182 + 0.67909 I, 0.376178 + 0.100881 I,
0.295465 + 0.798618 I, 0.675603 + 0.532656 I,
0.688044 + 0.389555 I, 0.768525 + 0.678484 I,
0.460119 + 0.337657 I, 0.177217 + 0.748684 I}

```
Table[Random[Complex, {2 + 3 I, 4 + 5 I}], {10}]
```

{2.41617 + 3.0807 I, 3.08038 + 3.19195 I
2.88143 + 4.76094 I, 3.40115 + 4.13645 I,
3.21484 + 3.81813 I, 3.81416 + 4.70834 I,
2.45585 + 3.1038 I, 3.18262 + 4.8596 I,
2.74898 + 3.82052 I, 2.6335 + 3.52903 I}

SeedRandom[n] resets the pseudorandom number generator, using the integer n as a seed. It allows us to get the same sequence of pseudorandom numbers on different occasions.

```
SeedRandom[123]
Table[Random[Integer, {1, 5}], {10}]
```

{3, 3, 3, 2, 5, 2, 2, 1, 2, 5}

```
SeedRandom[123]
Table[Random[Integer, {1, 5}], {10}]
```

{3, 3, 3, 2, 5, 2, 2, 1, 2, 5}

7.2 Evaluating π

Consider Figure 7.1 representing a unit square with inside a quarter of a disk of radius 1.

```
Show[Graphics[{
{RGBColor[0, 0, 1], Disk[{0, 0}, 1, {0, Pi / 2}]},
Line[{{0, 0}, {1, 0}, {1, 1}, {0, 1}, {0, 0}}]}],
AspectRatio → Automatic];
```

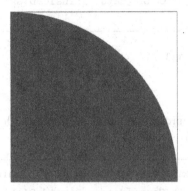

Fig. 7.1. *Quarter of a disk of radius 1 inside a unit square.*

To evaluate π, we select a sequence of random points distributed uniformly inside the unit square. The probability for a random point to lie inside the disk is proportional to its area, that is, equal to $\pi/4$. The function that generates a random point inside the unit square is

```
randomPoint := {Random[], Random[]}
```

```
randomPoint
```

{0.192417, 0.371977}

The function that tests if a point is inside the unit disk is

```
insideDisk[pt_] := Total[pt^2] ≤ 1
```

```
{insideDisk[{0.9, 0.7}], insideDisk[{0.4, 0.6}]}
```

{False, True}

To evaluate π we, therefore, generate a sequence of random points inside the unit square and determine the fraction of these points lying inside the unit disk quarter. The approximate value of π is obtained multiplying this fraction by 4.

```
data = Table[randomPoint, {1000000}]; // Timing
approximatePi = N[4* Count[data, _?insideDisk] /
Length[data]]
```

{4.54608 Second, Null}

3.13943

To obtain a better approximate value, we increase the number of random points from 10^6 to 10^7.

```
data = Table[randomPoint, {10000000}]; // Timing
approximatePi = N[4* Count[data, _?insideDisk] /
Length[data]]
```

{47.1171 Second, Null}

3.14204

As shown below, this approximate value is quite good; it differs from the exact value by less than 0.015%.

```
Abs[N[Pi]-3.14204] / N[Pi]
```

0.000142395

7.3 Probability Distributions

We present just a few of them. We access the most common discrete or continuous statistical distributions loading either the package Statistics'Discrete-

Distributions' or the package Statistics'ContinuousDistributions'. As illustrated below, we can find the properties of a given distribution using different functions that take as argument a symbolic representation of the distribution.

```
<<Statistics'DiscreteDistributions'
```

We can get the list of these discrete probability distributions entering the command:

```
?Statistics'DiscreteDistributions'*
```

7.3.1 Binomial Distribution

In the command BinomialDistribution[n, p], n is the number of independent trials and p the probability of a success in a trial.

```
bDist = BinomialDistribution[n, p]
```

BinomialDistribution[n, p]

We can ask *Mathematica* to give us the expression of the probability density function,

```
PDF[bDist, x]
```

$(1 - p)^{n - x} \, p^x \, \text{Binomial}[n, x]$

the characteristic function, which is the Fourier transform of the probability density function [5],

```
CharacteristicFunction[bDist, t]
```

$(1 - p + E^{I \, t} \, p)^n$

the average value, the variance, and the standard deviation.

```
Mean[bDist]
```

n p

```
Variance[bDist]
```

n (1 − p) p

```
StandardDeviation[bDist]
```

Sqrt[n (1 - p) p]

A discrete random variable distributed according to the binomial distribution is defined in a domain given by

```
Domain[bDist]
```

Range[0, n]

We can generate random numbers distributed according to the binomial distribution

```
Table[Random[BinomialDistribution[10, 0. 4]], {10}]
```

{6, 5, 3, 5, 5, 6, 3, 3, 4, 5}

and we can evaluate expected values.

```
ExpectedValue[x^4, BinomialDistribution[10, 0. 4], x]
```

510.304

7.3.2 Poisson Distribution

The command PoissonDistribution[λ] generates the Poisson distribution where λ represents the mean of the distribution.

```
pDist = PoissonDistribution[λ]
```

PoissonDistribution[λ]

As in the previous section, we can ask *Mathematica* to give us all the essential characteristics of this probability distribution.

```
PDF[pDist, x]
```

$$\frac{\lambda^x}{E^\lambda \, x!}$$

```
CharacteristicFunction[pDist, t]
```

$$E^{(-1 + E^{I\ t})\ \lambda}$$

```
Mean[pDist]
```

λ

```
Variance[pDist]
```

λ

```
StandardDeviation[pDist]
```

Sqrt$[\lambda]$

```
Domain[pDist]
```

Range[0, Infinity]

```
Table[Random[PoissonDistribution[3]], {10}]
```

$\{4,\ 4,\ 4,\ 0,\ 4,\ 2,\ 4,\ 4,\ 1,\ 1\}$

```
ExpectedValue[x^4, PoissonDistribution[3], x]
```

309

```
<<Statistics`ContinuousDistributions`
```

7.3.3 Normal Distribution

In the command `NormalDistribution[`μ`,` σ`]`, μ is the mean and σ is the standard deviation.

```
nDist = NormalDistribution[μ, σ]
```

NormalDistribution[μ, σ]

```
PDF[nDist, x]
```

$$\frac{1}{E^{(x-\mu)^2/(2\ \sigma^2)}\ \text{Sqrt}[2\ \text{Pi}]\ \sigma}$$

```
CharacteristicFunction[nDist, t]
```

$E^{I\ t\ \mu - (t^2\sigma^2)/2}$

```
Mean[nDist]
```

μ

```
Variance[nDist]
```

σ^2

```
StandardDeviation[nDist]
```

σ

```
Domain[nDist]
```

```
Interval[{- Infinity, Infinity}]
```

```
Plot[PDF[ NormalDistribution[0, 1], x], {x, -3, 3}];
```

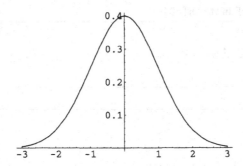

Fig. 7.2. *Probability density function of the normal distribution for $\mu = 0$ and $\sigma = 1$ in the interval $[-3, 3]$.*

```
Table[Random[NormalDistribution[0, 1]], {10}]
```

$\{-0.407312, -0.178324, -0.863422, 0.0359524, -0.24181,$

$-1.15673, 0.488383, 0.342282, 0.259808, -1.67701\}$

```
ExpectedValue[x^4, NormalDistribution[0, 1], x]
```

3

7.3.4 Cauchy Distribution

In the command CauchyDistribution[a,b], a is the location parameter and b the scale parameter.

```
cDist = CauchyDistribution[a, b]
```

CauchyDistribution[a, b]

```
PDF[cDist, x]
```

$$\frac{1}{b \; Pi \; (1 + \dfrac{(-a+x)^2}{b^2})}$$

```
CharacteristicFunction[cDist, t]
```

$E^{I \; a \; t - b \; t \; Sign[t]}$

```
Mean[cDist]
```

Indeterminate

```
Domain[cDist]
```

Interval[{−Infinity, Infinity}]

```
Plot[PDF[CauchyDistribution[0, 1], x], {x, - 3, 3}];
```

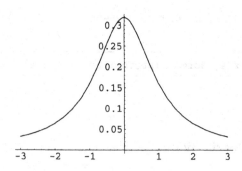

Fig. 7.3. *Probability density function of the Cauchy distribution for a = 0 and b = 1 in the interval [−3,3].*

```
Table[Random[CauchyDistribution[0,1]], {10}]
```

$\{-5.29566, -2.68135, 2.9326, -0.0757705, -1.10613,$
$0.996368, 0.345278, -1.30723, 0.142766, -1.00022\}$

7.4 Descriptive Statistics

We first have to load the packages:

```
<<Statistics'DescriptiveStatistics'
<<Statistics'DataManipulation'
<<Graphics'Graphics'
```

The functions of these packages compute the descriptive statistics of lists of data of the most common probability distributions.

7.4.1 Poisson Distribution

```
poissonData =
Table[Random[PoissonDistribution[2.5]], {5000}];
```

```
N[Mean[poissonData]]
```

2.5234

```
N[Variance[poissonData]]
```

2.52316

`Frequencies[list]` gives the distinct elements in a list paired with their frequencies.

```
poissonDataFrequencies = Frequencies[poissonData]
```

$\{\{431, 0\}, \{1032, 1\}, \{1315, 2\}, \{1029, 3\}, \{659, 4\},$
$\{335, 5\}, \{139, 6\}, \{43, 7\}, \{14, 8\}, \{2, 9\}, \{1, 10\}\}$

```
BarChart[poissonDataFrequencies];
```

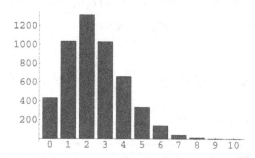

Fig. 7.4. *Bar chart of 5000 Poisson distributed random numbers.*

7.4.2 Normal Distribution

```
gaussData = Table[Random[NormalDistribution[2, 3]], {10000}];
```

```
N[Mean[gaussData]]
```

1.98548

```
N[Variance[gaussData]]
```

8.78256

```
gaussDataHistogram = Histogram[gaussData,
HistogramCategories → Table[- 10 + k, {k, 0, 24}],
HistogramScale → 1];
```

```
gaussPDF = Plot[PDF[NormalDistribution[2, 3], x],
{x, - 10, 12}, PlotStyle → {RGBColor[0, 0, 1],
AbsoluteThickness[2]}];
```

We can compare the histogram of Figure 7.5 with the exact probability density function

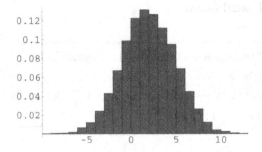

Fig. 7.5. *Histogram of 10,000 normally distributed random numbers.*

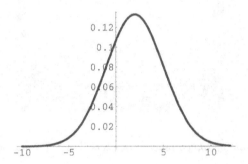

Fig. 7.6. *Probability density function of the normal distribution for $\mu = 2$ and $\sigma = 3$.*

```
Show[{gaussDataHistogram, gaussPDF}];
```

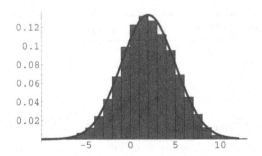

Fig. 7.7. *Comparing the histogram above with the exact probability density function.*

7.4.3 Cosine Distribution

```
cosData = N[Table[Cos[Pi Random[]], {10000}]];
Histogram[cosData,
HistogramCategories → Table[- 1+ k / 10, {k, 0, 20}],
HistogramScale → 1];
```

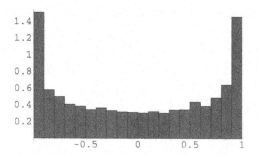

Fig. 7.8. *Histogram of 10,000 random numbers distributed according to the cosine distribution.*

```
{Mean[cosData], Variance[cosData]}
```

{−0.000358633, 0.50641}

```
Integrate[(Cos[x])^2 / Pi, {x, 0, Pi}]
```

$$\frac{1}{2}$$

7.4.4 Uniform Distribution

```
uniformData = Table[Random[], {10000}];
Histogram[uniformData,
HistogramCategories → Table[0 + k / 20, {k, 0, 20}],
HistogramScale → 1];
```

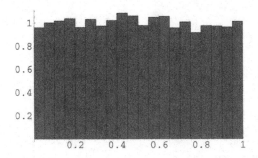

Fig. 7.9. *Histogram of 10,000 uniformly distributed random numbers in the interval* [0, 1].

```
{Mean[uniformData], Variance[uniformData]}
```

{0.49736, 0.082058}

The exact value of the variance is

```
Integrate[(x - 0.5)^2, {x, 0, 1}]
```

0.0833333

8

Basic Programming

8.1 The Mathematica Language

Everything you type in *Mathematica* is an expression. An expression is of the
form f[arguments] where f is the Head of the expression which identifies its
type. Expressions look like functions (or functions are expressions).

```
{Head[3], Head[3 / 4], Head[5.2], Head["Hello"]}
```

```
{Integer, Rational, Real, String}
```

```
{Head[Subtract], Head[Times], Head[a + b], Head[a b]}
```

```
{Symbol, Symbol, Plus,Times}
```

Expressions are represented in a uniform way that can be accessed using
FullForm. Some built-in functions are actually redundant and are translated
into basic forms such as Plus, Times, and Power.

```
{FullForm[Divide[a, b]], FullForm[Sqrt[a]], FullForm[a - b],
FullForm[a^b]}
```

```
{Times[a, Power[b, -1]], Power[a, Rational[1,2]],
Plus[a, Times[-1, b]], Power[a, b]}
```

```
{FullForm[a + b I], FullForm[(a + b)^2], FullForm[{a, b}],
FullForm[{a, b} c]}
```

{Plus[a, Times[Complex[0, 1], b]],

Power[Plus[a, b], 2], List[a, b], List[Times[a, c], Times[b,c]]}

```
{FullForm[a → b], FullForm[a == b], FullForm[a ≤ b],
FullForm[x_Integer]}
```

{Rule[a, b], Equal[a, b], LessEqual[a, b],

Pattern[x, Blank[Integer]]}

Consider the expression:

```
expr = 7 + (a - x)^2
```

$7 + (a - x)^2$

Its different parts are

```
expr[[0]]
```

Plus

```
expr[[1]]
```

7

```
expr[[2]]
```

$(a - x)^2$

```
expr[[2, 1]]
```

a — x

```
expr[[2,2]]
```

2

```
expr[[2,1,1]]
```

a

```
expr[[2, 1, 2]]
```

—x

8.2 Functional Programming

Mathematica is essentially a functional programming language that emphasizes rules and pattern matching.

8.2.1 Applying Functions to Values

```
Clear[f]
f[x_] := x^3
```

```
{f[2], f @ 2, 2 // f}
```

{8, 8, 8}

f @ x is the prefix form for f[x] and x//f is its postfix form.

```
Clear[f]
f[{a, b, c}]
f[x + y + z]
```

```
f[{a, b, c}]
f[x + y + z]
```

Map[f, expression] applies f to each element on the first level in expression.

```
Map[f, {a, b, c}]
Map[f, x + y + z]
```

```
{f[a], f[b], f[c]}
f[x] + f[y] + f[z]
```

MapThread[f, {{a1, a2, ...}, {b1, b2, ...}}] gives
f[a1, b1], f[a2,b2], ...

```
Clear[f]
MapThread[f, {{x, y, z}, {a, b, c}}]
MapThread[Rule, {{x, y, z}, {a, b, c}}]
MapThread[Plus, {{x, y, z}, {a, b, c}}]
```

```
{f[x,a], f[y,b], f[z,c]}
{x → a, y → b,z → c}
{a + x, b + y, c + z}
```

Apply[f, expression] replaces the head of expression by f.

```
Apply[Plus, {a, b, c}]
Apply[Times, {a, b, c}]
```

```
a + b + c
a b c
```

```
Clear[f]
Apply[f, {{a, b}, {c, d}}, 2]
Apply[Plus, {{a, b}, {c, d}}]
```

{f[a, b], f[c, d]}

{a + c, b + d}

8.2.2 Defining Functions

```
cube1[x_] := x^3 (* named function *)
cube2 = Function[{x}, x^3]; (* pure function: long form *)
cube3 = #^3 &; (* pure function: short form *)
{cube1[3], cube2[3], cube3[3]}
```

{27, 27, 27}

A function can be used to check if a condition is True or False. For example:

```
greaterThan3[n_] := n > 3
{greaterThan3[7.5], greaterThan3[2.9]}
```

{True, False}

Blank (_), BlankSequence (__), and BlankNullSequence (___) are pattern objects that can, respectively, stand for any *Mathematica* expression, any sequence of one or more *Mathematica* expressions, or any sequence of zero or more *Mathematica* expressions (see also below).

```
f1[x_, y__] := x + y
f2[x_, y___] := x + y
{f1[a, b, c, d], f1[a]}
{f2[a, b, c, d], f2[a]}
```

{a + b + c + d, f1[a]

{a + b + c + d}, a}

8.2.3 Iterations

In many programs we need to apply a function repeatedly. Here are various useful commands to perform such a task.

Nest[f, expression, n] generates f[f[f[...[expression]...]]] where f is nested n times.

NestList[f, expression, n] generates the list {x, f[x], f[f[x]], ...} where the last term of the list is f nested n times.

FixedPoint[f,x] applies f repeatedly until the result no longer changes.

FixedPoint[f, x, n] applies f repeatedly but stops after at most n steps.

FixedPointList[f, x] generates the list {x, f[x], f[f[x]], f[f[f[x]]], ...} until the elements no longer change.

NestWhile[f, expression, test] applies f repeatedly until applying test to the result no longer yields True.

NestWhileList[f, expression, test] generates the list of applying f repeatedly until applying test to the result no longer yields True.

```
Nest[Cos, 0.5, 7]
```

0.752356

```
NestList[Cos, 0.5, 7]
```

{0.5, 0.877583, 0.639012, 0.802685, 0.694778,
0.768196, 0.719165, 0.752356}

```
FixedPoint[Cos, 0.5]
```

0.739085

```
FixedPointList[Cos, 0.5]
```

{0.5, 0.877583, 0.639012, 0.802685, 0.694778, 0.768196, 0.719165,
0.752356, 0.730081, 0.74512, 0.735006, 0.741827, 0.737236,
0.74033, 0.738246, 0.73965, 0.738705, 0.739341, 0.738912,
0.739201, 0.739007, 0.739138, 0.73905, 0.739109, 0.739069,
0.739096, 0.739078, 0.73909, 0.739082, 0.739087, 0.739084,
0.739086, 0.739084, 0.739086, 0.739085, 0.739085, 0.739085,

```
0.739085, 0.739085, 0.739085, 0.739085, 0.739085, 0.739085,
0.739085, 0.739085, 0.739085, 0.739085, 0.739085, 0.739085,
0.739085, 0.739085, 0.739085, 0.739085, 0.739085, 0.739085,
0.739085, 0.739085, 0.739085, 0.739085, 0.739085, 0.739085,
0.739085, 0.739085, 0.739085, 0.739085, 0.739085, 0.739085,
0.739085, 0.739085, 0.739085, 0.739085, 0.739085, 0.739085,
0.739085, 0.739085, 0.739085, 0.739085, 0.739085, 0.739085,
0.739085, 0.739085, 0.739085, 0.739085, 0.739085, 0.739085,
0.739085, 0.739085, 0.739085, 0.739085, 0.739085, 0.739085,
0.739085}
```

The following function gives the smallest prime greater than x. It uses a pure function. The *Mathematica* command NextPrime[n] gives the smallest prime larger than n, but we have to first load the package NumberTheory'Number-TheoryFunctions'.

```
<<NumberTheory'NumberTheoryFunctions'
NextPrime[100]
```

101

```
firstPrimeAfter[x_] :=
NestWhile[(# + 1) &, x, !(PrimeQ[#]) &]
```

```
firstPrimeAfter[100]
```

101

Checking:

```
PrimeQ[101]
```

True

8.2.4 A Functional Program

Let us write a function giving the position of a sequence of k or more identical digits in the first n digits in the decimal part of an irrational number x.

We progress from small units to larger ones (bottom-up programming).

IntegerDigits[int] gives a list of the decimal digits in the integer int.

```
x = Pi; n = 30;
IntegerDigits[Floor[N[FractionalPart[x], n] 10^n]]
```

{1, 4, 1, 5, 9, 2, 6, 5, 3, 5, 8, 9, 7, 9, 3, 2, 3, 8, 4, 6,
2, 6, 4, 3, 3, 8, 3, 2, 7, 9}

Partition[list, k, d] generates sublists of list of length k with offset d.

```
x = Pi; n = 30; k = 4;
Partition[IntegerDigits[Floor[N[FractionalPart[x], n]
10^n]], k, 1]
```

{{1, 4, 1, 5}, {4, 1, 5, 9}, {1, 5, 9, 2}, {5, 9, 2, 6},
{9, 2, 6, 5}, {2, 6, 5, 3}, {6, 5, 3, 5}, {5, 3, 5, 8},
{3, 5, 8, 9}, {5, 8, 9, 7}, {8, 9, 7, 9}, {9, 7, 9, 3},
{7, 9, 3, 2}, {9, 3, 2, 3}, {3, 2, 3, 8}, {2, 3, 8, 4},
{3, 8, 4, 6}, {8, 4, 6, 2}, {4, 6, 2, 6}, {6, 2, 6, 4},
{2, 6, 4, 3}, {6, 4, 3, 3}, {4, 3, 3, 8}, {3, 3, 8, 3},
{3, 8, 3, 2}, {8, 3, 2, 7}, {3, 2, 7, 9}}

Among all the sublists of length 4 ($k = 4$), we search the position of the sublists equal to $\{9, 9, 9, 9\}$ ($d = 9$).

```
x = Pi; n = 10000; k = 4; d = 9;
Position[Partition[IntegerDigits[Floor[N[FractionalPart[x],
n] 10^n]], k,1], Table[d,k]]
```

{{762}, {763}, {764}}

We can now group all these elementary steps into a final function pos[x, k, d, n], where x is the irrational number, k the length of the sequence of consecutive digits d, and n the number of digits of the fractional part of x.

```
pos[x_, k_, d_, n_] := Position[Partition[IntegerDigits[Floor[
N[FractionalPart[x], n] 10^n]], k, 1], Table[d, {k}]]
```

```
pos[Pi, 4, 9,10000]
```

{{762}, {763}, {764}}

Here is a table of sequences of length 4 for all digits 0, 1, 2, 3, ... , 9 in the decimal expansion of π considering 10,000 digits.

```
Table[{m, pos[Pi, 4, m, 10000]}, {m, 0, 9}]//MatrixForm
```

$$
\begin{pmatrix}
0 & \{\} \\
1 & \{\} \\
2 & \{\{4902\}, \{7964\}\} \\
3 & \{\} \\
4 & \{\} \\
5 & \{\} \\
6 & \{\} \\
7 & \{\{1589\}, \{5241\}, \{5322\}, \{5863\}\} \\
8 & \{\{4751\}\} \\
9 & \{\{762\}, \{763\}, \{764\}\}
\end{pmatrix}
$$

Table of sequences of length 4 for all digits 0, 1, 2, 3, ... , 9 in the decimal expansion of e considering 10,000 digits.

```
Table[{m, pos[E, 4, m, 10000]}, {m, 0, 9}] // MatrixForm
```

$$\begin{pmatrix}
0 & \{\{7688\}\} \\
1 & \{\} \\
2 & \{\} \\
3 & \{\{3354\}\} \\
4 & \{\} \\
5 & \{\{3620\},\{8905\}\} \\
6 & \{\{2175\},\{4992\},\{4993\}\} \\
7 & \{\{1071\},\{5040\}\} \\
8 & \{\{723\}\} \\
9 & \{\}
\end{pmatrix}$$

Table of sequences of length 4 for all digits 0, 1, 2, 3, ... , 9 in the decimal expansion of $\sqrt{2}$ considering 10,000 digits.

```
Table[{m, pos[Sqrt[2], 4, m, 10000]}, {m, 0, 9}] /
/MatrixForm
```

$$\begin{pmatrix}
0 & \{\} \\
1 & \{\{952\}\} \\
2 & \{\{4701\}\} \\
3 & \{\{1481\}\} \\
4 & \{\{3308\}\} \\
5 & \{\{2016\}\} \\
6 & \{\} \\
7 & \{\{1559\}\} \\
8 & \{\{4214\}\} \\
9 & \{\{2515\},\{2707\},\{2708\},\{7326\}\}
\end{pmatrix}$$

Table of sequences of length 6 for all digits 0, 1, 2, 3, ... , 9 in the decimal expansion of π considering 1,000,000 digits. The computation takes less than three minutes.

```
Timing[Table[{m, pos[Pi, 6, m, 1000000]}, {m, 0, 9}] //
MatrixForm]
```

$$\left\{ 158.29 \text{ Second}, \begin{pmatrix} 0 & \{\} \\ 1 & \{\{255945\}\} \\ 2 & \{\{963024\}\} \\ 3 & \{\{710100\}, \{710101\}\} \\ 4 & \{\{828499\}\} \\ 5 & \{\{244453\}, \{253209\}, \{419997\}\} \\ 6 & \{\{252499\}\} \\ 7 & \{\{399579\}, \{452071\}\} \\ 8 & \{\{222299\}\} \\ 9 & \{\{762\}, \{193034\}\} \end{pmatrix} \right\}$$

We can solve the same problem in any base b.

```
BaseForm[[Sqrt[2], 20], 2]
```

1.011010100000100111100110011001111111100111011110011001001000\
10001$_2$

```
BaseForm[N[FractionalPart[Sqrt[2]], 20], 2]
```

0.011010100000100111100110011001111111100111011110011001001000\
1000110$_2$

```
BaseForm[N[FractionalPart[Sqrt[2]], 20] 10^20, 2]
```

1.000111110110101100011100000001001101111111101111110011011010111\
00000$_2 \times 10^{65}$

```
BaseForm[Floor[N[FractionalPart[Sqrt[2]], 20] 10^20], 2]
```

1000111110110101100011100000001001101111111101111110011011010111\
0000$_2$

```
IntegerDigits[Floor[N[FractionalPart[Sqrt[2]], 20] 10^20], 2]
```

{1, 0, 0, 0, 1, 1, 1, 1, 1, 0, 1, 1, 0, 1, 0, 1, 1, 0, 0, 0, 1,
1, 1, 0, 0, 0, 0, 0, 0, 0, 1, 0, 0, 1, 1, 0, 1, 1, 1, 1, 1, 1,
1, 0, 1, 1, 1, 1, 1, 0, 0, 1, 1, 0, 1, 1, 0, 1, 0, 1,1, 1, 0,
0, 0, 0}

```
Length[%]
```

66

Grouping all these elementary steps into a final function basePos[x, k, d, n, b], where x is the irrational number, k the length of the sequence of consecutive digits d, n the number of digits of the fractional part of x in base 10, and b the base yields

```
basePos[x_, k_, d_, n_, b_] :=
Position[Partition[IntegerDigits[Floor[N[FractionalPart[x],
n] 10^n], b], k, 1], Table[d, {k}]]
```

We can use this function to find sequences of 16 identical digits in the base 2 expansion of π.

```
Table[{m, basePos[Pi,16, m, 10000, 2]}, {m, 0, 1}] //
MatrixForm
```

$$\begin{pmatrix} 0 & \{\{7802\}, \{10558\}\} \\ 1 & \{\{24831\}\} \end{pmatrix}$$

The length of the sequence of digits in base 2 of the fractional part of π when we consider 10,000 digits in base 10 is given by

```
Length[IntegerDigits[Floor[N[FractionalPart[Pi], 10000]
10^10000], 2]]
```

33217

When, in base 10, we consider a fractional part of 10^k digits, the lengths of digit lists in the expansion of an irrational number in any base can be found by adapting the function above. For example, for π, and e in base 3, we find:

```
Table[Length[IntegerDigits[Floor[N[FractionalPart[Pi], 10^k]
10^(10^k)]], 3]], {k, 1, 4}]
```

{20, 208, 2095, 20958}

```
Table[Length[IntegerDigits[Floor[N[FractionalPart[E], 10^k]
10^(10^k)]], 3]], {k, 1, 4}]
```

{21, 210, 2096, 20959}

8.3 Replacement Rules

8.3.1 The Two Kinds of Rewrite Global Rules

They use either = or :=. In the first case the rule is applied immediately whereas in the second one it is applied only when it is used (i.e., when it is called).

```
x = Random[];
y := Random[];
Table[x, {5}]
Table[y, {5}]
```

{0.348571, 0.348571, 0.348571, 0.348571, 0.348571}
{0.25592, 0.216969, 0.465141, 0.0332237, 0.888415}

```
Clear[x, y]
a = Expand[(1 + x)^3];
b := Expand[(1 + x)^3]
{a, b}
x = y + z;
{a, b}
```

$\{1 + 3 x + 3 x^2 + x^3, 1 + 3 x + 3 x^2 + x^3\}$
$\{1 + 3 (y + z) + 3 (y + z)^2 + (y + z)^3,$

$$1 + 3\ y + 3\ y^2 + y^3 + 3\ z + 6\ y\ z + 3\ y^2\ z +$$
$$3\ z^2 + 3\ y\ z^2 + z^3\}$$

8.3.2 Local Rules

They use either $\rightarrow$ or $:\rightarrow$ (entered respectively as `->` and `:>`). As above, the first rule is immediately applied and the second one is applied only when it is used. These rules are applied using the operator `/.`.

```
x^2 - 3 x + 2 /. {{x -> 1}, {x -> 2}}
```

{0, 0}

```
f[a + x] /. f[X_] -> Expand[(1 + X)^3]
f[a + x] /. f[X_] :> Expand[(1 + X)^3]
```

$$1 + 3\ (a + x) + 3\ (a + x)^2 + (a + x)^3$$
$$1 + 3\ a + 3\ a^2 + a^3 + 3\ x + 6\ a\ x + 3\ a^2\ x + 3\ x^2 + 3\ a\ x^2 + x^3$$

```
rndList = Table[Random[], {10}]
```

{0.214286, 0.35825, 0.694216, 0.642354, 0.694934,

0.269416, 0.592734, 0.558784, 0.360521, 0.541512}

```
rndList /. x_ -> - Log[x]
```

{1.54044, 1.02652, 0.364971, 0.442615, 0.363939,

1.3115, 0.52301, 0.581992, 1.0202, 0.61339}

We can also add a condition. In the example below, it is only if $0.4 < x < 0.6$ that x is replaced by $-\log[x]$, which is the case for the seventh, eighth, and tenth elements of **rndList**. The other elements are left unchanged.

```
rndList /. x_ /; 0.4 < x < 0.6 -> - Log[x]
```

{0.214286, 0.35825, 0.694216, 0.642354, 0.694934, 0.269416, 0.52301, 0.581992, 0.360521, 0.61339}

Replacement rules are not commutative.

```
a + b /. a -> 3 b /. b -> 2 a
a + b /. b -> 2 a /. a -> 3 b
```

8 a

9 b

8.3.3 The Operators /. and //.

The following example illustrates the difference between these two operators. With /. rules are applied only if they match the lhs whereas with //. rules are applied recursively. That is, //. means keep applying the rules until no further substitution is possible.

```
fib[5] /. {fib[1] -> 1, fib[2] -> 1,
fib[n_] -> fib[n - 2] + fib[n - 1]}
fib[5] //. {fib[1] -> 1, fib[2] -> 1,
fib[n_] -> fib[n - 2] + fib[n - 1]}
```

fib[3] + fib[4]

5

In the first case, the rules fib[1] -> 1, fib[2] -> 1 do not match anything in fib[5]. Only the rule fib[n_] -> fib[n - 2] + fib[n - 1] can be applied.

8.3.4 Patterns

```
Clear[f]
f[x_] := x^2
{f[a], f[a + b], f[a, b], f[{a, b}]}
```

$\{a^2, (a + b)^2, f[a, b], \{a^2, b^2\}\}$

That is, a, a + b, and {a, b} are matching the pattern x_ and applying the rule we find a^2, $(a + b)^2$, and $\{a, b\}^2$. On the contrary a,b does not match the pattern x_.

We can specify the type of the argument x with a **Head**. For example,

```
Clear[f]
f[x_Integer] := x^2
{f[4], f[3.2]}
```

{16, f[3.2]}

x has to be an integer. Here are more examples.

```
Clear[f]
f[x_Real] := x^2
{f[3], f[3.], f[2.5]}
```

{f[3], 9., 6.25}

We can restrict patterns using /;.

```
Clear[f]
f[x_Integer] := x^2 /; x > 5
{f[4], f[6]}
```

{f[4], 36}

Other examples are

```
Clear[f]
f[x_Integer /; OddQ[x]] := x
f[x_Integer /; EvenQ[x]] := x^2
{f[1], f[2], f[3], f[4]}
```

{1, 4, 3, 16}

```
DeleteCases[Range[10], x_ /; OddQ[x]]
DeleteCases[Range[10], x_/ ; EvenQ[x]]
```

{2, 4, 6, 8, 10}
{1, 3, 5, 7, 9}

```
Cases[Range[- 5, 5], _?Positive]
Cases[Range[- 5, 5], x_ /; x > 0]
```

{1, 2, 3, 4, 5}
{1, 2, 3, 4, 5}

```
f[n_Integer?Positive /; IntegerQ[n/3]] := n / 3
{f[27], f[- 6], f[5]}
```

{9, f[- 6], f[5]}

As mentioned above:

x__ means a sequence of one or more expressions named x (two underscores).

x___ means zero or more expressions named x (three underscores)

```
Clear[f]
f[x_] := Length[x]  (* one underscore *)
{f[], f[a], f[{a}], f[a, b], f[{a, b}]}
```

{f[], 0, 1, f[a,b], 2}

```
Clear[f]
f[x__] := Length[{x}]  (* two underscores *)
{f[], f[a], f[{a}], f[a, b]}
```

{f[], 1, 1, 2}

```
Clear[f]
f[x___] := Length[{x}]   (* three underscores *)
{f[], f[a], f[a,b]}
```

{0, 1, 2}

8.3.5 Example: the Fibonacci Numbers

We use the recursive definition of the Fibonacci numbers[1] to determine the
nth Fibonacci number. Note that `Fibonacci[n]` is a *Mathematica* built-in
function.

```
Fibonacci[500]
```

1394232245616978801397243828704072839500702565876973072641\

089629483255716228632906915576588876222521294125

The symbol \ at the end of the first line of the output cell indicates that the
sequence of digits, being too long, includes the next line.

Ignoring the built-in function, we first build up a particularly inefficient program.

```
Clear[F]
F[1] = 1; F[2] = 1;
F[n_Integer /; n > 0] := F[n - 1] + F[n - 2];
F[15]
```

610

Let us evaluate the CPU time needed to compute the first Fibonacci numbers
using the function `F[n]` defined above.

---

[1]Fibonacci is the nickname of Leonardo Pisano (*circa* 1170–1250). With his
father he lived in the Mediterranean town Bejaïa in Kabylia where he studied arith-
metics. He later traveled widely to improve his knowledge in the art of manipulating
numbers. In 1202, he published his famous *Liber abaci* (Book of the Abacus), the
first European work on Indian and Arabian mathematics, introducing the positional
number system.

```
Clear[F]
F[1] = 1; F[2] = 1;
F[n_Integer /; n > 0] := F[n - 1] + F[n - 2];
timeCompute = Table[{2 k, Timing[F[2 k]][[1]] / Second},
{k, 1, 15}]
```

$\{\{2, 0.\}, \{4, 0.\}, \{6, 0.\}, \{8, 0.\}, \{10, 0.\},$

$\{12, 0.\}, \{14, 0.01\}, \{16, 0.01\}, \{18, 0.03\}, \{20, 0.08\},$

$\{22, 0.24\}, \{24, 0.6\}, \{26, 1.57\}, \{28, 4.12\}, \{30, 10.75\}\}$

To show the inefficiency of this method, let us estimate how long it would take to compute F[100].

```
plTime = ListPlot[timeCompute, PlotRange → All,
PlotStyle → {RGBColor[0,0,1], PointSize[0.02]}];
```

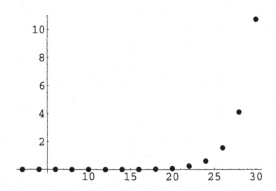

Fig. 8.1. *List plot of the CPU time to compute the first Fibonacci numbers using the inefficient method described above.*

```
Fit[timeCompute, {1, x, x^2, x^3, x^4}, x]
```

$1.9795 - 1.1436\, x + 0.181665\, x^2 - 0.0105719\, x^3 + 0.00020295\, x^4$

```
functionFit[x_] := 1.9795 - 1.1436 x + 0.181665 x^ 2 -
0.0105719 x^3 + 0.00020295 x^4

plFit = Plot[functionFit[x], {x, 0, 30},
PlotStyle → {RGBColor[1, 0, 0]}, PlotRange → All];
```

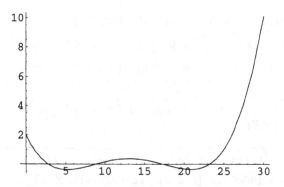

Fig. 8.2. *Plot of* functionFit *that fits the list of CPU times.*

```
Show[plFit,plTime];
```

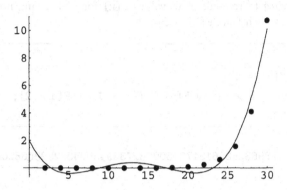

Fig. 8.3. *Plotting together* functionFit *the list plot of CPU times.*

It is not such a good fit but we can use it to find a rough estimate of the computing time.

```
Plot[functionFit[x] / 60, {x, 30, 100}];
```

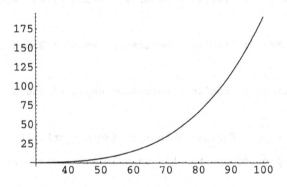

Fig. 8.4. *Plot* functionFit *in order to estimate the CPU time to evaluate the 100th Fibonacci number using the first inefficient method.*

The estimated CPU time is given in hours by

```
functionFit[100] / (60 60)
```

3.17425

More than three hours!!

It is much faster to use dynamic programming. That is, when evaluating F[n], ask *Mathematica* to remember all values F[k] for k < n, and not reevaluate all these values each time F[n] is called.

```
Clear[F]
F[1] = 1; F[2] = 1;
F[n_Integer /; n > 0] := F[n] = F[n - 1] + F[n - 2];
Timing[F[250]]
```

{0. Second, 7896325582613173050928273894363433289368686268675876375}

This is a much faster method. But let us try to evaluate F[330].

```
Clear[F]
F[1] = 1; F[2] = 1;
F[n_Integer /; n > 0] := F[n] = F[n - 1] + F[n - 2];
F[300];
```

$RecursionLimit :: reclim : Recursion depth of 256 exceeded.

More . . .

$RecursionLimit :: reclim : Recursion depth of 256 exceeded.

More . . .

$RecursionLimit :: reclim : Recursion depth of 256 exceeded.

More . . .

General :: stop : Further output of $RecursionLimit :: reclim :

will be suppressed during this calculation. More . . .

To evaluate higher order Fibonacci numbers, we have to increase Recursion-Limit. Note that dynamic programming is, however, much faster than the first method we used.

```
$RecursionLimit = 10000;
Clear[F]
F[1] = 1; F[2] = 1;
F[n_Integer /; n > 0] := F[n] = F[n - 1] + F[n - 2];
Timing[F[500]]
Table[{1000 k, Timing[F[1000 k]][[1]] / Second}, {k, 1, 20}]
```

{0.01 Second, \

13942322456169788013972438287040728395007025658 7697\

3072641089629483255716228632906915576588762225212941 25}

{{1000, 0.01}, {2000, 0.03}, {3000, 0.03}, {4000, 0.04},
{5000, 0.03}, {6000, 0.03}, {7000, 0.04}, {8000, 0.03},
{9000, 0.03}, {10000, 0.03}, {11000, 0.04}, {12000, 0.04},
{13000, 0.04}, {14000, 0.04}, {15000, 0.05},{16000, 0.04},
{17000, 0.04}, {18000, 0.04}, {19000, 0.05}, {20000, 0.04}}

The CPU time is greatly reduced. but we show below that it can be further reduced.

8.4 Control Structures

8.4.1 Conditional Operations

They are If, Which, and Switch.

The command If[test,then,else] gives then if test evaluates to True and else if it evaluates to False.

```
{If[3 > 2, 1, 0], If[3 < 2, 1, 0]}
```

{1, 0}

```
Clear[f]
f[x_] := If[x  0, 1, - 1]
Plot[f[x], {x, - 1, 1}];
```

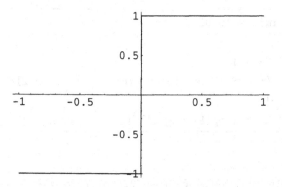

Fig. 8.5. *Plot of the function defined above.*

Note the difference with conditional rewrite rules.

```
Clear[f, g]
f[x_] := If[x ≥ 0, x^2, - x^2]
g[x_] := x^2 /; x ≥ 0
g[x_] := - x^2 /; x < 0
```

```
{D[f[x], x], D[g[x], x]}
```

$\{If[x >= 0, 2 x, - 2 x], g'[x]\}$

The command Which[test1, expression1, test2, expression2, ...] for $k = 1, 2, \ldots$, gives expressionk if testk evaluates to True.

```
Which[3 < 0, 0, 3 < 1, 1, 3 < 2, 2, 3 < 3, 3, 3 < 4, 4]
```

4

```
f[x_] := Which[0 < x ≤ 1, x, 1 < x ≤ 2, 2 x ,
2 < x ≤ 3, 3 x]
Plot[f[x], {x, 0, 3}];
```

The command Switch[expression, pattern1, value1, pattern2, value2, ...] compares expression to each patternj in turn and gives valuek if patternk is the first match found.

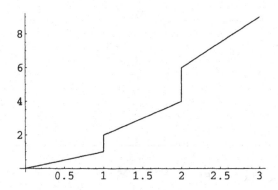

Fig. 8.6. *Plot of the function defined above.*

```
numberType[x_] := Switch[x, _Integer, ''this is an integer'',
_Rational, ''this is a rational'',
_Real, ''this is a real'',
_Complex, ''this is a complex'']
{numberType[6], numberType[3/2], numberType[2.45],
numberType[2.1 + 3.6 I]}
```

```
{this is an integer, this is a rational, this is a real,
this is a complex}
```

Note that **expression** is first evaluated before being compared to the different patterns.

```
numberType[6/3]
```

```
this is an integer
```

Evaluating 6/3 *Mathematica* finds 2 which is an integer.

8.4.2 Loops

There exist three built-in looping functions Do, While, and For.

The command Do[expression, {k, k1, k2}] evaluates expression starting from k = k1 to k2.

```
Do[Print[5^k], {k, 3, 7}]
```

125

625

3125

15625

78125

```
prod=1;
Do[prod *= k, {k, 1, 5}]
prod (* gives Factorial[5] *)
```

120

x *= c multiplies x by c and returns the new value of x.

```
Do[Plot[Cos[n x], {x, 0, 2 Pi}], {n, 1, 10, 0.25}];
```

The command above generates a sequence of plots of cos(nx) in the interval $[0, 2\pi]$ for n varying from 1 to 10 by steps 0.25. Below only the last plot of the sequence is represented.

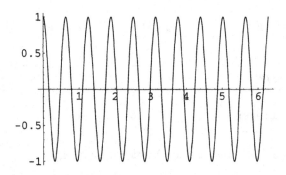

Fig. 8.7. *Plot of* cos(nx) *for* $n = 10$ *in the interval* $[0, 2\pi]$.

Coming back to Fibonacci numbers, we examine if we can build up more efficient program generating these numbers.

1. Generating a Fibonacci sequence with Do.

```
k = 0; a = 1; b = 1; FibonacciSequence = {1, 1};
Do[c = a + b;
FibonacciSequence = Join[FibonacciSequence, {c}];
a = b; b = c; c = 0, {k, 1, 10}]
FibonacciSequence
```

{1, 1, 2, 3, 5, 8, 13, 21, 34, 55, 89, 144}

The command While[test, expression] evaluates expression until test fails to give True.

```
s = 0; n = 0;
While[n   10, s += n; n++]
s (* yields Sum[n, {n, 1, 10}] *)
```

55

x += c adds c to x and returns the new value of x. n++ increases the value of n by 1 returning the old value of n.

2. Generating a Fibonacci sequence with While.

```
k = 0; a = 1; b = 1; FibonacciSequence = {1, 1};
While[k < 10, k = k + 1; c = a + b;
FibonacciSequence = Join[FibonacciSequence, {c}];
a = b; b = c; c = 0]
FibonacciSequence
```

{1, 1, 2, 3, 5, 8, 13, 21, 34, 55, 89, 144}

The command For[start, test, increment, expression] executes start, then repeatedly evaluates expression and increment until test fails to give True.

```
For[i = 1, i < 5, i++, Print[i]]
```

1

2

3

4

3. Generating a Fibonacci sequence with For.

```
a = 1; b = 1; FibonacciSequence = {1, 1};
For[k = 1, k ≤ 10, k += 1; c = a + b;
FibonacciSequence=Join[ FibonacciSequence, {c}];
a = b; b = c; c= 0]
FibonacciSequence
```

{1, 1, 2, 3, 5, 8, 13, 21, 34, 55, 89, 144}

8.5 Modules

In the command Module[{a, b, ...}, expression] the variables a,b,...
in expression are local. These variables may be initialized (a = a0, b =
b0).

```
{x = 10, Module[{x = 5}, x = x + 1; x], x}
```

{10, 6, 10}

8.5.1 Example 1

Find the first prime number greater than n.

```
firstPrimeGreater[n_Integer?Positive] := Module[{k},
k = n + 1;
While[Not[PrimeQ[k]], k++]; k]
```

```
Table[firstPrimeLarger[10 k], {k, 1, 10}]
```

{11, 23, 31, 41, 53, 61, 71, 83, 97, 101}

As mentioned above when we build up the function firstPrimeAfter, the *Mathematica* command NextPrime[n] gives the smallest prime larger than n. We have to first load the package NumberTheory'NumberTheoryFunctions'.

8.5.2 Example 2

Given a list {x1, x2, x3, ...}, generate the list of partial sums {x1, x1 + x2, x1 + x2 + x3, ...}.

```
partialSumList[lis_List] := Module[{sumList = {lis[[1]]}},
Do[sumList = Join[sumList, {sumList[[i - 1]] + lis[[i]]}],
{i, 2, Length[lis], 1}]; sumList]
```

```
lis = Table[2k, {k, 1, 10}]
partialSumList[lis]
```

{2, 4, 6, 8, 10, 12, 14, 16, 18, 20}
{2, 6, 12, 20, 30, 42, 56, 72, 90, 110}

8.5.3 Example 3

Determine the nth Fibonacci number without resetting RecursionLimit using a While loop.

```
Clear[fibonacciNumber]
fibonacciNumber[n_Integer /; n > 0] := Module[{k = 1, a = 0,
b = 1},
While[!(k == n), {a, b} = {b, a + b}; k++]; b]
```

```
Timing[fibonacciNumber[5000]]
```

{0.07244 Second, 38789684543883256337019163083259053120821277146\
4624510616059721489555013904403709701082291646221066947929345\
2858882973813483102008954982940361430156911478938364216563944\
10691021450563413370655865562382546567007125259299038549338139\
2883637834751890876297071203333705292310769300851809384980180\
3847813996748881765554653788291644268912980384613778969021502\

2930824756663462249230718833248032803750391303529033045058427\
0114763524227021093463769910400671417488329842289149127310405\
4328753298044273676822977244987749874555691907703880637046832\
7948113589737399931101062193081490185708153978543791953056175\
1076105307568878376603366735544525884488624161921055345749367\
5897849027988234351023599844663934853256411952221859563060475\
3646454707603309024208063825849291564528762915757591423438091\
4230291749108898415520985443248659407979357131684169286803954\
5309545388698114665082066862897420639323438488465240988742395\
8738019769938203171742089322654688793640026307977800587591296\
7138963421425257911687275560036031137054775472460463998758804\
6985178408674382863125}

The same problem using a Do loop is as follows.

```
Clear[fibonacciNumber]
fibonacciNumber[n_Integer?Positive] := Module[ {a = 0, b =1},
Do[{a, b} = {b, a + b}, {n - 1}]; b]
```

```
Timing[fibonacciNumber[5000]]
```

{0.03 Second, 3878968454388325633701916308325905312082127714\6\
4624510616059721489555013904403709701082291646221066947929345\
2858882973813483102008954982940361430156911478938364216563944\
1069102145056341337065586562382546567007125259299038549338139\
2883637834751890876297071203333705292310769300851809384980180\
3847813996748881765554653788291644268912980384613778969021502\
2930824756663462249230718833248032803750391303529033045058427\
0114763524227021093463769910400671417488329842289149127310405\
4328753298044273676822977244987749874555691907703880637046832\
7948113589737399931101062193081490185708153978543791953056175\
1076105307568878376603366735544525884488624161921055345749367\
5897849027988234351023599844663934853256411952221859563060475\

364645470760330902420806382584929156452876291575759142343809\
42302917491088984155209854432486594079793571316841692868039544\
530954538869811466508206686289742063932343848846524098874239\
873801976993820317174208932265468879364002630797780058759129\
713896342142525791168727556003603113705477547246046399875880\
6985178408674382863125}

The same problem using a For loop follows.

```
Clear[fibonacciNumber]
fibonacciNumber[n_Integer?Positive] := Module[{a = 0, b = 1},
For[k = 1, k < n, k++, {a, b} = {b, a + b}]; b]
```

```
Timing[fibonacciNumber[5000]]
```

{0.06 Second, 3878968454388325633701916308325905312082127714\
4624510616059721489555013904403709701082291646221066947929345\
2858882973813483102008954982940361430156911478938364216563944\
1069102145056341337065586562382546567007125259299038549338139\
2883637834751890876297071203333705292310769300851809384980180\
3847813996748881765554653788291644268912980384613778969021502\
2930824756663462249230718833248032803750391303529033045058427\
0114763524227021093463769910400671417488329842289149127310405\
4328753298044273676822977244987749874555691907703880637046832\
7948113589737399931101062193081490185708153978543791953056175\
1076105307568878376603366735544525884488624161921055345749367\
5897849027988234351023599844663934853256411952221859563060475\
3646454707603309024208063825849291564528762915757591423438091\
4230291749108898415520985443248659407979357131684169286803954\
5309545388698114665082066862897420639323438488465240988742395\
8738019769938203171742089322654688793640026307977800587591296\
7138963421425257911687275560036031137054775472460463998758804\
6985178408674382863125}

The Do loop method is the fastest. Actually an even faster program uses the function MatrixPower, which is the method used by the *Mathematica* built-in function Fibonacci[n] (refer to S. Wagon [64]). Note first that the vector {a, b} used in the above programs evolves according to

```
{b, a+b} == {a, b}. {{0, 1}, {1, 1}}
```

True

The successive powers of the matrix {{0, 1}, {1, 1}} are shown below:

```
Table[MatrixPower[{{0, 1}, {1, 1}}, k] // MatrixForm,
{k, 1, 10}]
```

$$\left\{\begin{pmatrix} 0 & 1 \\ 1 & 1 \end{pmatrix}, \begin{pmatrix} 1 & 1 \\ 1 & 2 \end{pmatrix}, \begin{pmatrix} 1 & 2 \\ 2 & 3 \end{pmatrix}, \begin{pmatrix} 2 & 3 \\ 3 & 5 \end{pmatrix}, \begin{pmatrix} 3 & 5 \\ 5 & 8 \end{pmatrix},\right.$$

$$\left.\begin{pmatrix} 5 & 8 \\ 8 & 13 \end{pmatrix}, \begin{pmatrix} 8 & 13 \\ 13 & 21 \end{pmatrix}, \begin{pmatrix} 13 & 21 \\ 21 & 34 \end{pmatrix}, \begin{pmatrix} 21 & 34 \\ 34 & 55 \end{pmatrix}, \begin{pmatrix} 34 & 55 \\ 55 & 89 \end{pmatrix}\right\}$$

The sequence of Fibonacci numbers is the sequence of the [[1,2]] elements of the matrices MatrixPower[{{0, 1}, {1, 1}}, n] for n = 1, 2, ... Hence:

```
Clear[fibonacciNumber]
fibonacciNumber[n_Integer?Positive] :=
MatrixPower[{{0, 1}, {1, 1}}, n][[1,2]]
```

```
Timing[fibonacciNumber[10000]]
```

{0. Second,

336447648764317832666216120051075433103021484606800639065\

647699746800814421666623681555955136337340255820653326808\

361593737347904838652682630408924630564318873545443695598\

274916066020998841839338646527313000888302692356736131351\

175792974378544137521305205043477016022647583189065278908\

551543661595829872796829875106312005754287834532155151038\

7081829896791613127856265033195487140214287532698187962\0

4693609787990035096230229102636813149319527563022783762\84\

415403605844025721143349611800230912082870460889239623288\

354615057765832712525460935911282039252853934346209042452\

489294039017062338889910858410651831733604374707379085526\

317643257339937128719375877468974799263058370657428301616\

374089691784263786242128352581128205163702980893320999057\

079200643674262023897831114700540749984592503606335609338\

838319233867830561364353518921332797329081337326426526339\

897639227234078829281779535805709936910491754708089318410\

561463223382174656373212482263830921032977016480547262438\

423748624114530938122065649140327510866433945175121615265\

453613331113140424368548051067658434935238369596534280717\

687753283482343455573667197313927462736291082106792807847\

180353291311767789246590899386354593278945237776744061922\

403376386740040213303432974969020283281459334188268176838\

930720036347956231171031012919531697946076327375892535307\

725523759437884345040677155557790564504430166401194625809\

722167297586150269684431469520346149322911059706762432685\

159928347098912847067408620085871350162603120719031720860\

940812983215810772820763531866246112782455372085323653057\

759564300725177443150515396009051686032203491632226408852\

488524331580515348496224348482993809050704834824493274537\

326245677558790891871908036620580095947431500524025327097\

469953187707243768259074199396322659841474981936092852239\

450397071654431564213281576889080587831834049174345562705\

202235648464951961124602683139709750693826487066132645076\

650746115126775227486215986425307112984411826226610571635\

150692600298617049454250474913781151541399415506712562711\

971332527636319396069028956502882686083622410820505624307\

01794976171121233066073310059947366875}

This Fibonacci number has 2090 digits!

```
Length[IntegerDigits[fibonacciNumber[10000]]]
```

2090

Using the *Mathematica* built-in function, we can check the result above

```
fibonacciNumber[10000] == Fibonacci[10000]
```

True

Part II

Applications

This second part presents a variety of examples from mathematics and physics illustrating how *Mathematica* can be used to study and solve problems. For each example we tried to build up simple-to-understand short programs. In most cases, we adopted the so-called bottom-up technique, progressing step by step from small units to the final program, making every effort to give a similar structure to all programs. Most problems can be tackled in many different ways. When a solution is found, it is good practice to try to find a different one and compare the different solutions. Many *Mathematica* applications can be found on the Web; the Wolfram site: http://mathworld.wolfram.com/, created and maintained by Eric Weisstein, is a particularly extensive source of interesting applications.

No subtle criterion has been used to order the list of the different applications. They are just listed in the alphabetical order of their TeX file names.

Mathematicians and physicists are often involved in time-consuming calculations. Using *Mathematica* to carry the bulk of the computational burden and check the results should enable them to devote more time to think about ideas. *Mathematica* computational capabilities are particularly helpful when one tries to discover a pattern. We illustrated this feature at the end of the first chapter when we tried to generalize the Collatz conjecture.

Discovering a pattern is often the first step towards the proof of a theoretical result. This has not been the case for the Collatz problem whose solution seems for the moment beyond the reach of today's mathematics but there exist many examples showing that "computational experimentation" does suggest a theoretical result.

Here is, for instance, a very simple example. There exists an extensive literature dedicated to *happy numbers*. If f is a function, defined on the set of positive integers, such that the image $f(n)$ of any integer n is equal to the sum of the squares of the digits of n, then, a number n is said to be happy if there exists a finite integer k such that the kth iterate $f^k(n) = 1$.

Experimenting with *Mathematica*, one soon discovers that either a number is happy or the sequence of iterates of f ends in the same eight-cycle: $(4, 16, 37, 58, 89, 145, 42, 20)$. Because the `ListPlot` of $f(n)$ shows that $f(n) < n$ for $n > f(99) = 162$, one can then easily prove that the iterates of f either converge to the fixed point 1 or to the eight-cycle above. This simple conclusion leads immediately to the question: what about functions f_k also defined on the set of positive integers, such that the image $f_k(n)$ of any integer n is equal to the sum of the powers k of the digits of n. Here again, experimenting with *Mathematica* shows that f_3 has only a few fixed points, namely $\{1, 153, 370, 371, 407\}$ and, because $f_3(n) < n$ for $n > f(999) = 2187$, it is sufficient to restrict our search for fixed points in a rather small range, which can be done in a fraction of a second using *Mathematica*. The function

Axially Symmetric Electrostatic Potential

Consider a conducting sphere of radius R held at potential zero and placed in a uniform electric field $\mathbf{E}_0$ directed along the Oz-axis oriented upward. If we use spherical coordinates, the symmetry of the system implies that the potential V depends only upon r and θ but not on φ, and can, therefore, be written as a sum of Legendre polynomials: LegendreP[n,Cos[θ]]. We first briefly review how to solve the Laplace equation in this case (for more details see Vvedensky [63], Chapter 6).

Legendre polynomials are encountered in several problems of mathematical physics involving the eigenfunctions of the angular part of the Laplacian operator.

The Laplacian operator of the function V in spherical coordinates is

$$\nabla^2 V = \frac{1}{r}\frac{\partial^2}{\partial r^2}(rV) + \frac{1}{r^2\sin^2\theta}\frac{\partial^2 V}{\partial\varphi^2} + \frac{1}{r^2\sin\theta}\frac{\partial}{\partial\theta}\left(\sin\theta\frac{\partial V}{\partial\theta}\right).$$

If the electric potential does not depend upon φ, the Laplace equation reads

$$\nabla^2 V = \frac{1}{r}\frac{\partial^2}{\partial r^2}(rV) + \frac{1}{r^2\sin\theta}\frac{\partial}{\partial\theta}\left(\sin\theta\frac{\partial V}{\partial\theta}\right) = 0.$$

This equation can be solved using the method of separation of variables. That is, we can factor the solution writing that $V(r,\theta)$ is the product of a radial part $F_r(r)/r$ and an angular part $F_\theta(\theta)$. Substituting

$$V(r,\theta) = \frac{F_r(r)}{r}F_\theta(\theta)$$

in the expression of the Laplacian yields

$$F_\theta\frac{d^2 F_r}{dr^2} + \frac{F_r}{r^2\sin\theta}\frac{d}{d\theta}\left(\sin\theta\frac{dF_\theta}{d\theta}\right) = 0.$$

Multiplying by r^2 and dividing by $F_r F_\theta$ the left-hand side of this equation gives

$$\frac{r^2}{F_r} \frac{d^2 F_r}{dr^2} + \frac{1}{F_\theta \sin\theta} \frac{d}{d\theta}\left(\sin\theta \frac{dF_\theta}{d\theta}\right) = 0,$$

that is, the sum of a function of r only and a function of θ only. Because the sum of these two functions is zero, each function is necessarily a constant and we have, therefore, to solve the following two equations:

$$\frac{d^2 F_r}{dr^2} - k\frac{F_r}{r^2} = 0, \text{ and } \frac{1}{\sin\theta}\frac{d}{d\theta}\left(\sin\theta \frac{dF_\theta}{d\theta}\right) + kF_\theta = 0,$$

where k is a constant.

The radial equation has obvious solutions of the form r^a. Replacing $F_r(r)$ by r^a, we find that r^a is a solution provided $a(a-1) = k$. Writing k as $n(n+1)$, we find that a must be equal either to n or $-(n+1)$. Hence, the general solution of the radial equation is

```
radialPart[r_] := A[n] r^n + B[n] r^(- n - 1)
```

where A[n] and B[n] are constants.

The angular equation takes a simpler form if we introduce the new variable $x = \cos\theta$. Because the range of θ is $[0, \pi]$, the range of x is $[-1, 1]$. After this change of variable the angular part satisfies the ordinary Legendre equation:

$$\frac{d}{dx}\left((1 - x^2)\frac{dF_\theta}{dx}\right) + n(n+1) = 0.$$

This equation is usually solved by representing the solution as a power series about the origin. The series are converging for all values of x of the closed interval $[-1, 1]$ if, and only if, F_θ is a Legendre polynomial of degree n. These functions are *Mathematica* built-in functions.

The general solution of the Laplace's equation in the case of axial symmetry is therefore

```
Vpotential[r_, theta_] := Sum[(A[n] r^n + B[n] r^(1-n))
LegendreP[n, Cos[theta]]
```

Coming back to our problem of the grounded conducting sphere of radius R in a uniform electric field $\mathbf{E}_0$, the electric potential Vpotential[r,theta] can be simplified taking into account the boundary conditions.

For $r \to \infty$, the electric potential must behave as $-E_0 r \cos\theta$, where E_0 is the only nonzero component of $\mathbf{E}_0$ on the Oz-axis. This condition implies that all constants A[n] are zero except A[1] = - E0 R. The electric potential being equal to zero on the sphere implies that Vpotential[R, theta] = 0. Thus B[0] = 0, B[1] = - A[1] = E0 R, and B[n] = 0 for all $n \geq 2$. The final solution is therefore

```
Vpotential[r_, theta_] := (- E0 r + E0 R^3 r^2) Cos[theta]
```

because

```
LegendreP[1, Cos[theta]]
```

Cos[theta]

We can plot the equipotentials around the sphere in the plane $y = 0$, choosing the numerical values $E_0 = 1$, $R = 1$. But first we have to transform spherical coordinates to Cartesian coordinates. Because $y = 0$, the transformation rule is

```
{r → Sqrt[x^2 + z^2], Cos[theta] → z / Sqrt[x^2 + z^2]}
```

```
numVpotential[r_, theta_] := Vpotential[r, theta] /. {E0 → 1,
R → 1}
```

Hence,

```
numVpotential[r, theta]
```

$(r^{-2} - r)$ Cos[theta]

In a more complicated case we would have used *Mathematica* to obtain the transformation rule as shown below. First we load the package Calculus'VectorAnalysis'.

```
<<Calculus'VectorAnalysis'
```

Then using the command CoordinatesFromCartesian we get

```
{r, theta, phi} → CoordinatesFromCartesian[{x, y, z},
Spherical] // Thread
```

$\{r \to \text{Sqrt}[x^2 + y^2 + z^2], \text{theta} \to \text{ArcCos}[\dfrac{z}{\text{Sqrt}[x^2 + y^2 + z^2]}],$

```
phi → ArcTan[x, y]}
```

`ArcTan[x, y]` gives $\arctan(y/x)$ taking into account in which quadrant the point (x, y) is.

The 2D potential as a function of x and z is thus defined by

```
Vpotential2D[x_, z_] := numVpotential[r, theta] //.
{Cos[theta] → z / r, r → Sqrt[x^2 + z^2]} // Simplify
```

that is,

```
Vpotential2D[x,z]
```

$$z \, (-1 + (x^2 + z^2)^{-3/2})$$

and using `ContourPlot` we can plot a few equipotentials:

```
cpl = ContourPlot[Vpotential2D[x, z],
{x, - 2, 2}, {z, - 2, 2},
Contours → 30, ContourShading → False,
DisplayFunction → Identity];
sphere = Graphics[{GrayLevel[0.2], Disk[{0, 0}, 1]}];
Show[{cpl, sphere}, DisplayFunction → $DisplayFunction,
AspectRatio → Automatic, Frame → True];
```

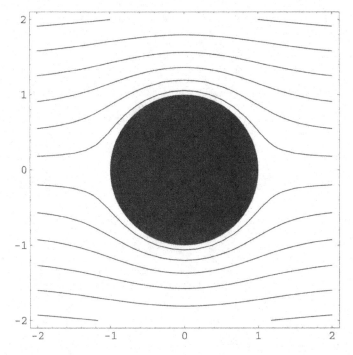

Fig. 9.1. *Equipotentials in the plane $y = 0$ in the vicinity of a grounded sphere placed in a uniform electric field directed along the Oz-axis.*

Motion of a Bead on a Rotating Circle

A bead of mass m is constrained to move without friction on a circular wire of radius R. The circular wire rotates with constant angular velocity ω around its vertical diameter. This system has two degrees of freedom. If, to describe the motion of the bead, we use the two spherical coordinates θ and φ, the Cartesian coordinates of the bead in terms of these generalized coordinates are $x = R\sin\theta\cos\varphi$, $y = R\sin\theta\sin\varphi$, and $z = R\cos\theta$, if we take $\theta = 0$ at the bottom of the circular wire. Let us draw the figure.

We load the package Graphics'Arrow' which implements a new graphics primitive to generate arrows.

```
<<Graphics'Arrow'
```

```
g = Graphics[{Circle[{0, 0}, 1],
Circle[{0, 1.2}, {0.15, 0.07}, {- 4Pi / 3, Pi / 3}],
Line[{{0, 1.3}, {0, - 1.1}}],
Line[{{0, 0}, {Cos[4 Pi / 3], Sin[4 Pi / 3]}}],
Circle[{0, 0}, 0.3, {4 Pi / 3, 3 Pi / 2}],
{PointSize[0.04], Point[{Cos[4 Pi / 3], Sin[4 Pi / 3]}]}}];
txt = Graphics[{ Text[''m'', {Cos[4 Pi / 3] - 0.15,
Sin[4 Pi / 3] - 0.05}],
Text[''g'', {1.3, - 0.1}],
Text[''R'', {0.5 Cos[4 Pi / 3] - 0.2,
0.5 Sin[4 Pi / 3] - 0.2}],
Text[''θ'', {0.3 Cos[17 Pi / 12] - 0.04,
0.3 Sin[17 Pi / 12] - 0.1}]}];
```

```
a = Graphics[{Arrow[{1.2, 0.2}, {1.2, - 0.5}],
Arrow[{- 0.05, 1.13}, {0.05, 1.13}]}];
fig = Show[{g, txt, a}, AspectRatio → Automatic,
TextStyle → {FontSlant → ''Italic'', FontSize → 16}];
```

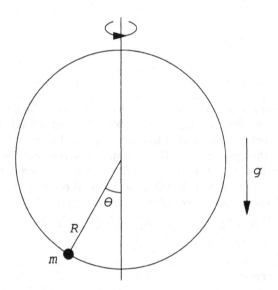

Fig. 10.1. *A bead on a rotating circle.*

Taking into account that $\varphi' = \omega$, the Cartesian components of the velocity vector are

$$x' = R\theta' \cos\theta \cos\varphi - R\omega \sin\theta \sin\varphi,$$
$$y' = R\theta' \cos\theta \sin\varphi + R\omega \sin\theta \cos\varphi,$$
$$z' = R\theta' \sin\theta,$$

and the kinetic energy is

```
K = (1 / 2) m R^2 (theta'[t]^2 + omega^2 Sin[theta[t]]^2);
```

The potential energy being

```
U = - m g R Cos[theta[t]];
```

the expression of the Lagrangian is

```
Lagrangian = K - U
```

$$g\ m\ R\ \text{Cos[theta[t]]} + \frac{m\ R^2\ (\text{omega}^2\ \text{Sin[theta[t]]}^2 + \text{theta}'[t]^2)}{2}$$

Loading the *Mathematica* package `Calculus`VariationalMethods`` that contains the command `EulerEquations` we can get Euler-Lagrange equations.

```
<<Calculus`VariationalMethods`
```

```
EulerEquations[Lagrangian, theta[t], t] // Simplify
```

$m\ R\ ((-\ g\ +\ R\ \text{omega}^2\ \text{Cos[theta[t]]})\ \text{Sin[theta[t]]}$
$-\ R\ \text{theta}''[t])\ ==\ 0$

The equation of motion is

```
eqn = theta"[t] + (g / R - omega^2 Cos[theta[t]])
Sin[theta[t]] == 0
```

Taking into account the Coriolis force, the effective potential is

```
Ueffective[theta_] := - m g R Cos[theta] -
(1 / 2) m omega^2 R^2 Sin[theta]^2
```

Its minimum gives the stable equilibrium position.

```
derTheta = D[Ueffective[theta], theta]
```

$g\ m\ R\ \text{Sin[theta]}\ -\ m\ R^2\ \text{omega}^2\ \text{Cos[theta]}\ \text{Sin[theta]}$

Define

```
Omega^2 = g / R
```

and replace in the expression of the derivative of the effective potential

```
derTheta = derTheta /. g -> Omega^2 R // Simplify
```

m R^2 (Omega^2 - omega^2 Cos[theta]) Sin[theta]

The equilibrium positions are

$$\theta = 0 \text{ and } \theta = \pm\theta_0 \text{ such that } \cos\theta_0 = \frac{\Omega^2}{\omega^2}.$$

If the oscillations around the stable equilibrium are small, from the expression of the second derivative of the effective potential, we find that the period T of these oscillations is

$$T = \frac{2\pi}{\sqrt{\Omega^2 - \omega^2}} \quad \text{when } \theta = 0 \text{ is stable for } \omega < \Omega, \text{ and}$$

$$T = \frac{2\pi}{\sqrt{\omega^4 - \Omega^4}} \quad \text{when } \theta = \pm\theta_0 \text{ is stable for } \omega > \Omega.$$

Here are two plots of the effective potential, one for $\omega < \Omega$, when $\theta = 0$ is stable, and the other one for $\omega > \Omega$, when $\theta = \theta_0$ is stable.

```
m = 1; g = 9.8; R = 2; Omega = g / R;
Clear[omega]
omega = 0.5;
pl1 = Plot[Ueffective[theta], {?, - Pi, Pi},
PlotStyle → {RGBColor[0, 0,1]},
DisplayFunction → Identity];
Clear[omega]
omega = 3.5;
pl2 = Plot[Ueffective[theta], {theta, - Pi, Pi},
PlotStyle → {RGBColor[1, 0, 0]},
DisplayFunction → Identity];
Show[{pl1, pl2}, DisplayFunction → $DisplayFunction];
```

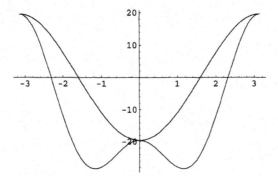

Fig. 10.2. *Effective potentials when either $\theta = 0$ is stable or $\theta = \theta_0 \neq 0$ is stable.*

The Brachistochrone

The brachistocrone (from the Greek, *brakhisto* meaning "shortest" and *chronos* meaning "time"), is the planar curve on which a body subjected only to the force of gravity slides without friction between two points in the least possible time. Finding the curve was a problem first posed by Galileo (1564-1642). In June 1696 the Swiss mathematician Johann Bernoulli (1667-1748), father of Daniel (1700-1782), another famous Bernoulli, who was the first to find the correct solution in 1696, challenged his contemporaries in *Acta Eruditorum* to find the solution. Correct solutions came from his brother Jakob (1654-1705), Gottfried Wilhelm von Leibniz (1646-1716), Guillaume de l'Hôpital (1661-1704), and Isaac Newton (1643-1727). The solution of this famous problem led to the development of the calculus of variations.

A bead of mass m slides without friction on a wire in a vertical plane. In the vertical plane, we choose the Ox-axis directed vertically downward. If the bead is released from the origin $(0,0)$ with an initial velocity equal to zero, the time to reach the point (x_0, y_0) is given by

$$t = \int_0^{x_0} \frac{ds}{v} = \int_0^{x_0} \frac{\sqrt{1 + (y')^2}}{v} \, dx$$

Conservation of energy implies that the velocity v along the wire is equal to $\sqrt{2gx}$, where g is the acceleration due to gravity. In order to find the equation of the curve representing the shape of the wire such that the time to reach a given point (x_0, y_0) is minimum, we have to minimize the integral

$$t = \int_0^{x_0} \sqrt{\frac{1 + (y')^2}{2gx}} \, dx.$$

The function $x \mapsto y(x)$ minimizing this integral is the solution of the Euler–Lagrange differential equation that could be obtained using the *Mathematica* command `EulerEquations` found in the package `Calculus'VariationalMeth-`

ods'. But it is simpler to first note that, because the integrand depends only upon y' and not y, the Euler–Lagrange equation is

$$\frac{d}{dx}\left(\frac{y'}{\sqrt{x(1+(y')^2}}\right) = 0.$$

That is,

```
eqn = (y')^2 / (x (1 + (y')^2)) == c;
```

where c is a positive constant less than 1. Solving for y' yields

```
Solve[eqn, y']
```

$$\left\{\left\{y' \rightarrow -\frac{\text{Sqrt}[c]\ \text{Sqrt}[x]}{\text{Sqrt}[1-cx]}\right\},\ \left\{y' \rightarrow \frac{\text{Sqrt}[c]\ \text{Sqrt}[x]}{\text{Sqrt}[1-cx]}\right\}\right\}$$

Here the function y is defined by

```
y[x_] := Integrate[Sqrt[c u] / Sqrt[1 - c u], {u, 0, x}]
```

Because $0 < cx < 1$, it is better to replace cx by $\sin^2\theta$; that is, $x(\theta) = (1/2c)(1 - \cos(2\theta))$. To obtain the expression of $y(\theta)$ we have just to write

$$\frac{dy}{d\theta} = \frac{dy}{dx}\frac{dx}{d\theta} = \frac{1}{c}\tan\theta\sin(2\theta).$$

Hence,

$$y(\theta) = \frac{1}{2c}(2\theta - \sin(2\theta)).$$

The parametric equation of the curve representing the wire shape is, therefore,

$$x(\theta) = \frac{1}{2c}(1 - \cos(2\theta))\ y(\theta) = \frac{1}{2c}(2\theta - \sin(2\theta)).$$

It is the equation of an inverted cycloid with its base along the Oy-axis and its cusp at the origin as shown in the figure below, where we took into account that the Ox-axis is directed vertically downward.

```
ParametricPlot[{(1 / 2) (2 theta - Sin[2 theta]),
- (1 / 2) (1 - Cos[2 theta])}, {theta, 0, Pi},
PlotStyle → {RGBColor[0, 0, 1]}, Axes → False,
Frame → True, FrameTicks → None,
PlotLabel → ''Brachistochrone'',
DefaultFont → {''Helvetica'', 14},
AspectRatio → Automatic];
```

Brachistochrone

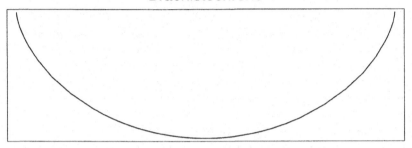

Fig. 11.1. *Brachistochrone.*

Negative and Complex Bases

It is common practice to represent numbers using positional number systems. In such a system, each number has a unique representation by an ordered sequence of symbols, the value of the number being determined by the position of the symbols and the base b of the system. If b is a positive integer we need b different symbols which are digits if $2 \leq b \leq 10$ and extra symbols if $b > 10$. A number N_b is then represented by the sequence of digits $d_n d_{n-1} \ldots d_1 d_0$ such that

$$N_b = d_n \times b_n + d_{n-1} \times b^{n-1} + \cdots + b_1 \times b + d_0.$$

The concept of base can be extended to negative and even complex bases.

12.1 Negative Bases

The representation of numbers in negative bases offers the advantage of not requiring the minus sign preceding a negative number. For example, in base -10, the negative number -253 is represented by 1867 because

```
- 253 == (-10)^3 + 8 (-10)^2 + 6 (-10) + 7
```

True

This representation, which can be shown to be unique, is more economical because it does not need the extra symbol "$-$" and the problem of $+0$ being equal to -0 does not exist.

Note that, of course, positive numbers can also be represented in base -10. For example, 253 is represented by 353.

```
253 == 3 (-10)^2 + 5 (-10) + 3
```

True

Formally, as for positive bases, a number N in base b, written $(d_n d_{n-1} \ldots d_1 d_0)_b$ is equal to $d_n b^n + d_{n-1} b^{n-1} + \cdots + b_1 b + d_0$ where, for all k, $0 \le d_k < |b|$. As for positive bases, the digits d_k are calculated using the usual division algorithm. The successive quotients are, therefore, given by $q_k = (q_{k+1} - d_k)/b$. Here is, for example, the sequence of operations done to find the representations of 253 and -253 in base -10 using *Mathematica*.

```
n = 253; b = -10;
d[1] = Mod[n, Abs[b]]
d[2] = Mod[(n - d[1]) / (b), Abs[b]]
d[3] = Mod[((n - d[1]) / (b) - d[2]) / (b), Abs[b]]
```

3

5

3

```
n = - 253; b = -10;
d[1] = Mod[n, Abs[b]]
d[2] = Mod[(n - d[1]) / (b), Abs[b]]
d[3] = Mod[((n - d[1]) / (b) - d[2]) / (b), Abs[b]]
d[4] = Mod[(((n - d[1]) / (b) - d[2]) / (b) - d[3]) / (b),
Abs[b]]
```

7

6

8

1

All arithmetic operations can be carried out in negative base systems. To add, subtract, and multiply using negative-base representations, we proceed as usual. The problem of carrying digits requires, however, more care. Because

```
{10 == 1 (-10)^2 + 9 (-10) + 0,
 20 == 1 (-10)^2 + 8 (-10) + 0, 30 == 1 (-10)^2 + 7 (-10) + 0,
 40 == 1 (-10)^2 + 6 (-10) + 0, 50 == 1 (-10)^2 + 5 (-10) + 0,
 60 == 1 (-10)^2 + 4 (-10) + 0, 70 == 1 (-10)^2 + 3 (-10) + 0,
 80 == 1 (-10)^2 + 2 (-10) + 0, 90 == 1 (-10)^2 + 1 (-10) + 0}
```

{True, True, True, True, True, True, True, True, True}

instead of carrying $1, 2, 3, \ldots$ we have to carry $19, 18, 17, \ldots$ which will affect the next two higher places. This can create some complications when the carry digits accumulate giving an infinite series of carry digits.

Here are a few simple examples with no accumulation of carry digits. In base -10,

$$
\begin{array}{r}
1\,9 \\
207 \\
+303 \\
\hline
690
\end{array}
$$

because $7 + 3 = 10$, we write 0 and carry 19, $9 + 0 + 0 = 9$, we write 9 and then add 1 (of 19), 2 and 3 to get 6, to obtain the final result 690. We verify that in base 10, 207 and 303 have the same representation, and that their sum, 510, is represented by 690 in base -10.

```
207 == 2 (-10)^2 + 0 (-10) + 7
303 == 3 (-10)^2 + 0 (-10) + 3
510 == 6 (-10)^2 + 9 (-10) + 0
```

True

True

True

When subtracting two numbers in base -10, we can borrow 10 in one column by adding one to the next higher column. For example, $353 - 187 = 386$,

$$
\begin{array}{r}
1\,1 \\
353 \\
-187 \\
\hline
386
\end{array}
$$

because 7 being greater than 3 we borrow 10 from the next left column, then 13 minus 7 equals 6, then 8 is greater than 6 (5 + 1) so we borrow again 10 from the next left column, 16 minus 8 is 8, finally 4 (3 + 1) minus 1 is 3. Subtracting 187 from 353 in base −10 gives 386. In base 10, 353 is 253 and 187 is 27, so 253 − 27 being equal to 226 we verify that 226 in base −10 is 386:

```
253 == 3 (-10)^2 + 5 (-10) + 3
27 == 1 (-10)^2 + 8 (-10) + 7
226 == 3 (-10)^2 + 8 (-10) + 6
```

True

True

True

In order to multiply 304 by 107 (note that these two numbers have the same representation in bases 10 and −10), we write

$$
\begin{array}{r}
1\,8 \\
1\,8 \\
304 \\
\times 107 \\
\hline
18288 \\
30400 \\
\hline
48688
\end{array}
$$

because we first multiply 304 by 7 which gives 18,288 (where instead of carrying 2 we carried 18, etc.) then multiplying 304 by 0 gives 0, multiplying 304 by 1 gives 304. Finally adding 18,288 + 30,400 = 48,688, which is equal in base 10 to 304 × 107 = 32,528.

```
304 == 3 (-10)^2 + 4
107 == 1 (-10)^2 + 7
304 107 == 32528 == 4 (-10)^4 + 8 (-10)^3 + 6 (-10)^2 +
8 (-10) + 8
```

True

True

True

As mentioned above, it often happens that carry digits accumulate. For instance, if we add the two negative numbers -44 and -13, represented, respectively, by 56 and 27 in base -10, because $6 + 7 = 13$, write 3 and carry 19, $9 + 5 + 2 = 16$, write 6 and carry 19, $9 + 1 = 10$, write 0, and carry 19, and so on. We thus obtain an infinite sequence of zeros. We just have to stop when obtaining the infinite sequence of zeros. If we do eliminate this infinite sequence of zeros we correctly obtain 63 representing $-57 = (-44) + (-13)$ in base -10.

```
- 44 == 5 (-10) + 6
- 13 == 2 (-10) + 7
- 57 == 6 (-10) + 3
```

True

True

True

Then,

$$
\begin{array}{r}
1\,9 \\
1\,9 \\
1\,9 \\
56 \\
+27 \\
\hline
0063
\end{array}
$$

Many other problems such as representations of rationals having either terminating or periodic expansions and reals that do not have periodic expansions can be studied. For more details, the interested reader should consult [21].

12.2 Complex Bases

12.2.1 Arithmetic in Complex Bases

A Gaussian integer $z = x + iy$ is said to be expressible in the complex base $b = \alpha + i\beta$ if it can be written in the form $z = \sum_{k=0}^{n} d_k b^k$ where, as for usual positive bases, the numbers d_k $(0 \le k \le n)$ are called digits. In the base $b = -\alpha + i\beta$ $(\alpha > 0)$, it can be shown that all Gaussian integers can be represented with the "digits" $0, 1, 2, \ldots, \alpha^2$. Thus, the base $b = -1 + i$ provides a binary representation of all Gaussian integers using only the digits 0 and 1.

For example, $7 + 6i = (1101001)_{-1+i}$ because

```
7 + 6 I == (-1 + I)^6 + (-1 + I)^5 + (-1 + I)^3 + 1
```

True

As with negative bases, all arithmetic operations can be performed in complex bases. But, here again, the problem of carry digits has to be handled with care. In base $-1+i$, each time a sum exceeds $(-1)^2 = 1$, we have to carry $(-1)^2 + 1$ or some multiple of it to the next left columns. Because

```
(-1)^2 + 1 == 2 == (-1+ I)^3 + (-1+ I)^2 + 0 (-1+ I) + 0
```

True

we have to write 0 and carry 110 to the next left columns. For example, the representations of $2 + 3i$ and $-1 - i$ in base $-1 + i$ are

```
2 + 3 I == 1 (-1 + I)^3 + 1 (-1+I) + 1
-1 - I == 1 (-1+I)^2 + 1 (-1 + I) + 0
```

True

True

Then,

$$
\begin{array}{r}
1\,1\,0 \\
1\,1\,0 \\
1011 \\
+1\ 10 \\
\hline
1110101
\end{array}
$$

To add 1011 and 110, in base $-1 + i$, add $1 + 0 = 1$, write 1, add $1 + 1 = 2$, write 0 and carry 110, $0 + 1 + 0$ (from carry 110) $= 1$, write $1 + 1$ (from previous carry 110) $= 2$, write 0 and carry 110, 1 (from first carry) $+0$ (from second carry) $= 1$, write 1, and because we have the remaining two 1s from the second carry, write 11. The result is, therefore, 1110101, which is the binary representation of $(2 + 3i) + (-1 - i) = 1 + 2i$. Asking *Mathematica* to check yields

```
(2 + 3 I) + (- 1 - I) == 1 + 2 I ==
(1 (- 1 + I)^3 + 1 (-1 + I) + 1) + (1 (-1+I)^2 +
1 (-1 + I) + 0) ==
1 (-1 + I)^6 + (-1 + I)^5 + (-1 + I)^4 + 1 (-1 + I)^2 + 1
```

True

12.2.2 Fractal Images

In what follows we generate fractal images plotting Gaussian integers as points of $\mathbb{R}^2$ using complex bases.

Consider the base $b = -1 + i$ with the digit set $\{0, 1\}$, and first build up the list of Gaussian integers defined recursively by

```
b = -1+I;
Gint1[0] = {0, 1}; (* set of digits *)
Gint1[1] = Join[Gint1[0], Gint1[0] + b];
Gint1[n_] := Join[Gint1[n - 1], Gint1[n - 1] + b^n]
```

For example, we find

```
Gint1[2]
```

$\{0, 1, -1 + I, I, -2I, 1 - 2I, -1 - I, -I\}$

In order to transform the list Gint1[n] of Gaussian integers in points in $\mathbb{R}^2$, we use the function complex2point and make it listable.

```
Attributes[complex2point] = Listable;
complex2point[z_] := {Re[z], Im[z]}
```

```
complex2point[Gint1[2]]
```

$\{\{0, 0\}, \{1, 0\}, \{-1, 1\}, \{0, 1\}, \{0, -2\},$
$\{1, -2\}, \{-1, -1\}, \{0, -1\}\}$

In order to represent the list of points we associate to each point of the list a rectangle using the function:

```
rectangleList[lis_List] :=
Table[Rectangle[lis[[i]], lis[[i]] + 1.],
{i, 1, Length[lis]}]
```

Here are the plots corresponding to the first 4 lists of points

```
d1 = Graphics[{Hue[1 / 15],
rectangleList[complex2point[Gint1[1]]]}];
d2 = Graphics[{Hue[2 / 15],
rectangleList[complex2point[Gint1[2]]]}];
d3 = Graphics[{Hue[3 / 15],
rectangleList[complex2point[Gint1[3]]]}];
d4 = Graphics[{Hue[4 / 15],
rectangleList[complex2point[Gint1[4]]]}];
Show[GraphicsArray[{{d1, d2}, {d3, d4}}],
AspectRatio → Automatic];
```

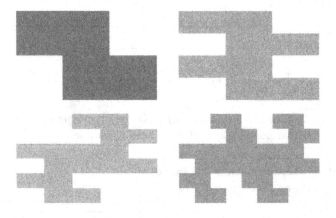

Fig. 12.1. *Images associated with lists* Gint1[[1]], Gint1[[2]], Gint1[[3]], *and* Gint1[[4]].

Increasing the list length generates a dragon-type fractal.

```
d14 = Show[Graphics[{Hue[14 / 15],
rectangleList[complex2point[Gint[14]]]}],
AspectRatio → Automatic];
```

Fig. 12.2. *Dragon-type fractal associated with list* Gint[[14]].

In the base $b = -2 + i$ the digit set is $\{0, 1, 2, 3, 4\}$, and, as above we build up the list of Gaussian integers defined recursively by

```
b = - 2 + I;
Gint2[0] = {0, 1, 2, 3, 4}; (* set of digits *)
Gint2[1] = Join[Gint2[0], Gint2[0] + b,
Gint2[0] + 2 b,Gint2[0] + 3 b, Gint2[0] + 4 b];
Gint2[n_] := Join[Gint2[n - 1], Gint2[n - 1] + b^(n-1),
Gint2[n - 1] + 2 b^(n-1), Gint2[n - 1] + 3 b^(n-1),
Gint2[n - 1] + 4 b^(n-1)]
```

For example, we find

Gint2[2]

{0, 1, 2, 3, 4, − 2 + I, − 1 + I, I, 1 + I, 2 + I, − 4 +2 I,
− 3 + 2 I, − 2 + 2 I, − 1 + 2 I, 2 I, − 6 + 3 I, − 5 + 3 I,
− 4 + 3 I, − 3 + 3 I, − 2 + 3 I, − 8 + 4 I, − 7 + 4 I,
− 6 + 4 I, − 5 + 4 I, − 4 + 4 I, − 2 + I, − 1 + I, I, 1 + I,
2 + I, − 4 + 2 I, − 3 + 2 I, − 2 + 2 I, − 1 + 2 I, 2 I,
− 6 + 3 I, − 5 + 3 I, − 4 + 3 I, − 3 + 3 I, − 2 + 3 I,
− 8 + 4 I, − 7 + 4 I, − 6 + 4 I, − 5 + 4 I, − 4 + 4 I,
− 10 + 5 I, − 9 + 5 I, − 8 + 5 I, − 7 + 5 I, − 6 + 5 I,
− 4 + 2 I, − 3 + 2 I, − 2 + 2 I, − 1 + 2 I, 2 I, − 6 + 3 I,
− 5 + 3 I, − 4 + 3 I, − 3 + 3 I, − 2 + 3 I, − 8 + 4 I,
− 7 + 4 I, − 6 + 4 I, − 5 + 4 I, − 4 + 4 I, − 10 + 5 I,
− 9 + 5 I, − 8 + 5 I,− 7 + 5 I, − 6 + 5 I, − 12 + 6 I,
− 11 + 6 I, − 10 + 6 I, − 9 + 6 I, − 8 + 6 I, − 6 + 3 I,
− 5 + 3 I, − 4 + 3 I, − 3 + 3 I, − 2 + 3 I, − 8 + 4 I,
− 7 + 4 I, − 6 + 4 I, − 5 + 4 I, − 4 + 4 I, − 10 + 5 I,
− 9 + 5 I, − 8 + 5 I, − 7 + 5 I, − 6 + 5 I, − 12 + 6 I,
− 11 + 6 I, − 10 + 6 I, − 9 + 6 I, − 8 + 6 I, − 14 + 7 I,
− 13 + 7 I, − 12 + 7 I, − 11 + 7 I, − 10 + 7 I, − 8 + 4 I,
− 7 + 4 I, − 6 + 4 I, − 5 + 4 I, − 4 + 4 I, − 10 + 5 I,
− 9 + 5 I, − 8 + 5 I, − 7 + 5 I, − 6 + 5 I, − 12 + 6 I,
− 11 + 6 I, − 10 + 6 I, − 9 + 6 I, − 8 + 6 I, − 14 + 7 I,
− 13 + 7 I, − 12 + 7 I, − 11 + 7 I, − 10 + 7 I, − 16 + 8 I,
− 15 + 8 I, − 14 + 8 I, − 13 + 8 I, − 12 + 8 I}

Below we represent the plots corresponding to the first four lists of points.

```
d1 = Graphics[{Hue[1 / 10],
rectangleList[complex2point[Gint2[1]]]}];
d2 = Graphics[{Hue[2 / 10],
rectangleList[complex2point[Gint2[2]]]}];
d3 = Graphics[{Hue[3 / 10],
rectangleList[complex2point[Gint2[3]]]}];
d4 = Graphics[{Hue[4 / 10],
rectangleList[complex2point[Gint2[4]]]}];
Show[GraphicsArray[{{d1, d2}, {d3, d4}}],
AspectRatio → Automatic];
```

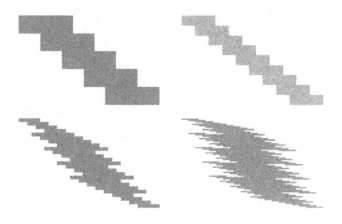

Fig. 12.3. *Images associated with lists* Gint2[[1]], Gint2[[2]], Gint2[[3]], *and* Gint2[[4]].

Increasing the list length generates a fractal image. The image associated to Gint2[8], obtained using the following command, has a size of more than 100 MB. It is not displayed.

```
d8 = Show[Graphics[{Hue[8 / 10],
rectangleList[complex2point[Gint2[8]]]}],
AspectRatio → Automatic];
```

13

Convolution and Laplace Transform

If f and g are two functions defined on the semi-infinite interval $[0, \infty[$, the convolution of these two functions is defined by

$$\int_0^t f(t-u)g(u)\,du.$$

As shown using the *Mathematica* command LaplaceTransform, the Laplace transform of the convolution of two functions is simply the product of their Laplace transforms:

```
LaplaceTransform[Integrate[f[t - u] g[u], {u, 0, t}], t, s]
```

```
LaplaceTransform[f[t], t, s] LaplaceTransform[g[t], t, s]
```

If F and G are, respectively, the Laplace transforms of f and g, the result above reads

$$\int_0^\infty e^{-ts}\,dt \int_0^t f(t-u)g(u)\,du = F(s)\,G(s).$$

In other words, the Laplace transform transforms a convolution into an ordinary product.

This theorem can be used to solve Volterra integral equations. For example, find f[t] such that

```
eqn = f[t] - Integrate[Exp[t - u] f[u], {u, 0, t}] == t
```

```
f[t] - Integrate[E^{t - u} f[u], {u, 0, t}] == t
```

```
Solve[LaplaceTransform[eqn, t, s],
LaplaceTransform[f[t], t, s]]
```

$$\{\{\text{LaplaceTransform}[f[t], t, s] \rightarrow \frac{-1+s}{(-2+s) s^2}\}\}$$

The solution is

```
f[t_] = InverseLaplaceTransform[(s - 1) / (s^2 (s - 2)),
s, t] // Simplify
```

$$\frac{-1+E^{2\ t}+2\ t}{4}$$

and it can be checked

```
Simplify[f[t] - Integrate[Exp[t - u] f[u], {u, 0, t}]] == t
```

True

More details on the Laplace transform and its use in solving convolution equations can be found in [4, 8].

Within the framework of distribution theory, the Laplace transform allows us to justify the so-called operational calculus. In most of the literature from the early twentieth century, Oliver Heaviside (1850–1925) is said to be the inventor of operational calculus. According to J. Lützen [33], "Today we know that this view is wrong, but it reflects the central role that Heaviside has played in the history of this branch of mathematics. His work became the starting point of the development of the operational calculus in this century, his predecessors apparently being for a period totally forgotten." On the history of operational calculus from Heaviside to Laurent Schwartz, see [8].

Double Pendulum

A double pendulum is a simple dynamical system that exhibits a complex dynamical behavior. It consists of two mass points at the end of rigid massless rods, one suspended from a fixed pivot and the second one suspended from the bob of the first. The system is free to oscillate in a vertical plane.

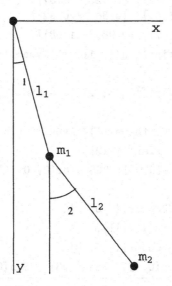

Fig. 14.1. *A double pendulum.*

Here is the list of graphics commands to draw the figure. Note the commands used to generate the symbols θ_1 and θ_2.

```
l1 = Graphics[Line[{{0, 0}, {Cos[1.3], - Sin[1.3]}}]];
l2 = Graphics[Line[{{Cos[1.3], - Sin[1.3]},
{Cos[1.3] + Cos[0.9], - Sin[1.3] - Sin[0.9]}}]];
X = Graphics[Line[{{0, 0}, {1.1, 0}}]];
Y = Graphics[Line[{{0, 0}, {0, - 1.8}}]];
U = Graphics[Line[{{Cos[1.3], - Sin[1.3]},
{Cos[1.3], - 1.8}}]];
theta = Graphics[{Circle[{0, 0}, 0.3,
{- 90 Degree, - 75 Degree}],
Circle[{Cos[1.3], - Sin[1.3]}, 0.3,
{- 90 Degree, - 50 Degree}]}];
T = Graphics[{Text[''x'', {1.05, - 0.05}],
Text[''y'', {0.05, - 1.75}],
Text[Subscript[''l'', 1], {0.22, - 0.5}],
Text[Subscript[''l'', 2], {0.62, - 1.3}],
Text[Subscript[''m'', 1], {0.38, - 0.92}],
Text[Subscript[''m", 2], {0.98, - 1.68}],
Text[FontForm[Subscript[''q'', 1], {''Symbol'', 12}],
{0.062, - 0.42}],
Text[FontForm[Subscript[''q'', 2],{''Symbol'', 12}],
{0.4, - 1.34}]}];
Show[{l1, l2, X, Y, U, theta, T}, AspectRatio -> Automatic,
DefaultFont -> {''Courier'', 12},
Epilog -> {{PointSize[0.05], RGBColor[0, 0, 1],
Point[{0, 0}]},
{PointSize[0.05], RGBColor[1, 0, 0],
Point[{Cos[1.3], - Sin[1.3]}]},
{PointSize[0.05], RGBColor[1, 0, 0],
Point[{Cos[1.3] + Cos[0.9], - Sin[1.3] - Sin[0.9]}]}}];
```

Symbol[''name''] refers to a symbol with a specified name.

If ℓ_1 and ℓ_2 are the respective lengths of the first and second pendulum strings, the bob positions are given by

```
bob[1] := l1 {Sin[theta1[t]], Cos[theta1[t]]}
bob[2] := bob[1] + l2 {Sin[theta2[t]], Cos[theta2[t]]}
```

If m_1 and m_2 are the respective masses of the bobs, the following commands will give the expression of the Lagrangian.

```
speed[bob_] := Sqrt[D[bob, t] . D[bob, t]];
v1 = speed[bob[1]] // Simplify
v2 = speed[bob[2]] // Simplify
```

Sqrt[l1^2 theta1'[t]^2]

Sqrt[l1^2 theta1'[t]^2 +

2 l1 l2 Cos[theta1[t] − theta2[t]] theta1'[t] theta2'[t] +

l2^2 theta2'[t]^2]

The kinetic energy K is

```
K = (1 / 2 m1 v1^2 + 1 / 2 m2 v2^2) // Simplify
```

(l1^2 (m1 + m2) theta1'[t]^2 +

2 l1 l2 m2 Cos[theta1[t] - theta2[t]] theta1'[t] theta2'[t] +

l2^2 m2 theta2'[t]^2) / 2

and, if g denotes the acceleration due to gravity, the potential energy U is given by

```
U = - m1 g bob[1][[2]] - m2 g bob[2][[2]]
```

− (g l1 m1 Cos[theta1[t]]) − g m2 (l1 Cos[theta1[t]] +

l2 Cos[theta2[t]])

Hence, the Lagrangian is given by

```
Lagrangian = (K - U) // Simplify
```

g (l1 (m1 + m2) Cos[theta1[t]] + l2 Cos[theta2[t]]) +

$\dfrac{\text{l1}^2\ (\text{m1}+\text{m2})\ \text{theta1}'[\text{t}]^2}{2}$ + l1 l2 m2 Cos[theta1[t] −

theta2[t]]   theta1′[t]   theta2′[t] $\dfrac{12^2 \ \text{m2 theta2}'[t]^2}{2}$

In order to write down the Euler–Lagrange's equations, we first have to load the package Calculus‘VariationalMethods‘.

```
<<Calculus‘VariationalMethods‘
```

```
eqns = EulerEquations[Lagrangian, {theta1[t], theta2[t]}, t]
```

{− (l1 (g m1 Sin[theta1[t]] + g m2 Sin[theta1[t]] +

l2 m2 Sin[theta1[t] − theta2[t]] theta2′[t]^2 +

l1 (m1 + m2) theta1″[t] + l2 m2 Cos[theta1[t] −

theta2[t]] theta2″[t])) == 0,

− (l2 m2 (g Sin[theta2[t]] − l1 Sin[theta1[t] −

theta2[t]] theta1′[t]^2 + l1 Cos[theta1[t] −

theta2[t]] theta1″[t] + l2 theta2″[t])) == 0}

Choosing the following numerical values

```
Clear[values]
values = {g → 9.8, l1 → 1, l2 → 1, m1 → 1, m2 → 1};
```

we obtain

```
numEqns = eqns /. values
```

{− 19.6 Sin[theta1[t]] − Sin[theta1[t] − theta2[t]] theta2′[t]^2

− 2 theta1″[t] − Cos[theta1[t] − theta2[t]] theta2″[t])) == 0,

− 9.8 Sin[theta2[t]] + Sin[theta1[t] − theta2[t]] theta1′[t]^2 −

Cos[theta1[t] − theta2[t]] theta1″[t] − theta2″[t])) == 0}

and choosing the initial conditions

```
Clear[initCond]
initCond = {theta1[0] == 1.5, theta2[0] == 3.0, theta1'[0] ==
0.0, theta2'[0] == 0.0};
```

we can solve numerically the equations of motion:

```
Clear[solution]
solution = NDSolve[Join[numEqns, initCond],
{theta1[t], theta2[t]}, {t, 0, 10}]
```

{{theta1[t] → InterpolatingFunction[{{0., 10.}}, <>][t],
theta2[t] → InterpolatingFunction[{{0., 10.}}, <>][t]}}

Plotting the solution reveals a fairly complicated behavior.

```
Plot[Evaluate[theta1[t] /. solution], {t, 0, 10}];
Plot[Evaluate[theta2[t] /. solution], {t, 0, 10}];
```

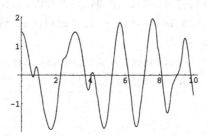

Fig. 14.2. *Variations of angle θ_1 as a function of time.*

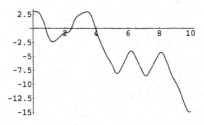

Fig. 14.3. *Variations of angle θ_2 as a function of time.*

The positions of the bobs are defined by rod1 and rod2 after having flipped the y-coordinate which pointed down.

```
rod1[t_] :=
Evaluate[l1 {Sin[theta1[t]], - Cos[theta1[t]]}] /. values
rod2[t_] := Evaluate[l1 {Sin[theta1[t]], - Cos[theta1[t]]} +
l2 {Sin[theta2[t]], - Cos[theta2[t]]}] /. values
```

Although the trajectory of bob[1] is the circle of radius l1 = 1, centered at the origin, the trajectory of bob[2] may be very complicated if the initial values of theta1 and theta2 are not small as illustrated below. To follow the trajectory of bob[2] a sequence of numbered points shows all points from 0 (initial point) to 20 (last point).

```
pl = ParametricPlot[Evaluate[rod2[t] /. solution],
{t, 0, 10}, Ticks → None, AspectRatio → 1,
DisplayFunction → Identity];
pts = Flatten[Table[Evaluate[rod2[t] /. solution /. values],
{t, 0, 10, 0.5}], 1];
numPts = Graphics[{{PointSize[0.04], CMYKColor[0, 0, 1, 0],
Map[Point, pts]}, Table[Text[i - 1, Part[pts, i]],
{i, 1, Length[pts]}]}];
Show[{pl, numPts}, PlotRange → All, AspectRatio → 1,
DisplayFunction → $DisplayFunction];
```

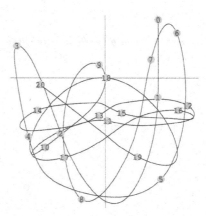

Fig. 14.4. *Trajectory of* bob[2].

The following animation better reveals the chaotic behavior of the double pendulum when the initial angles theta1[0] and theta2[0] are not small.

```
endPts1 = Flatten[Table[Evaluate[rod1[t] /.
solution/.values],
{t, 0, 10, 0.5}], 1];
endPts2 = Flatten[Table[Evaluate[rod2[t] /. solution /.
values],
{t, 0, 10, 0.5}], 1];
```

```
{Length[endPts1], Length[endPts2]}
```

{21, 21}

```
Table[Show[Graphics[{{RGBColor[1, 0, 0], Circle[{0, 0}, 1]},
Line[{{- 2, - 2}, {2, - 2}, {2, 2}, {- 2, 2}, {- 2,- 2}}],
Line[{{0, 0}, endPts1[[i]], endPts2[[i]]}],
{RGBColor[0, 0, 0], PointSize[0.04], Point[{0, 0}]},
{RGBColor[1, 0, 0], PointSize[0.04], Point[endPts1[[i]]]},
{RGBColor[0, 0, 1], PointSize[0.04], Point[endPts2[[i]]]}}],
AspectRatio -> 1], {i, 1, Length[endPts1]}];
```

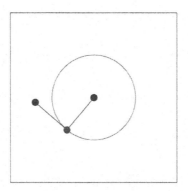

Fig. 14.5. *Last figure of the sequence generating the animation of the double pendulum.*

Duffing Oscillator

15.1 The Anharmonic Potential

The Duffing equation is

$$x'' - ax + bx^3 = 0,$$

where a and $b > 0$ are two parameters. Depending upon the sign of a we have either a single-well ($a < 0$) or a double-well ($a > 0$) anharmonic potential V defined by

$$V(x) = -\frac{a}{2}x^2 + \frac{b}{4}x^4.$$

We can ask *Mathematica* to plot the potential when a is either negative or positive.

```
pl1 = Plot[ - a x^2 / 2 + b x^4 / 4 /. {a → - 4, b → 0.05},
{x, - 15, 15}, AxesOrigin → {0, 0},
Ticks → {{- 15, - 10, - 5, 0, 5, 10, 15}, {300, 600, 900}},
PlotStyle → RGBColor[0, 0, 1],
PlotLabel → ''single-well potential'',
DefaultFont → {''Helvetica'', 12},
DisplayFunction → Identity];
pl2 = Plot[ - a x^2 / 2 + b x^4 / 4 /. {a → 4, b → 0.05},
{x, - 15, 15}, AxesOrigin → {0, - 80},
Ticks → {{- 15, - 10, - 5, 0, 5, 10, 15},
{- 20, 40, 100, 160}},
```

```
PlotStyle → {RGBColor[0, 0, 1]},
PlotLabel → ''double-well potential'',
DefaultFont → {''Helvetica'', 12},
DisplayFunction → Identity];
Show[GraphicsArray[{pl1, pl2}]];
```

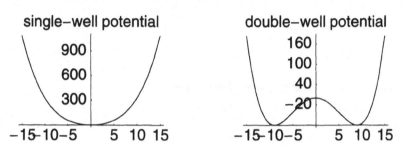

Fig. 15.1. *Anharmonic potential $V(x) = -(a/2)x^2 + (b/4)x^4$, for $a = -4$ (left figure) and $a = 4$ (right figure). In both cases $b = 0.05$.*

15.2 Solving Duffing Equations

15.2.1 Single-Well Potential

```
Clear[a, b]
a = - 4; b = 0.05;
sol1 = NDSolve[{x''[t] - a x[t] + b x[t]^3 == 0,
x[0] == - 10, x'[0] == 0}, x[t],
{t, 0, 30}]
```

```
{{x[t] → InterpolatingFunction[{{0.,30.}},<>][t]}}
```

```
Plot[Evaluate[x[t] /. sol1], {t, 0, 30},
PlotStyle → {RGBColor[0, 0, 1]},
DefaultFont → {''Helvetica'', 10},
AxesLabel → {''t'', ''x(t)''}];
```

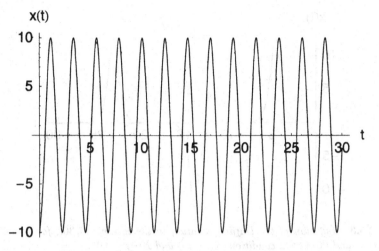

Fig. 15.2. *Solution of the Duffing equation in the interval* $[0, 30]$, *for* $a = -4$ *and* $b = 0.05$, *and the initial conditions* $x(0) = -10$ *and* $x'(0) = 0$.

15.2.2 Double-Well Potential

```
Clear[a, b]
a = 4; b = 0.05;
sol1 = NDSolve[{x''[t] - a x[t] + b x[t]^3 == 0,
x[0] == 0, x'[0] == 0.01}, x[t],
{t, 0, 30}]
```

$\{\{x[t] \rightarrow \text{InterpolatingFunction}[\{\{0., 30.\}\}, <>][t]\}\}$

```
Plot[Evaluate[x[t] /. sol2], {t, 0, 30},
PlotStyle → {RGBColor[0, 0, 1]},
DefaultFont → {''Helvetica'', 10},
AxesLabel → {''t'', ''x(t)''}];
```

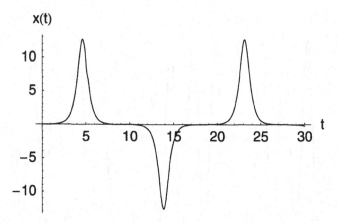

Fig. 15.3. *Solution of the Duffing equation in the interval* $[0, 30]$, *for* $a = 4$ *and* $b = 0.05$, *and the initial conditions* $x(0) = 0$ *and* $x'(0) = 0.01$.

15.3 Oscillations in a Potential Well

If $V(x)$ is the potential, the conservation of energy implies

$$E = \frac{1}{2}m \left(\frac{dx}{dt} \right)^2 + V(x),$$

where E is the total energy. From the equation above, we obtain

$$\frac{dx}{dt} = \pm \sqrt{\frac{2(E - V(x))}{m}}.$$

Thus, the period of oscillation T is given by

$$T = 2 \int_{x_1}^{x_2} \sqrt{\frac{m}{2(E - V(x))}} \, dx,$$

where x_1 and x_2 are such that $V(x_1) = V(x_2) = E$ for $x_1 < x < x_2$ and $V(x) < E$.

15.3.1 Single-Well Potential

Here $m = 1$, $x_1 = -10$, $x_2 = 10$, and $E = V(10)$.

```
energy1 = 2 x^2 + 0.0125 x^4 /. x → 10
```

```
period1 = 2 NIntegrate[ 1 / Sqrt[2 (energy1 - 2 x^2 -
0.0125 x^4)], {x, - 10, 10}]
```

$2.26816 - 7.29805 \times 10^{-13}$ I

Using Chop replaces the imaginary part close to zero by zero.

```
Chop[period1]
```

2.26816

We can also get rid of the spurious small imaginary part replacing the bounds −10 and 10 by -9.99999999999999, and 9.99999999999999.

```
period2 =
2 NIntegrate[ 1 / Sqrt[2(energy1- 2 x^2 - 0.0125 x^4)],
{x,-9.99999999999999, 9.99999999999999}]
```

2.26816

15.3.2 Double-Well Potential

Here $m = 1$ and $E = \frac{1}{2}(x'(0))^2$.

```
energy2 = (1/2) (0.01)^2
```

0.00005

Then x_1 and x_2 are the real solutions of

```
NSolve[energy2 == - 2 x^2 + 0.0125 x^4, x]
```

$\{\{x \to -12.6491\}, \{x \to 0. - 0.005 \text{ I}\}, \{x \to 0. + 0.005 \text{ I}\},$
$\{x \to 12.6491\}\}$

Hence,

```
period2 = 2 NIntegrate[ 1 / Sqrt[2(energy2 + 2 x^2 -
0.0125 x^4)], {x, - 12.6491, 12.6491}]
```

NIntegrate :: slwcon : Numerical integration converging
too slowly; suspect
one of the following: singularity, value of the
integration being 0, oscillatory integrand, or
insufficient WorkingPrecision. If your integrand is
oscillatory try using the option Method->Oscillatory
in NIntegrate. More

NIntegrate :: ncvb :
NIntegrate failed to converge to prescribed accuracy
after 7 recursive bisections in x near x = -0.0988211
. More

18. 3727

Let us try to include the approximative position of the singularity in the limits
of integration as suggested when we click More in the message above.

```
period2 = 2 NIntegrate[ 1/Sqrt[2(energy2 + 2 x^2 -
0.0125 x^4)], {x, - 12.6491, 0, 12.6491}]
```

18.436

15.4 Forced Duffing Oscillator with Damping

Adding a damping term and a harmonic forcing term to the Duffing equation
yields:
$$x'' + \gamma x' - ax + bx^3 = c\cos(\omega t).$$

If we choose $a = 0.4$ and $b = 0.5$ we have a double-well as shown below.

```
Clear[a, b]
pl3 = Plot[ - a x^2 / 2 + b x^4 / 4 /. {a → 0.4, b → 0.5},
{x, - 1.5, 1.5}, PlotStyle → {RGBColor[0, 0, 1]},
PlotRange → All, DefaultFont → {''Helvetica'', 12}];
```

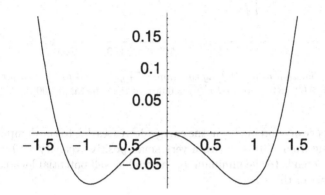

Fig. 15.4. *Double-well potential* $V(x) = -(1/2)ax^2 + (1/4)bx^4$ *for* $a = 0.4$ *and* $b = 0.5$.

15.4.1 No Forcing Term

```
Clear[a, b, g, omega, c]
a = 0.4; b = 0.5; g = 0.02; omega = 0.125; c = 0;
sol3 = NDSolve[{x''[t] + g x'[t] - a x[t] + b x[t]^3 ==
c Cos[omega t],
x[0] == 0, x'[0] == 0.001}, x[t], {t, 0, 200}]
```

```
{{x[t] → InterpolatingFunction[{{0., 200.}}, <>][t]}}
```

```
Plot[Evaluate[x[t] /. sol3], {t, 0, 200},
PlotStyle → {RGBColor[0, 0, 1]}, PlotRange → All,
DefaultFont → {''Helvetica'', 10},
AxesLabel → {''t'', ''x(t)''}];
```

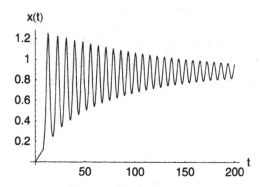

Fig. 15.5. *Solution of the Duffing equation:* $x'' + gx' - ax + bx^3 == 0$ *for* $a = 0.4$, $b = 0.5$. $g = 0.02$, $x(0) = 0$, *and* $x'(0) = 0.001$ *in the interval* $[0, 200]$.

In the absence of a forcing term, the energy decreases due to the damping term, and starting from $x(0) = 0$ with a very small positive velocity $x'(0) = 0.001$, the position tends to the minimum of the double-well potential located on the positive part of the x-axis.

15.4.2 With Forcing Term

In the presence of a forcing term, the solution has a more irregular behavior.

```
Clear[a, b, g, omega, c]
a = 0.4; b = 0.5; g = 0.02; omega = 0.125; c = 0.1;
sol4 = NDSolve[{x''[t] + g x'[t] - a x[t] + b x[t]^3 ==
c Cos[omega t],
x[0] == 0, x'[0] == 0.001}, x[t], {t, 0, 200}]
```

```
{{x[t] → InterpolatingFunction[{{0., 200.}}, <>][t]}}
```

```
Plot[Evaluate[x[t] /. sol4], {t, 0, 200},
PlotStyle → {RGBColor[0, 0,1]},
PlotRange → All, DefaultFont → {''Helvetica'', 10},
AxesLabel → {''t'', ''x(t)''}];
```

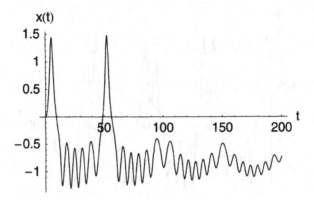

Fig. 15.6. *Solution of the Duffing equation:* $x'' + gx' - ax + bx^3 == c\cos(\omega t)$ *for* $a = 0.4$, $b = 0.5$. $g = 0.02$, $\omega = 0.125$, $c = 0.1$, $x(0) = 0$, *and* $x'(0) = 0.001$ *in the interval* $[0, 200]$.

```
Clear[a, b, g, omega, c, sol4]
a = 0.4; b = 0.5; g = 0.02; omega = 0.125; c = 0.1;
sol5 = NDSolve[{x''[t] + g x'[t] - a x[t] + b x[t]^3 ==
c Cos[omega t],
x[0] == 0.1, x'[0] == 0.001}, x[t], {t, 0, 200}]
```

```
{{x[t] → InterpolatingFunction[{{0., 200.}}, <>][t]}}
```

```
Plot[Evaluate[x[t] /. sol5], {t, 0, 200},
PlotStyle → {RGBColor[0, 0,1]}, PlotRange → All,
DefaultFont → {''Helvetica'', 10},
AxesLabel → {''t'', ''x(t)''}];
```

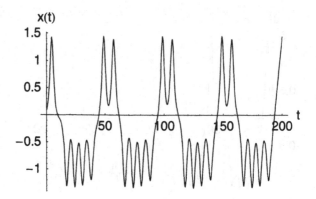

Fig. 15.7. *Same as above but with $x(0) = 0.1$ instead of $x(0) = 0$.*

A slight change in the initial position greatly modifies the solution suggesting a possible chaotic behavior of the forced oscillator (on chaos see Chapter 5 of [9]).

Egyptian Fractions

In 1858 the Scottish antiquarian Alexander Henry Rhind (1833–1863), traveling in Egypt, bought in Luxor an ancient scroll that has been the source of much information about Egyptian mathematics. This important document, known as the *Rhind Mathematical Papyrus*, contains, in particular, tables to help find a representation of rational numbers less than one as sums of different unit fractions, that is, with numerators equal to 1, as, for example,

$$\frac{2}{29} = \frac{1}{24} + \frac{1}{58} + \frac{1}{174} + \frac{1}{232}.$$

As illustrated below, this so-called Egyptian fraction representation is not unique.

$$\frac{2}{29} = \frac{1}{15} + \frac{1}{435} = \frac{1}{16} + \frac{1}{232} + \frac{1}{464}.$$

These results can be checked using *Mathematica*:

```
2/29 == 1/24 + 1/58 + 1/174 + 1/232 == 1/15 + 1/435
== 1/16 + 1/232 + 1/464
```

True

In fact, any fraction less than one has an infinite number of different representations because taking any representation we can replace the last fraction of the representation, that is, the unit fraction with the greatest denominator, by its Egyptian fraction representation. This result is true if we can first prove that any fraction has at least one Egyptian fraction representation. Let us first describe the well-known algorithm due to Fibonacci—and called the *greedy algorithm*—that, given a fraction a/b generates a strictly increasing sequence of integers $(n-1, n_2, \ldots)$ whose sum of reciprocals is equal to a/b.

The idea is to first find the greatest unit fraction $1/n_1$ that is less than or equal to a/b, then to find the greatest unit fraction $1/n_2$ less than or equal to the remainder $a/b - 1n_1$, and so on. The name of the algorithm comes from the fact that at each step we "greedily eat" as much as possible of the remainder. If we take, for example, the fraction $2/29$ of the Rhind papyrus, the greatest unit fraction less than $2/29$ is $1/15$ (note here something that will be useful to write our *Mathematica* program: 15 is the least integer greater than or equal to $29/2$, that is, $15 = \lceil 29\ 2 \rceil$, then the largest unit fraction less than or equal to $2/29 - 1/15$ is precisely $1/435$. Hence $2/29 = 1/15 + 1/435$, which is the second representation given above.

To prove the existence of at least one Egyptian representation we still have to prove that the sequence of unit fractions generated by the greedy algorithm is not infinite. This is not difficult if we look at the sequence of remainders.

The first remainder is $a/b - 1/n_1 = (an_1 - b)/bn_1$ but, because $1/(n_1 - 1) > a/b$, $b/(n_1 - 1) > a$, that is, the numerator $an_1 - b$ of the first remainder is strictly less than the numerator a and the denominator bn_1 of this first remainder is strictly greater than the denominator b of the fraction. More generally, after the kth step the numerator of the remainder $a_k n_{k+1} - b_k$ is strictly less than the numerator a_k, and it is also clear that the denominator $b_k n_{k+1}$ of the remainder is strictly greater than b_k. Hence, the sequence of remainders is a sequence of fractions whose denominators are strictly increasing while their numerators are strictly decreasing. Because numerators are nonnegative integers, this sequence converges to zero.

Finally, to prove that the number of Egyptian fraction representations of any rational number is infinite we have to show that we can replace the last fraction of the representation, that is, the unit fraction with the greatest denominator, which is already an Egyptian representation by a different representation. This new representation cannot of course be found using the greedy algorithm, but any unit fraction $1/n$ can always be written under the form $1/m + a_1/b_1$ where $m > n$ and use the greedy algorithm to find the Egyptian representation of the fraction a_1/b_1.

In order to write a program generating the sequence of unit fractions representing a given rational number $0 < r < 1$, we have, at each step, to determine the remainder using, for instance, the function:

```
remainder[r_Rational /; 0 < r < 1] :=
r - Rational[1, Ceiling[1 / r]]
```

For example, the sequence of remainders of $7/11$ is

```
remainder[7 / 11]
```

$$\frac{3}{22}$$

```
remainder[3 / 22]
```

$$\frac{1}{88}$$

```
remainder[1 / 88]
```

0

When the remainder is either zero or a unit fraction the program should stop. We, therefore, slightly modify the function remainder as follows.

```
remainder[r_Rational/; 0 < r < 1] :=
If[r ∈ Integers || Numerator[r] == 1,
0, r - Rational[1,Ceiling[1/r]]]
```

Let us now look at the sequence of remainders using the function FixedPoint-List[f, expression] that gives the list of repeatedly applying f to expression until the result does not change. In our case the last two elements of the list generated by this function will be two zeros.

```
FixedPointList[remainder, 7 / 11]
```

$$\left\{\frac{7}{11}, \frac{3}{22}, \frac{1}{88}, 0, 0\right\}$$

The Egyptian representation of 7/11 is then simply given by

$$\left\{\frac{7}{11}, \frac{3}{22}, \frac{1}{88}\right\} - \left\{\frac{3}{22}, \frac{1}{88}, 0\right\} = \left\{\frac{1}{2}, \frac{1}{8}, \frac{1}{88}\right\}.$$

This result can be obtained using the function Take[list, {m, n}] which gives the elements m through n of list as shown below.

```
Take[{7 / 11, 3 / 22, 1 / 88, 0, 0}, {1, - 3}] -
Take[{7 / 11, 3 / 22, 1 / 88, 0, 0}, {2, - 2}]
```

$$\left\{\frac{1}{2}, \frac{1}{8}, \frac{1}{88}\right\}$$

This translated difference can be written for any list as

```
translatedDifference[lis_List] := Take[lis, {1, - 3}] -
Take[lis, {2, - 2}]
```

that we can verify

```
translatedDifference[{7 / 11, 3 / 22, 1 / 88, 0, 0}]
```

$$\left\{\frac{1}{2}, \frac{1}{8}, \frac{1}{88}\right\}$$

Finally, we can find the Egyptian representation of any rational number using the following function,

```
greedyEgyptianSequence[r_Rational /; 0 < r < 1] :=
Module[{remainder, translatedDifference},
remainder[x_] := If[x == 0 || Numerator[x] == 1,
0, x - Rational[1, Ceiling[1 / x]]];
translatedDifference[lis_List] := Take[lis, {1, - 3}] -
Take[lis, {2, - 2}];
translatedDifference[FixedPointList[remainder,x]]]
```

and we make this function Listable.

```
SetAttributes[greedyEgyptianSequence, Listable];
```

Here are a few examples.

```
greedyEgyptianSequence[2 / 9]
```

$$\left\{\frac{1}{5}, \frac{1}{45}\right\}$$

```
greedyEgyptianSequence[7 / 11]
```

$$\left\{\frac{1}{2}, \frac{1}{8}, \frac{1}{88}\right\}$$

```
greedyEgyptianSequence[131 / 263]
```

$$\left\{\frac{1}{3}, \frac{1}{7}, \frac{1}{46}, \frac{1}{5909}, \frac{1}{51766508}, \frac{1}{7771336864724278},\right.$$
$$\left.\frac{1}{1509841916625564200758955415415818828932}\right\}$$

Because the numerators of the sequence of remainders strictly decrease, the number of terms of the Egyptian representation of the rational number a/b is at most equal to a. The sequence of denominators is strictly increasing and, as shown in the last example, may become quite large.

Here is a very simple recursive program taken from S. Wagon's book [64] which gives the sequence of denominators of the unit fractions.

```
EgyptianFraction[0] := {}
EgyptianFraction[q_] := Prepend[EgyptianFraction[q - 1 /
Ceiling[1 / q]], Ceiling[1 / q]]
Attributes[EgyptianFraction] = Listable;
```

```
EgyptianFraction[131 / 263]
```

{3, 7, 46, 5909, 51766508, 7771336864724278,

1509841916625564200758955415818828932}

Egyptian representations of rational numbers given by the greedy algorithm often involve unit fractions with very large denominators. It is possible to find simpler representations combining the representations of two fractions whose sum is equal to the original one. For example, because $62 + 69 = 131$, we obtain a much simpler representation using the command

```
Sort[Flatten[EgyptianFraction[{62 / 263, 69 / 263}]]]
```

{4, 5, 28, 81, 36820, 85212}

And we verify that the sum of the corresponding unit fractions is

```
Total[1 / %]
```

$$\frac{131}{263}$$

Electrostatics

Electostatics is the study of time-independent distributions of electric charges.

17.1 Potential and Field

Given a spatial distribution of electric charge $\rho(x, y, z)$, the laws of electrostatics make it possible to calculate the electric potential $V(x, y, z)$ and the electric field $\mathbf{E}(x, y, z)$. These physical quantities are given by

$$V(x, y, z) = \frac{1}{4\pi} \int \int \int \frac{\rho(x', y', z')}{\sqrt{(x - x')^2 + (y - y')^2 + (z - z')^2}} \, dx' \, dy' \, dz',$$

$$\mathbf{E}(x, y, z) = -\nabla V(x, y, z),$$

$$= -\left(\frac{\partial V(x, y, z)}{\partial x}, \frac{\partial V(x, y, z)}{\partial y}, \frac{\partial V(x, y, z)}{\partial z}, \right).$$

17.1.1 Useful Packages

The commands in the package `Graphics'PlotField'` can be used to draw arrows representing vectors, the direction of the arrow indicating the direction of the vector field at its base point, and its magnitude being proportional to the magnitude of the vector field. The package `Calculus'VectorAnalysis'` offers a variety of tools for doing calculus in various three-dimensional coordinate systems.

```
<<Graphics'PlotField'
<<Calculus'VectorAnalysis'
```

As usual, we can obtain the list of all commands provided by these packages entering

```
?Graphics'PlotField'*
```

and

```
?Calculus'VectorAnalysis'*
```

17.1.2 Point Charge

The electric potential at point (x, y, z) created by a point **charge** located at (x_0, y_0, z_0) is given by

```
monopolePotential[charge_, {x0_, y0_, z0_}, {x_, y_, z_}] :=
charge / (4 Pi Sqrt[(x - x0)^2 + (y - y0)^2 + (z - z0)^2])
```

The potential and the field at (x, y, z) created by a point charge q located at $(x_0, y_0, z_0) = (0, 0, 0)$ are

```
monopolePotential[q, {0, 0, 0}, {x, y, z}]
```

$$\frac{q}{4 \text{ Pi Sqrt}[x^2 + x^2 + x^2]}$$

```
monopoleField = - Grad[monopolePotential[q, {0, 0, 0},
{x, y, z}], Cartesian[x, y, z]]
```

$$\left\{ \frac{q \, x}{4 \text{ Pi } (x^2 + y^2 + z^2)^{3/2}}, \frac{q \, y}{4 \text{ Pi } (x^2 + y^2 + z^2)^{3/2}}, \frac{q \, z}{4 \text{ Pi } (x^2 + y^2 + z^2)^{3/2}} \right\}$$

The potential and the field in the plane $z = 0$ are represented below. We take $z_0 \neq 0$ to avoid having an infinite expression at the origin (see below: electricField).

```
unitMonopoleV[x_, y_] := monopolePotential[1, {0, 0, 0},
{x, y, 0.01}]
```

```
unitMonopoleV[x, y]
```

$$\frac{1}{4\ \text{Pi Sqrt}[0.0001 + x^2 + y^2]}$$

```
equiPotentials = ContourPlot[unitMonopoleV[x, y],
{x, - 2, 2}, {y, - 2, 2}, PlotPoints → 60,
ColorFunction → Hue, ContourSmoothing → True];
```

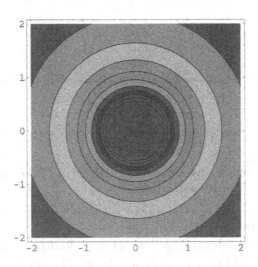

Fig. 17.1. *Equipotentials, in the plane z = 0.01, of a unit electric charge located at the origin.*

The option ContourSmoothing specifies what smoothing to use for contour lines.

```
unitCharge = {{RGBColor[1, 1, 0], AbsolutePointSize[20],
Point[ {0, 0}]}, {Text[''+1'', {0, 0}]}};
```

In the following command, using the option ScaleFunction → (1&), all arrows have the same unit length.

```
electricField = PlotGradientField[- unitMonopoleV[x, y],
{x, - 1, 1}, {y, - 1, 1}, ScaleFunction → (1&),
Epilog → unitCharge];
```

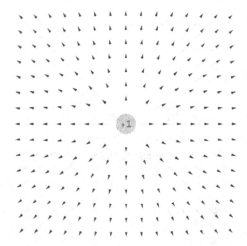

Fig. 17.2. *Electric field created by a unit electric charge located at the origin.*

17.1.3 Dipole

The following function gives the electric potential of a dipole $(p, 0, 0)$ localized at the origin and directed along the x-axis.

```
dipolePotential[p_, {x_, y_, z_}] :=
Limit[(monopolePotential[p / a, {a / 2, 0, 0}, {x, y, z}]
- monopolePotential[p / a, {- a / 2 ,0, 0}, {x, y, z}]),
a -> 0]
```

```
unitDipoleV[x_, y_] := dipolePotential[1, {x, y, 0.01}]
```

We load the package `Graphics'Arrow'` to be able to use the command `Arrow` to draw arrows.

```
<<Graphics'Arrow'
```

```
unitDipole = Arrow[{0. 1, 0}, {- 0.1, 0},
HeadScaling -> Absolute];
```

```
PlotGradientField[unitDipoleV[x,y], {x, - 1, 1}, {y, - 1, 1},
ScaleFunction -> (1&), Epilog -> unitDipole];
```

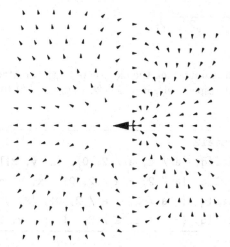

Fig. 17.3. *Electric field created by a unit dipole, represented by a bigger arrow, located at the origin.*

The option `HeadScaling → Absolute` makes the head of the arrow, representing the dipole, slightly bigger (see Figure 17.3).

17.1.4 Quadrupoles

We consider different functions that give the electric potential of three or four electric charges whose sum is equal to zero. In the next section we define a function to plot the corresponding equipotentials and field lines.

We first define the command `quadrupolePotential1` generating the electric potential created by three charges $2q, -q, -q$ localized, respectively, at

$$(0,0,0), \ \left(\frac{a}{2},0,0\right), \ \text{and} \ \left(-\frac{a}{2},0,0\right).$$

```
quadrupolePotential1 =
monopolePotential[2 q, {0, 0, 0}, {x, y, z}] -
monopolePotential[q, {a / 2, 0, 0}, {x, y, z}] -
monopolePotential[q, {- a / 2, 0, 0}, {x, y ,z}]
```

$$\frac{q}{2 \ \text{Pi Sqrt}[x^2 + y^2 + z^2]} - \frac{q}{4 \ \text{Pi Sqrt}[(\frac{-a}{2} + x)^2 + y^2 + z^2]} -$$

$$\frac{q}{4 \; \text{Pi} \; \text{Sqrt}[(\frac{a}{2}+x)^2+y^2+z^2]}$$

Here is another command giving the electric potential created by four charges $q, -q, q, -q$ localized, respectively, at $(a/2, a/2, 0)$, $(a/2, -a/2, 0)$, $(-a/2, -a/2, 0)$, and $(a/2, a/2, 0)$.

```
quadrupolePotential2 =
monopolePotential[q, {a / 2, a / 2, 0}, {x, y, z}] -
monopolePotential[q, {a / 2, - a / 2, 0}, {x, y, z}] +
monopolePotential[q, {- a / 2, - a / 2, 0}, {x, y, z}] -
monopolePotential[q, {- a / 2, a / 2, 0}, {x, y, z}]
```

$$\frac{q}{4 \; \text{Pi} \; \text{Sqrt}[(\frac{-a}{2}+x)^2+(\frac{-a}{2}+y)^2+z^2]} -$$

$$\frac{q}{4 \; \text{Pi} \; \text{Sqrt}[(\frac{a}{2}+x)^2+(\frac{-a}{2}+y)^2+z^2]} -$$

$$\frac{q}{4 \; \text{Pi} \; \text{Sqrt}[(\frac{-a}{2}+x)^2+(\frac{a}{2}+y)^2+z^2]} +$$

$$\frac{q}{4 \; \text{Pi} \; \text{Sqrt}[(\frac{a}{2}+x)^2+(\frac{a}{2}+y)^2+z^2]}$$

We again consider the electric potential created by three charges $2q, -q, -q$ but localized, respectively, at

$$(0,0,0), \quad \left(\frac{a}{2}, -\frac{a}{2}, 0\right) \quad \text{and} \quad \left(-\frac{a}{2}, -\frac{a}{2}, 0\right).$$

Their electric potential is given by

```
quadrupolePotential3 =
monopolePotential[2 q, {0, 0, 0}, {x, y, z}] -
monopolePotential[q, {a / 2, - a / 2, 0}, {x, y, z}] -
monopolePotential[q, {- a / 2, - a / 2, 0}, {x, y, z}]
```

$$\frac{q}{2 \; \text{Pi} \; \text{Sqrt}[x^2+y^2+z^2]} - \frac{q}{4 \; \text{Pi} \; \text{Sqrt}[(\frac{-a}{2}+x)^2+(\frac{a}{2}+y)^2+z^2]} -$$

$$\frac{q}{4 \ \text{Pi} \ \text{Sqrt}[(\frac{a}{2}+x)^2 + (\frac{a}{2}+y)^2 + z^2]}$$

17.1.5 Plots

We define the function `equipotentialFieldPlot` that plots (in the xOy-plane) the equipotentials and the field, and we use it to plot the equipotentials and the electric field lines of the quadrupoles defined above. `optionList__` stands for zero or more options.

```
equipotentialFieldPlot[potential_, xRange_, yRange_,
optionList___] :=
Module[{equiPotentials, fieldLines},
equiPotentials = ContourPlot[potential, xRange, yRange,
ContourShading -> False, ContourSmoothing -> True,
PlotPoints -> 60, DisplayFunction -> Identity];
fieldLines = PlotGradientField[- potential, xRange, yRange,
ScaleFunction -> (1 &), DisplayFunction -> Identity];
Show[{equiPotentials, fieldLines}, optionList,
DisplayFunction -> $DisplayFunction]];
```

```
equipotentialFieldPlot[quadrupolePotential1 /.
{q -> 1,a -> 1, z -> 0.001}, {x, - 2, 2}, {y, - 2, 2},
Epilog -> {{RGBColor[1, 1, 0], AbsolutePointSize[20],
Point[{0, 0}]}, {RGBColor[1, 1, 0], AbsolutePointSize[20],
Point[{0.5, 0}]}, {RGBColor[1, 1, 0],
AbsolutePointSize[20], Point[{-0.5, 0}]},
{Text["+ 2", {0, 0}], Text["-1", {0.5, 0}],
Text["-1", {-0.5, 0}]}}];
```

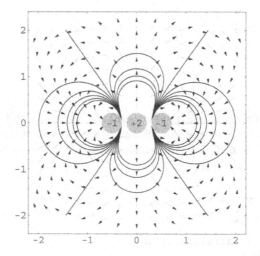

Fig. 17.4. *Equipotentials and electric field lines created by three charges respectively equal to +2 localized at the origin and −1 localized on the Ox-axis at a distance −1/2 and 1/2 from the origin.*

```
equipotentialFieldPlot[quadrupolePotential2 /.
{q → 1,a → 1, z → 0.001}, {x, - 2, 2}, {y, - 2, 2},
Epilog → {{RGBColor[1,1,0], AbsolutePointSize[20],
Point[{0.5, 0.5}]}, {RGBColor[1, 1, 0],
AbsolutePointSize[20], Point[{0.5, - 0.5}]},
{RGBColor[1, 1, 0], AbsolutePointSize[20],
Point[{- 0.5, - 0.5}]}, {RGBColor[1,1,0],
AbsolutePointSize[20], Point[{- 0.5, 0.5}]},
{Text[''+1'', {0.5, 0.5}], Text[''-1'', {0.5, - 0.5}],
Text[''+1'', {- 0.5, - 0.5}], Text[''-1'', {- 0.5, 0.5}]}}];
```

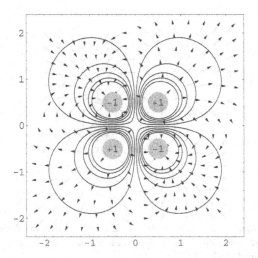

Fig. 17.5. *Equipotentials and electric field lines created by four charges respectively equal to* −1, +1, −1 *and* +1 *localized at the vertices of a unit square centered at the origin.*

```
equipotentialFieldPlot[quadrupolePotential3 /.
{q → 1,a → 1, z → 0.001}, {x, - 2, 2}, {y, - 2, 2},
Epilog → {{RGBColor[1, 1, 0], AbsolutePointSize[20],
Point[{0, 0}]}, {RGBColor[1,1,0], AbsolutePointSize[20],
Point[{0.5, - 0.5}]}, {RGBColor[1, 1, 0],
AbsolutePointSize[20], Point[{- 0.5, - 0.5}]},
{Text[''+2'', {0, 0}], Text[''-1'', {0.5, - 0.5}],
Text[''-1'', {- 0.5, - 0.5}]}}];
```

Output represented in Figure 17.6.

17.1.6 Uniformly Charged Sphere

Let R be the radius of a uniformly charged sphere whose center coincides with the origin , and let ρ be the charge density. Spherical symmetry implies that the electric field is radial and depends only upon the distance r to the origin. This field is easily determined using Gauss law.

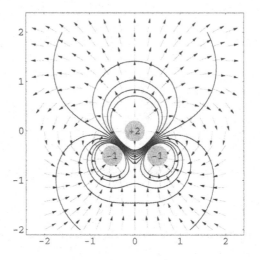

Fig. 17.6. *Equipotentials and electric field lines created by three charges respectively equal to +2 localized at the origin and two negative unit charges localized at* $(-1/2, -1/2, 0)$ *and* $(1/2, -1/2, 0)$.

```
eqn = {4 Pi r^2 internalElectricField == 4 Pi r^3 rho
/ 3, 4 Pi r^2 externalElectricField == 4 Pi R^3 rho /
3}; sol = Flatten[Solve[eqn, {internalElectricField,
externalElectricField}]]
```

$$\{\text{internalElectricField} \rightarrow \frac{r\ \text{rho}}{3}, \text{externalElectricField} \rightarrow \frac{R^3\ \text{rho}}{3\ r^2}\}$$

```
Clear[electricField]
electricField[r_] := If[r < R, r rho / 3, R^3 rho / (3 r^2)]
plE = Plot[electricField[r] /. {R → 1,rho → 1}, {r, 0, 5},
PlotStyle → {RGBColor[0,0,1]}, DisplayFunction → Identity];
```

```
t1 = Graphics[Text["inside", {0.5, 0.33}]];
t2 = Graphics[Text["outside", {3, 0.33}]];
r1 = Graphics[{RGBColor[0.5, 0.4, 0],
Rectangle[{0, 0}, {1, 0.36}]}];
r2 = Graphics[{RGBColor[0.8, 0.8, 0],
Rectangle[{1, 0}, {5, 0.36}]}];
Show[{r1, r2, plE, t1, t2}, Axes → False, Frame → True,
TextStyle → {FontSlant → "Italic", FontSize → 12},
FrameLabel → {"r", "E(r)"}, RotateLabel → False,
FrameTicks → {{0, 1, 2, 3, 4, 5}, {0. 05, 0.15, 0.25, 0.35},
{}, {}}, DisplayFunction → $DisplayFunction];
```

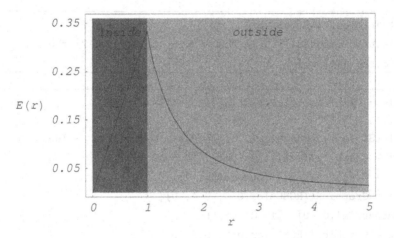

Fig. 17.7. *Electric field created by a uniformly charged sphere as a function of the distance r from the sphere center.*

RotateLabel specifies whether labels on vertical frame axes should be rotated to be vertical.

The order of the graphics objects in Show is important, The order {plE, t1, t2, r1, r2} would mask the plot and the text. The list of graphics objects has to start with the colored rectangles.

The electric potential can be derived integrating the radial electric field component.

```
externalElectricPotential[r_] = - Integrate[R^3 rho /(3 r^2),
r]
```

$$\frac{R^3 \ rho}{3 \ r}$$

```
internalElectricPotential[r_] = externalElectricPotential[R]
-
Integrate[r rho / 3, {r, R, r}] // Simplify
```

$$\frac{-(r^2 - 3 \ R^2) \ rho}{6}$$

```
Clear[electricPotential]
electricPotential[r_] := If[r < R, - (r^2 - 3 R^2) rho / 6,
R^3 rho / (3 r)]
```

```
plV = Plot[electricPotential[r] /. {R → 1,rho → 1}, {r, 0,
5},
PlotStyle → {RGBColor[0,0,1]}, DisplayFunction → Identity];
t1 = Graphics[Text[''inside'', {0.5, 0.3}]];
t2 = Graphics[Text[''outside'', {3, 0.3}]];
r1 = Graphics[{RGBColor[0.4, 0.4, 0],
Rectangle[{0, 0}, {1, 0. 5}]}];
r2 = Graphics[{RGBColor[0.8, 0.8, 0],
Rectangle[{1, 0}, {5, 0.5}]}];
Show[{r1, r2, plV, t1, t2}, Axes → False, Frame → True,
TextStyle → {FontSlant → ''Italic'', FontSize → 12},
FrameLabel → {''r'', ''V(r)''}, RotateLabel → False,
DisplayFunction → $DisplayFunction];
```

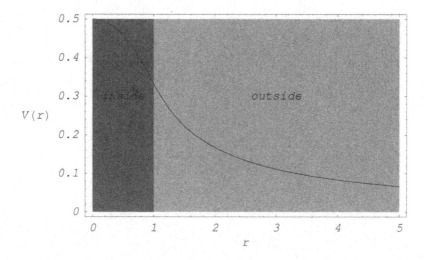

Fig. 17.8. *Electric potential created by a uniformly charged sphere as a function of the distance r from the sphere center.*

Foucault Pendulum

The stars appear to move in circles about a line through the poles of the earth as if they were attached to a sphere rotating about the earth.

Aristarchus of Samos (310–230 BC) was the first astronomer who explained the apparent motion of the stars and planets assuming that the earth turns on its own axis and also travels around the sun. This theory was not accepted by the Greeks.

Around 1514 Mikolaj Kopernik (1473–1543), better known as Nicolaus Copernicus, distributed a little handwritten book, called the *Little Commentary*, to a few of his friends, in which he stated, in particular, that the center of the universe is close to the sun and that the rotation of the earth accounts for the apparent daily rotation of the stars.

At the close of the 16th Century, Filippo Bruno (1548–1600), who took the name Giordano in 1565 when he entered the Dominican convent of San Domenico Maggiore in Naples, was soon suspected of heresy for his unorthodox ideas. He was nevertheless ordained as a priest in 1572. He came to the attention of the Inquisition in Naples and, in 1576, left the city to escape prosecution. For 14 years he wandered about Europe defending, in particular, Copernicus' heliocentric theory. Probably believing that the Inquisition had lost some of its strength, he thought that he might safely return to Italy. Betrayed by Giovanni Mocenigo, a Venetian nobleman who invited him in August 1591, he was denounced to the Inquisition and arrested on May 22, 1592. At the request of the Holy Office, he was transferred to Rome and arrived at the prison of the Holy Office near St. Peters on February 27, 1593. After seven years, on February 19, 1600, refusing to renounce his beliefs. he was brought to the Campo de' Fiori, his tongue in a gag, and burned alive.

In 1633, after the publication of *Dialogo dei due massimi sistemi del mondo*, Galileo Galilei (1564–1642) was convicted of heresy by the Inquisition and forced to recant his support of Copernicus. He confessed to having erred in writing his book, and asked for mercy. Sentenced to life imprisonment, he was allowed to live in his villa in Arcetri close to Florence.

In 1851, the French physicist Jean, Bernard, Léon Foucault (1819–1868), devised an experiment to demonstrate the rotation of the earth. Inside the Panthéon in Paris, he suspended from the dome a 67-meter, 28-kg pendulum. To show the motion of the plane of oscillation, he attached a stylus to the ball and placed a ring of damp sand on the floor below. It was observed that swing after swing this plane rotated slowly clockwise with respect to the earth.

In order to study the motion of the Foucault pendulum we first briefly review the equation of motion of a particle in a moving frame of reference.

Assuming that the moving frame is just rotating, the time derivatives of a vector quantity $\mathbf{A}$ with respect to a fixed frame and a rotating frame of reference satisfy the relation

$$\left(\frac{d\mathbf{A}}{dt}\right)_{\text{fixed}} = \left(\frac{d\mathbf{A}}{dt}\right)_{\text{rotating}} + \boldsymbol{\omega} \times \mathbf{A},$$

where $\boldsymbol{\omega}$ is the angular velocity vector. In particular, for the time derivative $\mathbf{v}$ of the position vector $\mathbf{r}$ we have

$$\mathbf{v}_{\text{fixed}} = \mathbf{v}_{\text{rotating}} + \boldsymbol{\omega} \times \mathbf{r}_{\text{rotating}}$$

Similarly, for the time derivative $\mathbf{a}$ of the velocity vector $\mathbf{v}$ we have

$$\mathbf{a}_{\text{fixed}} = \mathbf{a}_{\text{rotating}} + \boldsymbol{\omega}' \times \mathbf{r}_{\text{rotating}} + 2\,\boldsymbol{\omega} \times \mathbf{v}_{\text{rotating}} + \boldsymbol{\omega} \times (\boldsymbol{\omega} \times \mathbf{r}_{\text{rotating}})$$

- $\boldsymbol{\omega}' \times \mathbf{r}_{\text{rotating}}$ is the azimuthal or transverse acceleration,

- $2\,\boldsymbol{\omega} \times \mathbf{v}_{\text{rotating}}$ is the Coriolis acceleration, and

- $\boldsymbol{\omega} \times (\boldsymbol{\omega} \times \mathbf{r}_{\text{rotating}})$ is the centripetal acceleration.

In a local frame fixed to the earth, neglecting air resistance, the equation of motion is

$$m\,\mathbf{a}_{\text{rotating}} = \mathbf{F} - m\,\boldsymbol{\omega}' \times \mathbf{r}_{\text{rotating}} - 2\,m\,\boldsymbol{\omega} \times \mathbf{v}_{\text{rotating}} - m\,\boldsymbol{\omega} \times (\boldsymbol{\omega} \times \mathbf{r}_{\text{rotating}}).$$

In a rotating frame, to the real physical force $\mathbf{F}$ we have, therefore, to add three inertial forces, namely, the azimuthal, Coriolis, and centripetal forces.

Hence, in a local frame fixed to the earth, neglecting air resistance, the equation of motion becomes

$$m\mathbf{r}''_{\text{rotating}} = \mathbf{F} + m\mathbf{g} - 2\,m\,\boldsymbol{\omega} \times \mathbf{r}'_{\text{rotating}} - m\,\boldsymbol{\omega} \times (\boldsymbol{\omega} \times \mathbf{r}_{\text{rotating}})$$

where $\mathbf{r}'$ and $\mathbf{r}''$ are, respectively, the first and second time derivative of the vector $\mathbf{r}$.

The numerical value of the earth's angular velocity ω is small. It is equal to:

$$\frac{2\pi}{24} \times 3600 = 0.727 \times 10^{-4} \text{ rad/s}.$$

We may, therefore, neglect the centripetal force and write the equation of motion as

$$m\, \mathbf{r}''_{\text{rotating}} = \mathbf{F} + m\mathbf{g} - 2\, m\, \omega \times \mathbf{r}'_{\text{rotating}}.$$

The Foucault pendulum illustrates the earth's rotation through the Coriolis force. We treat the earth as rotating about its axis with an angular velocity ω with respect to an inertial frame and use the equation above to study the motion of the pendulum.

Let us choose the local Cartesian coordinate system such that its origin O is the equilibrium position of the bob. The Ox- and Oy-axes are in the horizontal plane pointing, respectively, south and east, and the vertical Oz-axis points up towards the point of suspension of the string.

If $\mathbf{r}$ represents the position of the bob in the rotating frame,

```
r[t_] := {x[t], y[t], z[t]}
```

the equation of motion of the bob in the rotating frame is

$$m\, \mathbf{r}''_{\text{rotating}} = \mathbf{T} + m\mathbf{g} - 2\, m\, \omega \times \mathbf{r}'_{\text{rotating}},$$

where $\mathbf{T}$ is the tension of the string. The acceleration due to gravity $\mathbf{g}$ is directed along the Oz-axis and points down, and the coordinates in the rotating frame of angular velocity vector ω are

```
omega[t_] := {- omega0 Cos[theta], 0, omega0 Sin[theta]}
```

where omega0 is the norm of omega and theta the latitude. Hence, in the rotating frame, the Coriolis force is

```
CoriolisForce = - 2 m Cross[omega[t], r'[t]]
```

```
{2 m Sin[theta] omega0 y'[t],

- 2 m (Sin[theta]  omega0 x'[t] + Cos[theta] omega0 z'[t]),

2 m Cos[theta] omega0 y'[t]}
```

If the amplitude of the oscillations is small, the tension **T** is nearly constant and equal to mg. Its coordinates in the rotating frame are

```
T = {- m g x[t] / l, - m g y[t] / l, 0}                              ;
```

where l is the string length.

Because $m \neq 0$, we use the command `Assuming[assumption, expression]` to tell *Mathematica* that `assumption` should be appended to the default assumptions when evaluating `expression`. Hence, the equation of motion of the bob is

```
eqn = Assuming[m != 0, Simplify[Thread[m r''[t] == T +
CoriolisForce]]]
```

$$\left\{\frac{g\ x[t]}{l} + x'' == 2\ \text{Sin[theta]}\ \text{omega0}\ y'[t] == 0,\right.$$

$$\frac{g\ y[t]}{l} + 2\ \text{omega0}\ (\text{Sin[theta]}\ x'[t] + \text{Cos[theta]}\ z'[t]) +$$

$$y''[t] == 0,$$

$$\left.2\ \text{Cos[theta]}\ \text{omega0}\ y'[t] == z''[t]\right\}$$

Because we cannot directly solve the vector differential equation, we used the command `Thread[f[arguments]]` that threads f over any lists that appear in `arguments`.

Neglecting z'[t] compared to y'[t], the equations of motion in the xOy-plane are

```
eqn[[1]]
eqn[[2]] /. z'[t] → 0
```

$$\frac{g\ x[t]}{l} + x'' == 2\ \text{Sin[theta]}\ \text{omega0}\ y'[t],$$

$$\frac{g\ y[t]}{l} + 2\ \text{omega0}\ (\text{Sin[theta]}\ x'[t] + y''[t] == 0$$

Let $\Omega = \omega_0 \sin\theta$ denote the vertical component of the angular velocity vector ω. Then, the equations of motion in the xOy-plane are

```
equations = x''[t] == - (g / l) x[t] + 2 Omega y'[t],
y''[t] == - (g / l) y[t] - 2 Omega x'[t];
```

Consider a new rotating frame that rotates around the Oz-axis with an angular velocity Ω, and denote $O\xi$ and $O\eta$ the new horizontal axes. We have

```
x[t] = xi[t] Cos[Omega t] + eta[t] Sin[Omega t];
y[t] = - xi[t] Sin[Omega t] + eta[t] Cos[Omega t];
```

Because

```
x'[t] = D[xi[t] Cos[Omega t] + eta[t] Sin[Omega t], t]
x''[t] = D[xi[t] Cos[Omega t] + eta[t] Sin[Omega t], {t, 2}]
y'[t] = D[- xi[t] Sin[Omega t] + eta[t] Cos[Omega t], t]
y''[t] = D[- xi[t] Sin[Omega t] + eta[t] Cos[Omega t], {t, 2}]
```

```
Omega Cos[t Omega] eta[t] − Omega Sin[t Omega] xi[t] +
Sin[t Omega] eta'[t] + Cos[t Omega] xi'[t]

− (Omega^2 Sin[t Omega] eta[t]) − Omega^2 Cos[t Omega] xi[t] +
2 Omega Cos[t Omega] eta'[t] − 2 Omega Sin[t Omega] xi'[t] +
Sin[t Omega] eta''[t] + Cos[t Omega] xi''[t]

− (Omega Sin[t Omega] eta[t]) − Omega Cos[t Omega] xi[t] +
Cos[t Omega] eta'[t] − Sin[t Omega] xi'[t]

− (Omega^2 Cos[t Omega] eta[t]) + Omega^2 Sin[t Omega] xi[t] −
2 Omega Sin[t Omega] eta'[t] − 2 Omega Cos[t Omega] xi'[t] +
Cos[t Omega] eta''[t] − Sin[t Omega] xi''[t]
```

```
equations = equations /. {x[t] → xi[t] Cos[Omega t] +
eta[t] Sin[Omega t],
x'[t] → D[x[t], t], x''[t] → D[x[t],t,2],
y[t] → - xi[t] Sin[Omega t] + eta[t] Cos[Omega t],
y'[t] → D[y[t], t], y'''[t] → D[y[t], {t, 2}]} //
FullSimplify
```

```
{(Sin[t Omega] ((g + l Omega^2) eta[t] + l eta''[t]) +
Cos[t Omega] ((g + l Omega^2) xi[t] + l xi''[t])) / l == 0,
(Cos[t Omega] ((g + l Omega^2) eta[t] + l eta''[t]) -
Sin[t Omega] ((g + l Omega^2 ) xi[t] + l xi''[t])) / l == 0}
```

Neglecting terms in Ω^2, which are very small, yields

```
equations = equations /. Omega^2 → 0
```

```
{(Sin[t Omega] (g  eta[t] + l eta''[t]) +
Cos[t Omega] (g xi[t] + l xi''[t])) / l == 0,
(Cos[t Omega] (g eta[t] + l eta''[t]) -
Sin[t Omega] (g xi[t] + l xi''[t]))/ l == 0}
```

These equations are obviously satisfied if

$$\xi'' + (g/l)\,\xi == 0 \text{ and } \eta'' + (g/l)\,\xi = 0.$$

These are the equations of a two-dimensional harmonic oscillator. The path in the $\xi O \eta$-plane is an ellipse. Hence in the rotating xOy-plane linked to the earth, the path is an ellipse that undergoes a steady precession with angular velocity Ω. For a fixed observer in the plane linked to the earth, the vertical plane containing the major axis of the ellipse turns clockwise in the northern hemisphere and counterclockwise in the southern with a period

$$T = \frac{2\pi}{\Omega} = \frac{24}{\sin\theta} \text{ hours.}$$

The precession vanishes at the equator and is maximum at the north pole. The 24-hour period has been checked at the south pole during the winter 2001. Details on the experiment can be found at www.phys-astro.sonoma.edu/people/students/baker/SouthPoleFoucault.html.

Fractals

Fractals are exotic sets that first appeared in the mathematical literature at the end of the 19th century. They were devised by Georg Ferdinand Ludwig Philipp Cantor (1845–1918), Giuseppe Peano (1858–1932), David Hilbert (1862–1943), Henri Léon Lebesgue (1875–1941), Arnaud Denjoy (1884–1974), George Pólya (1887–1985), Wacław Sierpiński (1882–1969), and many others. There is no precise definition but most authors agree to call fractals sets possessing certain characteristic properties such as self-similarity illustrated in the examples presented below. The idea of self-similarity originated implicitly in a paper of Niels Fabian Helge von Koch (1870–1924) (see the von Koch curve below), and was formulated explicitly by Ernesto Cesàro (1859–1906). The word fractal was coined by Benoît Mandelbrot (born 1924) who wrote a few books [34, 35] and many articles on fractal geometry, drawing attention to its relevance in such diverse fields as fluid mechanics, geomorphology, economics, and linguistics.

In order to better characterize fractals it is useful to introduce the notions of Hausdorff measure and Hausdorff dimension [25]. Let A be a compact subset of a metric space and $\{U_j \mid j \in J\}$ a countable cover of A by a family of open sets; the Hausdorff outer measure of A is

$$H_d^*(A) = \lim_{\varepsilon \to 0} \inf_{\{U_j \mid j \in J\}} \left\{ \sum_{j \in J} (\delta(U_j))^d \mid \forall j \in J, \delta(U_j) < \varepsilon \right\},$$

where $\delta(U_j)$ is the diameter of the open set U_j.

It can be shown (see p. 31 of [6]) that

If $H_{d_1}^*(A) < \infty$, then for $d_2 > d_1$, $H_{d_2}^*(A) = 0$, and
if $0 < H_{d_1}^*(A) < \infty$, then for $d_2 < d_1$, $H_{d_2}^*(A) = \infty$.

Let A be a compact subset of a metric space; the number

$$d_H(A) = \inf\{d \mid H_d^*(A) = 0\}$$

is the *Hausdorff dimension* of the set A.

In what follows, we will determine the Hausdorff dimension of various fractals.

19.1 Triadic Cantor Set

Georg Cantor is the founder of set theory. In 1878, he proved, in particular, that the sets $[0,1]$ and $[0,1] \times [0,1]$ have the same cardinality. This very surprising result even surprised Cantor himself.

All countable sets such as the set of integers or the set of rational numbers have a zero Lebesgue measure. Thus, all sets whose Lebesgue measure is not equal to zero are necessarily noncountable. The converse is wrong: there exist noncountable sets whose Lebesgue measure is zero. The triadic Cantor set is a classical example.

Let J_n be the union of 2^n disjoint closed sets of length 3^{-n} obtained from J_{n-1} by removing from each closed interval of length $3^{-(n-1)}$ the middle open interval of length 3^{-n}. Starting from $J_0 = [0,1]$, we obtain the sequence:

$$J_0 = [0,1]$$
$$J_1 = \left[0, \frac{1}{3}\right] \cup \left[\frac{2}{3}, 1\right]$$
$$J_2 = \left[0, \frac{1}{9}\right] \cup \left[\frac{2}{9}, \frac{1}{3}\right] \cup \left[\frac{2}{3}, \frac{7}{9}\right] \cup \left[\frac{8}{9}, 1\right]$$
$$J_3 = \cdots .$$

To build up a function generating the sets J_n we first define a function removing the middle third open interval of a given closed interval $[a, b]$ whose limit points a and b are rational numbers.

```
remainingIntervals[a_,b_] := Module[{m = (b - a) / 3},
{{a, a + m}, {b - m, b}}]
```

```
remainingIntervals[{0, 1}]
```

$$\left\{\{0, \frac{1}{3}\}, \{\frac{2}{3}, 1\}\right\}$$

We then define a similar function whose argument is a list of intervals.

```
remainingIntervalsList[intervals_List] :=
Flatten[Map[remainingIntervals, intervals], 1]
```

```
remainingIntervalsList[{{0, 1}}]
```

$$\{\{0, \frac{1}{3}\}, \{\frac{2}{3}, 1\}\}$$

```
remainingIntervalsList[{{0, 1 / 3}, {2 / 3, 1}}]
```

$$\{\{0, \frac{1}{9}\}, \{\frac{2}{9}, \frac{1}{3}\}, \{\frac{2}{3}, \frac{7}{9}\}, \{\frac{8}{9}, 1\}\}$$

We can now generate the sets of intervals J_n. Define

```
J[0] = {0, 1};
J[n_Integer] := Nest[remainingIntervalsList, {{0, 1}}, n]
```

For example,

```
J[3]
```

$$\{\{0, \frac{1}{27}\}, \{\frac{2}{27}, \frac{1}{9}\}, \{\frac{2}{9}, \frac{7}{27}\}, \{\frac{8}{27}, \frac{1}{3}\},$$
$$\{\frac{2}{3}, \frac{19}{27}\}, \{\frac{20}{27}, \frac{7}{9}\}, \{\frac{8}{9}, \frac{25}{27}\}, \{\frac{26}{27}, 1\}\}$$

The triadic Cantor set C is the intersection of all sets of intervals J_n:

$$C = \bigcap_{n \in \mathbb{N}} J_n.$$

Because it is the intersection of closed sets, C is closed. It contains no interval, so its interior is empty, which implies that all its points are boundary points. At each stage of its construction, the open intervals that constitute the middle thirds of the closed intervals left at the previous stage are removed, thus, any elements x of C can be written:

$$x = \sum_{n=1}^{\infty} \frac{x_n}{3^n},$$

where, for all positive integers n, $x_n = 0$ or 2. In other words, the ternary expansion, (i.e., the expansion to base 3), of an element of C does not contain the digit 1. We can use *Mathematica* to verify this characteristic property by writing in base 3 the list of remaining intervals at stage n, for $n = 1, 2, 3, \ldots$.

```
BaseForm[J[1] // N, 3]
```

$\{\{0._3 \ , \ 0.1_3\}, \quad 0.2 \ _3, \ 1._3$

```
BaseForm[J[2] // N, 3]
```

$\{\{0._3 \ , \ 0.01_3 \ \}, \ \{0.02_3 \ , \ 0.1_3\}, \ \{0.2_3 \ , \ 0.21_3\}, \ \{0.22_3 \ , \ 1._3\}\}$

```
BaseForm[J[3] // N, 3]
```

$\{\{0._3 \ , \ 0.001_3\}, \ \{0.002_3 \ , \ 0.01_3\}, \ \{0.02_3 \ , \ 0.021_3\},$
$\{0.022_3 \ , \ 0.1_3\}, \ \{0.2_3 \ , \ 0.201_3\}, \ \{0.202_3 \ , \ 0.21_3\},$
$\{0.22_3 \ , \ 0.221_3\}, \ \{0.222_3 \ , \ 1._3\}\}$

Observing the expressions of the endpoints of these intervals (which belong to the Cantor set because only the middle third open intervals are removed at each stage) we can note that the digit 1 can only appear as the last digit of a terminating ternary expansion. But, such digits can be replaced by a nonterminating sequence of digits 2; that is, in base 3, we have:

$$1.0 = 0.222\ldots, \qquad 0.1 = 0.0222\ldots, \qquad 0.01 = 0.00222\ldots,$$
$$0.001 = 0.000222\ldots, \quad 0.0001 = 0.0000222\ldots, \ 0.00001 = 0.00000222\ldots.$$

These two different representations (which exist for all bases) is a consequence of the relation

```
Sum[2 (1 / 3)^n, {n, 1, Infinity}]
```

1

The ternary representation shows that C is not countable because it is equipotent to the closed interval $[0, 1]$. To prove it we just have to exhibit a bijection

φ from $[0,1]$ onto C. If we write $\xi \in [0,1]$ under the form $\sum_{n=1}^{\infty} \xi_n/2^n$ where all ξ_n are either equal to 0 or 1, and define φ by

$$\varphi(\xi) = \sum_{n=1}^{\infty} \frac{2\xi_n}{3^n},$$

the bijection φ is known as the Cantor function. Because for all positive integers $n = 1, 2, \ldots$, $2\xi_n$ is either equal to 0 or 2, then the range of φ lies in C. To complete the proof, we have to show that the range of φ coincides with C. Let y be any element in C, and let its ternary expansion be $0.y_1y_2\ldots$ where for all positive integers n, y_n is equal to 0 or 2; then $x = \varphi^{-1}(y)$ exists and is unique. It is determined by its binary expansion $0.x_1x_2\ldots$, where, for all n, $x_n = y_n/2$.

Because J_n is obtained from J_{n-1} by removing 2^n open intervals of length $3^{-(n+1)}$, the Lebesgue measure m of C i.e., its length) is given by

$$m(C) = 1 - \sum_{n=0}^{\infty} \frac{2^n}{3^{n+1}} = 0.$$

That is, the Lebesgue measure of the Cantor set is zero but it has, however, the same cardinality as the interval $[0,1]$.

Taking into account the self-similar structure of the Cantor set, its Hausdorff dimension is easy to determine i9n as much as

$$3\left(C \cap \left[0, \frac{1}{3}\right]\right) = C \text{ and } 3\left(C \cap \left[\frac{2}{3}, 1\right]\right) - 2 = C,$$

but

$$H_d^*(C) = H_d^*\left(C \cap \left[0, \frac{1}{3}\right]\right) + H_d^*\left(C \cap \left[\frac{2}{3}, 1\right]\right),$$

and

$$H_d^*(C) = H_d^*\left(C \cap \left[0, \frac{1}{3}\right]\right) = H_d^*\left(C \cap \left[\frac{2}{3}, 1\right]\right) = \frac{1}{3^d} H_d^*(C).$$

Hence

$$H_d^*(C) = \frac{2}{3^d} H_d^*(C),$$

which is true only if $2/3^d = 1$, that is, if

$$d_H(D) = \frac{\log 2}{\log 3} \approx 0.63093 \ldots .$$

This result was obtained by Hausdorff and appears in his seminal paper [25].

There exists an interesting function related to the Cantor set called the *Lebesgue function*. Although it is fairly exotic, it has been recently rediscovered by physicists, and called the *Devil's staircase*. It is probably best viewed as the limit of a sequence L_n defined as follows. For each positive integer n, let L_n be the nondecreasing continuous function on $[0,1]$ such that $L(0) = 0$, $L(1) = 1$, and linear and increasing by 2^{-n} on each closed interval whose union is J_n and constant on the removed middle third intervals. If $n > m$, then, for any $x \in [0,1]$, $|L_n(x) - L_m(x)| < 2^{-m}$, that is, for all x, the sequence $(L_n(x))_{n \in \mathbb{N}}$ is a Cauchy sequence. The limit $L(x) = \lim_{n \to \infty} L_n(x)$ is, therefore, well defined for all $x \in [0,1]$, and from $|L_n(x) - L_m(x)| < 2^{-m}$ it follows that the convergence is uniform. The Lebesgue function is, consequently, a nondecreasing continuous function on $[0,1]$ such that $L(0) = 0$ and $L(1) = 1$. Because L is constant on each middle third removed interval, it is constant almost everywhere.

In order to define a *Mathematica* function **Lebesgue[n]** generating a plot of the function L_n we first define the following function that transforms a list of two points with nonidentical ordinates into a list of four points.

```
newSegments[{pt1_, pt2_}] :=
Module[{m = (pt2[[1]] - pt1[[1]]) / 3,
r = (pt2[[2]] - pt1[[2]]) / 2},
If[pt1 != pt2, {pt1, {pt1[[1]] +m, pt1[[2]] + r},
{pt2[[1]] - m, pt2[[2]] - r}, pt2},
{pt1, pt2}]]
```

If we apply this function to the list {{0,0},{1,1}}, we obtain

```
l1 = newSegments[{{0, 0}, {1, 1}}]
```

$$\{\{0, 0\}, \{\tfrac{1}{3}, \tfrac{1}{2}\}, \{\tfrac{2}{3}, \tfrac{1}{2}\}, \{1,1\}\}$$

From the list **l1** it is easy to obtain a list **l2** such that **ListPlot[l1]** would represent the graph of the function L_2.

```
l2 = Flatten[Map[newSegments, Partition[l1, 2]], 1]
```

$$\{\{0, 0\}, \{\tfrac{1}{9}, \tfrac{1}{4}\}, \{\tfrac{2}{9}, \tfrac{1}{4}\}, \{\tfrac{1}{3}, \tfrac{1}{2}\}, \{\tfrac{2}{3}, \tfrac{1}{2}\},$$
$$\{\tfrac{7}{9}, \tfrac{3}{4}\}, \{\tfrac{8}{9}, \tfrac{3}{4}\}, \{1,1\}\}$$

Plotting the lists of points **l1** and **l2** give the graphs of the functions L_1 and L_2

```
lpl1 = ListPlot[l1, PlotJoined → True, Frame → True,
AspectRatio → Automatic, DisplayFunction → Identity];
lpl2 = ListPlot[l2, PlotJoined → True, Frame → True,
AspectRatio → Automatic, DisplayFunction → Identity];
Show[GraphicsArray[{lpl1, lpl2}],
DisplayFunction → $DisplayFunction];
```

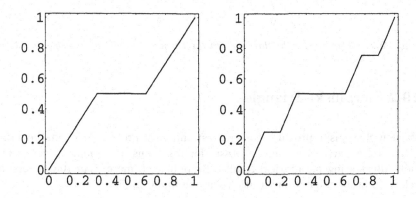

Fig. 19.1. *Graphs of L_1 and L_2, the first two steps in the construction of the Lebesgue function L.*

Generalizing these results we define the function LebesgueFunction[{pt1_, pt2_}, n] that plots the function:

```
LebesgueFunction[{pt1_, pt2_}, n_Integer] :=
ListPlot[Nest[Flatten[Map[newSegments,
Partition[#, 2]], 1] &, {pt1, pt2}, n],
TextStyle → {FontSize → 16}, PlotJoined → True,
Frame → True, AspectRatio → Automatic]
```

```
LebesgueFunction[{{0, 0}, {1, 1}}, 3];
```

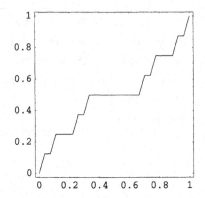

Fig. 19.2. *Graph of L_3 the third step in the construction of the Lebesgue function L.*

19.2 Sierpiński Triangle

Wacław Sierpiński, probably the greatest and most prolific Polish mathematician, had a particularly marked taste for ingenious mathematical constructions illustrating paradoxical results. He published more than 700 papers; a selection of his original publications can be found in [52].

The Sierpiński triangle, also called the Sierpiński gasket, is a bounded connected subset of $\mathbb{R}^2$ whose recursive construction is similar to the Cantor set construction. Starting from the closed equilateral triangle whose vertices are respectively located at $\{A_1, A_2, A_3\}$, we remove the open equilateral triangle whose vertices are $\{(A_1 + A_2)/2, (A_2 + A_3)/2, (A_3 + A_1)/2\}$ to obtain four equilateral triangles. We then repeat this operation on each remaining triangle.

We proceed as above and first define a function **remainingTriangles** similar to the function **remainingIntervals**.

```
remainingTriangles[{pt1_, pt2_, pt3_}] :=
Module[{pt12 = (pt1 + pt2) / 2, pt23 = (pt2 + pt3) / 2,
pt31 = (pt3 + pt1) / 2},
{{pt1, pt12, pt31}, {pt12, pt2, pt23}, {pt31, pt23, pt3}}]
```

```
A1= {0., 0.}; A2 = {1.,0.}; A3 = {1 / 2, Sqrt[3] / 2} //N;
```

```
rT = remainingTriangles[{A1, A2, A3}]
```

{{{0., 0.}, {0.5, 0.}, {0.25, 0.433013}},
{{0.5, 0.}, {1., 0.}, {0.75, 0.433013}},
{{0.25, 0.433013}, {0.75, 0.433013}, {0.5, 0.866025}}}

```
Show[Graphics[{RGBColor[0,0,1], Map[Polygon, rT]}],
AspectRatio → Automatic];
```

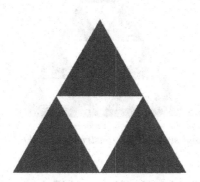

Fig. 19.3. *First stage in the construction of the Sierpiński triangle.*

As for the Cantor set, we define the function **remainingTrianglesList**.

```
remainingTrianglesList[triangles_List] :=
Flatten[Map[remainingTriangles, triangles], 1]
```

```
rtl = remainingTrianglesList[rT]
```

{{{0., 0.}, {0.25, 0.}, {0.125, 0.216506}},
{{0.25, 0.}, {0.5, 0.}, {0.375, 0.216506}},
{{0.125, 0.216506}, {0.375, 0.216506}, {0.25, 0.433013}},
{{0.5, 0.}, {0.75, 0.}, {0.625, 0.216506}},
{{0.75, 0.}, {1., 0.}, {0.875, 0.216506}},
{{0.625, 0.216506}, {0.875, 0.216506}, {0.75, 0.433013}},
{{0.25, 0.433013}, {0.5, 0.433013}, {0.375, 0.649519}},
{{0.5, 0.433013}, {0.75, 0.433013}, {0.625, 0.649519}},

$\{\{0.375, 0.649519\}, \{0.625, 0.649519\}, \{0.5, 0.866025\}\}\}$

```
Show[Graphics[{RGBColor[0,0,1], Map[Polygon, rtl]}],
AspectRatio → Automatic];
```

Fig. 19.4. *Second stage in the construction of the Sierpiński triangle.*

We can now build up the function `SierpinskiTriangle`, which depends upon the integer **n**, generating the nth iterate in the construction of the Sierpiński triangle corresponding to the limit $n \to \infty$.

```
SierpinskiTriangle[n_Integer] :=
Nest[remainingTrianglesList, {{A1, A2, A3}}, n];
```

```
Show[Graphics[{RGBColor[0, 0, 1],
Map[Polygon, SierpinskiTriangle[5]]}],
AspectRatio → Automatic];
```

Output represented in Figure 19.5.

Proceeding as we did for the Cantor set, we can determine the Hausdorff dimension of the Sierpiński triangle. We find

$$d_H(S_\triangle) = \frac{\log 3}{\log 2} \approx 1.58496\ldots,$$

where $S_\triangle$ denotes the Sierpiński triangle This result shows that the area of the Sierpiński triangle must be zero. We can check this result as follows. The area of an equilateral triangle with sides of length a is equal to $(\sqrt{3}/4)a^2$. Thus, starting with an equilateral triangle of side 1 and area $\sqrt{3}/4$, the first iteration

Fig. 19.5. *Fifth stage in the construction of the Sierpiński triangle.*

removes one equilateral triangle of side $\frac{1}{2}$; that is, we decrease the area of the original triangle by $(\sqrt{3}/4)\left(\frac{1}{2}\right)^2$, the second iteration removes three triangles of sides $\frac{1}{4}$; that is, we again decrease the area by $3(\sqrt{3}/4\left(\frac{1}{4}\right)^2$. More generally, the nth iteration removes an area equal to $3^{n-1}(\sqrt{3}/4)\left(\frac{1}{4}\right)^n$ which is equal to $(\sqrt{3}/12)\left(\frac{3}{4}\right)^n$. Therefore the total area removed from the original triangle in the limit $n \to \infty$ is

```
removedArea = (Sqrt[3] / 12) Sum[(3 / 4)^n, {n, 1, Infinity}]
```

$$\frac{\text{Sqrt}[3]}{4}$$

that is, the area of the original triangle.

19.3 Sierpiński Square

Also called the *Sierpiński carpet* (this fractal set was, according to Sierpiński, first found by Stefan Mazurkiewicz (1888–1945) who never published it), it is a bounded connected subset of $\mathbb{R}^2$ whose recursive construction is similar to the Sierpiński triangle construction. Starting from the closed square, whose vertices are respectively located at $(0,0)$, $(0,1)$, $(1,1)$, $(1,0)$, we first divide it in nine equal squares and then remove the open square whose vertices are $\left(\frac{1}{3},\frac{1}{3}\right)$, $\left(\frac{1}{3},\frac{2}{3}\right)$, $\left(\frac{2}{3},\frac{2}{3}\right)$, $\left(\frac{2}{3},\frac{1}{3}\right)$, to obtain eight closed squares. We then repeat this operation on each remaining square.

We again proceed as above and start defining the function `remainingSquares` similar to the function `remainingTriangles`.

```
remainingSquares[Rectangle[pt1_List, pt2_List]] :=
Module[{pt = Abs[pt1 - pt2] / 3, rectangleList},
rectangleList = {Rectangle[pt1, pt1 + pt],
Rectangle[{pt1[[1]],pt1[[2]]+ pt[[2]]},
{pt1[[1]]+ pt[[1]], pt2[[2]] - pt[[2]]}],
Rectangle[{pt1[[1]], pt2[[2]] - pt[[2]]},
{pt1[[1]]+ pt[[1]], pt2[[2]]}],
Rectangle[{pt1[[1]] + pt[[1]], pt1[[2]]},
{pt1[[1]] + 2 pt[[1]], pt1[[2]] + pt[[2]]}],
Rectangle[{pt1[[1]] + pt[[1]], pt2[[2]] - pt[[2]]},
{pt2[[1]] - pt[[1]], pt2[[2]]}],
Rectangle[{pt2[[1]] - pt[[1]], pt1[[2]]},
{pt2[[1]], pt1[[2]]+ pt[[2]]}],
Rectangle[{pt2[[1]] - pt[[1]], pt1[[2]] + pt[[2]]},
{pt2[[1]], pt2[[2]] - pt[[2]]}],
Rectangle[{pt2[[1]] - pt[[1]], pt2[[2]]- pt[[2]]},
{pt2[[1]], pt2[[2]]}]}]; rectangleList]
```

```
rS1 = remainingSquares[Rectangle[{0, 0}, {1, 1}]]
```

{Rectangle[{0, 0}, {$\frac{1}{3}$, $\frac{1}{3}$}], Rectangle[{0, $\frac{1}{3}$}, {$\frac{1}{3}$, $\frac{2}{3}$}],

Rectangle[{0, $\frac{2}{3}$}, {$\frac{1}{3}$, 1}], Rectangle[{$\frac{1}{3}$, 0}, {$\frac{2}{3}$, $\frac{1}{3}$}],

Rectangle[{$\frac{1}{3}$, $\frac{2}{3}$}, {$\frac{2}{3}$, 1}], Rectangle[{$\frac{2}{3}$, 0}, {1, $\frac{1}{3}$}],

Rectangle[{$\frac{2}{3}$, $\frac{1}{3}$}, {1, $\frac{2}{3}$}], Rectangle[{$\frac{2}{3}$, $\frac{2}{3}$}, {1,1}]}

```
Show[Graphics[{RGBColor[0, 0, 1], rS1}],
AspectRatio → Automatic];
```

Making the function remainingSquares listable, we can define the function SierpinskiCarpet that generates all the successive steps of the construction of the Sierpiński square.

```
Attributes[remainingSquares] = Listable;
```

Fig. 19.6. *First stage in the construction of the Sierpiński square.*

```
SierpinskiCarpet[n_Integer] :=
Nest[remainingSquares, Rectangle[{0, 0}, {1,1}], n]
```

```
Show[Graphics[{RGBColor[0, 0, 1], SierpinskiCarpet[5]}],
AspectRatio → Automatic];
```

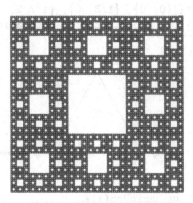

Fig. 19.7. *Fifth stage in the construction of the Sierpiński square.*

Proceeding as we did above for the Sierpiński triangle, we can determine the Hausdorff dimension of the Sierpiński square. We find

$$d_H(S_\Box) = \frac{\log 8}{\log 3} \approx 1.89279\ldots,$$

where $S_\square$ denotes the Sierpiński square. This result shows that the area of the Sierpiński square must be zero. We could check this result as we did for the Sierpinski triangle.

19.4 von Koch Curve

Helge von Koch was a student of Mittag-Leffler at Stockholm University. He is famous for the self-similar curve presented in his 1904 paper [28] entitled *Sur une courbe continue sans tangente, obtenue par une construction géométrique élémentaire* (On a continuous curve without tangents constructible from elementary geometry).

The von Koch curve is constructed by first dividing a segment of unit length into three segments of equal length and replacing the middle segment by the two sides of an equilateral triangle of the same length as the segment being removed. Repeat this process on the four resulting segments, dividing them into three equal parts and replacing each of the middle segments by two sides of an equilateral triangle. The von Koch curve is the limit curve obtained by repeating indefinitely this construction on each new generated segment.

Here is the first stage of the construction.

```
Show[Graphics[Line[{{0, 0}, {1/3, 0}, {1 / 2, Sqrt[3] / 6},
{2 / 3, 0}, {1, 0}}]], AspectRatio → Automatic];
```

Fig. 19.8. *First stage of the construction of the von Koch curve.*

We call this structure the basicProfile.

```
basicProfile = {{0, 0}, {1 / 3, 0}, {1 / 2, Sqrt[3] / 6},
{2 / 3, 0}, {1, 0}};
```

At each stage of the construction, we have to replace each segment by the line defined by basicProfile, correctly oriented and scaled; each segment being defined by the coordinates of its initial and final point denoted {x1, y1} and {x2, y2}.

The function `nextProfile[{{x1, y1}, {x2, y2}}]`, defined below, generates the next profile of a segment.

```
nextProfile[{{x1_, y1_}, {x2_, y2_}}] :=
Module[{rotation2D, x, y, basicProfile, newProfile},
rotation2D[x_, y_] = Module[{c, s},
c = x / Sqrt[x^2 + y^2]; s = y / Sqrt[x^2 + y^2];
{{c, - s},{s, c}}];
{x, y} = {x2 - x1, y2 - y1};
r = Sqrt[x^2 + y^2];
basicProfile = {{0, 0}, {1 / 3, 0}, {1 / 2, Sqrt[3] / 6},
{2 / 3, 0}, {1, 0}};
newProfile = Map[{x1, y1} + r rotation2D[x, y]. # &,
basicProfile] // N; newProfile]
```

```
nextProfile[{{0, 0}, {1, 0}}]
```

{{0., 0.}, {0.333333, 0.}, {0.5, 0.288675},
{0.666667, 0.}, {1., 0.}}

In order to generate the whole curve at a given stage we have to define the next profile of a line. Because the arguments of the function `nextProfile` are a pair of points, given a line given by a sequence of points we have to generate the sequence of segments that constitute this line. This is done using the function **Partition** with an offset equal to 1.

```
Partition[{{0, 0}, {1 / 3, 0}, {1 / 2, Sqrt[3] / 6},
{2 / 3, 0}, {1, 0}}, 2, 1]
```

$$\{\{\{0, 0\}, \{\tfrac{1}{3}, 0\}\}, \{\{\tfrac{1}{3}, 0\}, \{\tfrac{1}{2}, \tfrac{1}{2\,\mathrm{Sqrt}[3]}\}\},$$
$$\{\{\tfrac{1}{2}, \tfrac{1}{2\,\mathrm{Sqrt}[3]}\}, \{\tfrac{2}{3}, 0\}\}, \{\{\tfrac{1}{2\,\mathrm{Sqrt}[3]}, 0\}, \{1, 0\}\}\}$$

We then use the **Map** function to make the function `nextProfile` listable.

```
nextProfile[lis_List] := Map[nextProfile, lis]
```

```
nP = Flatten[nextProfile[Partition[{{0, 0}, {1 / 3, 0},
{1 / 2, Sqrt[3] / 6}, {2 / 3, 0}, {1, 0}}, 2, 1]], 1]
```

{{0., 0.}, {0.111111, 0.}, {0.166667, 0.096225},
{0.222222, 0.}, {0.333333, 0.}, {0.333333, 0.},
{0.388889, 0.096225}, {0.333333, 0.19245},
{0.444444, 0.19245}, {0.5, 0.288675}, {0.5, 0.288675},
{0.555556, 0.19245}, {0.666667, 0.19245},
{0.611111, 0.096225}, {0.666667, 0.}, {0.666667, 0.},
{0.777778, 0.}, {0.833333, 0.096225}, {0.888889, 0.},
{1., 0.}}

```
Show[Graphics[Line[nP]],
AspectRatio → Automatic];
```

Fig. 19.9. *Second stage of the construction of the von Koch curve.*

Apparently, everything is fine but looking carefully at the list nP we discover
that the fifth and sixth points are identical. This is due to the fact that
the nP line consists of four **basicProfiles** and that the endpoint of a given
basic profile is identical to the initial point of the following one. Because
basicProfile consists of five points this implies that the pairs (5,6), (10,11),
and (15,16) are pairs of identical points. This does not cause any problem
when visualizing the line because the three segments of zero length do not
appear on the visualization. But applying the function **nextProfile** to a pair
of identical points returns an error message caused by the indeterminate value
of the parameters c and s. We can eliminate from the list the extra points
using the following function.

```
lineSequence[lis_List] := Module[{seq = {}},
For[k = 1, k < Length[lis], k++,
If[lis[[k]] != lis[[k + 1]], seq = Append[seq, {lis[[k]]}]]];
seq = Flatten[Append[seq, {lis[[Length[lis]]]}], 1];
seq]
```

```
lineSequence[nP]
```

{{0., 0.}, {0.111111, 0.}, {0.166667, 0.096225},
{0.222222, 0.}, {0.333333, 0.}, {0.388889, 0.096225},
{0.333333, 0.19245}, {0.444444, 0.19245}, {0.5, 0.288675},
{0.555556, 0.19245}, {0.666667, 0.19245},
{0.611111, 0.096225}, {0.666667, 0.}, {0.777778, 0.},
{0.833333, 0.096225}, {0.888889, 0.}, {1., 0.}}

```
Show[Graphics[Line[lineSequence[nP]]],
AspectRatio → Automatic];
```

Fig. 19.10. *Second stage of the construction of the von Koch curve using*
lineSequence *instead of the listable version of the function* nextProfile.

We now have to group all these steps to build up the function KochCurve[{{x1, y1}, {x2, y2}}, n] that, starting from an initial segment {{x1, y1}, {x2, y2}} iterates n times the construction described above.

```
KochCurve[{{x1_, y1_}, {x2_, y2_}}, n_Integer] :=
Nest[lineSequence[Flatten[nextProfile[Partition[#, 2,
1]], 1]]&, {{x1, y1}, {x2, y2}}, n]
```

```
Show[Graphics[Line[KochCurve[{{0, 0}, {1,0}}, 4]]],
AspectRatio → Automatic];
```

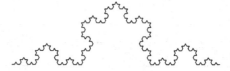

Fig. 19.11. *Fourth stage of the construction of the von Koch curve.*

Note that we could also have defined a slightly more general function.

```
KochCurve[pointsList_List, n_Integer] :=
Nest[lineSequence[Flatten[nextProfile[Partition[#, 2,
1]], 1]] &, pointsList, n]
```

where pointsList is a list of points such as {{0, 0}, {1, 0}} or any of
its iterates such as {{0, 0}, {1 / 3, 0}, {1 / 2, Sqrt[3] / 6}, {2 /
3, 0}, {1, 0}}.

```
Show[Graphics[Line[KochCurve[{{0, 0}, {1, 0}}, 5]]],
AspectRatio → Automatic];
```

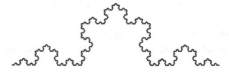

Fig. 19.12. *Fifth stage of the construction of the von Koch curve.*

```
Show[Graphics[Line[KochCurve[{{0, 0}, {1 / 3, 0},
{1 / 2, Sqrt[3] / 6}, {2 / 3, 0}, {1, 0}}, 4]]],
AspectRatio → Automatic];
```

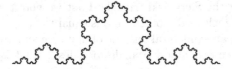

Fig. 19.13. *Same as above but starting from a different set of points.*

Taking into account the self-similar structure of the von Koch curve, we can determine its Hausdorff dimension. We find

$$d_H(K) = \frac{\log 4}{\log 3} \approx 1.266186\ldots,$$

where K denotes the von Koch curve. This Hausdorff dimension, greater than 1, implies that the length of the von Koch curve is infinite which is, by the way, obvious because at each step the length is multiplied by $4/3$.

We can even start from a list of points that are the vertices of a regular polygon. Starting from an equilateral triangle we generate the so-called *von Koch island* also called the *von Koch triangle*.

```
Show[Graphics[Line[KochCurve[{{0, 0}, {1 / 2, Sqrt[3] / 2},
{1, 0}, {0, 0}}, 5]]], AspectRatio → Automatic];
```

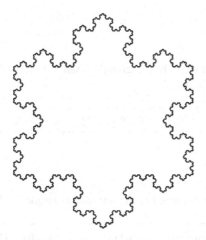

Fig. 19.14. *Fifth stage of the construction of the von Koch triangle.*

It is clear that although the length of the perimeter of the von Koch triangle is infinite, its area is finite. To find its value we proceed as we did above to obtain the area of the Sierpiński triangle. First we note that the area of an equilateral triangle with sides of length a is equal to $(\sqrt{3}/4)a^2$. Next we note that starting from an equilateral triangle of unit side and area $\sqrt{3}/4$, the first iteration adds three equilateral triangles of sides $\frac{1}{3}$, that is, we increase the area of the original triangle by $3(\sqrt{3}/4)\left(\frac{1}{3}\right)^2$. More generally, the nth iteration adds an area equal to

$$3\frac{\sqrt{3}}{4}4^{n-1}\left(\frac{1}{3}\right)^{2n} = \frac{3\sqrt{3}}{16}\left(\frac{4}{9}\right)^n.$$

Therefore, the total area added to the original triangle, in the limit b$n \to \infty$, is

```
addedArea = (3 Sqrt[3] / 16) Sum[(4 / 9)^n, {n, 1, Infinity}]
```

$$\frac{3\ \mathrm{Sqrt}[3]}{20}$$

The area of a von Koch triangle is larger than the area of the original equilateral triangle by a factor

```
addedArea / (Sqrt[3] / 4)
```

$$\frac{3}{5}$$

That is, the area of the von Koch triangle when the original equilateral triangle sides are equal to 1 is equal to

```
areaKochTriangle = (8 / 5) (Sqrt[3] / 4) // N
```

0.69282

Starting from a unit square, here is another example of the von Koch island.

```
Show[Graphics[Line[KochCurve[{{0, 0}, {1, 0}, {1, 1}, {0, 1},
{0, 0}}, 4]]], AspectRatio -> Automatic];
```

Fig. 19.15. *Fourth stage of the construction of the von Koch square.*

Iterated Function Systems

Let $\{f_i \mid 1 \leq i \leq n\}$ be a finite set of n mappings from a complete metric space (E, d) onto itself; $\{p_i \mid 1 \leq i \leq n\}$ a discrete probability distribution, that is, a set of nonnegative real numbers such that $\sum_{i=1}^n p_i = 1$; and $F : E \mapsto E$ the mapping defined by $x \mapsto F(x) = f_i(x)$ with probability p_i. The dynamical system $((E, d), F)$ is called an *iterated function system* or *IFS*.

A mapping f of a metric space onto itself is a contraction if there exists a constant $0 \leq s < 1$ such that for all pairs (x, y) of points of the metric space, $d(f(x), f(y)) \leq s d(x, y)$, where d is the distance defined on the metric space. If f, defined on a complete metric space, is a contraction, the sequence of iterates $(f^n(x))$ of any point x always converges to the same fixed point. If all the mappings f_i of an IFS $\{f_i \mid 1 \leq i \leq n\}$ are contractions, then the system has an attractor.

For example, $\{f_i \mid 1 \leq i \leq n\}$ can be a set of two-dimensional affine transformations. That is, each mapping is of the form $f_i(x) = A_i x + b_i$, where A_i is a 2×2 matrix and b_i a two-dimensional vector. We can verify that, if $|\det A_i| < 1$, f_i is a contraction.

20.1 Chaos Game

Iterated function systems are widely used to generate computer images that are approximations of the attractor of the IFS. These attractors are often fractals. In his book, M. Barnsley [1] uses IFS to give a detailed treatment of fractal images. The traditional example of IFS is the so-called chaos game. Start with an equilateral triangle with vertices at $(0, 0)$, $(1, 0)$, and $(0.5, \sqrt{3}/2)$, respectively, labeled A_1, A_2, and A_3. Select a random point P_1 and define a sequence of points $(P_1, P_2, P_3, \ldots)$ such that P_2 is the midpoint of the segment $A_i P_1$, where A_i is one of the vertices selected at random, P_3 the midpoint of the segment $A_j P_2$, where A_j is one of the vertices selected at random, and so on.

In order to build up a small program generating the sequence of successive iterates of an initial point P_1, we first define the vertices A1, A2, A3 of the triangle and the three functions f[1], f[2], and f[3] that transform a point {x, y} in the midpoint of the segment joining the point {x, y} to, respectively, A1, A2, and A3.

```
A1 = {0, 0}; A2 = {1, 0}; A3 = {0.5, Sqrt[3.0] / 2};
f[1][{x_, y_}] := 0.5 ({x, y} + A1);
f[2][{x_, y_}] := 0.5 ({x, y} + A2);
f[3][{x_, y_}] := 0.5 ({x, y} + A3);
```

We then define the function chaosGame[init, n] which generates a numerated sequence starting from init followed by its n iterates and draw the equilateral triangle.

```
chaosGame[init_List, n_Integer] :=
Module[{F, pts, ptList, triangle, trajectory, image},
F[x_] := f[Random[Integer, {1, 3}]][x];
pts = NestList[F, init, n];
ptList = Graphics[{{PointSize[0.05], CMYKColor[0, 0, 1, 0],
Map[Point, pts]}, Table[Text[i, Part[pts, i]],
{i, 1, Length[pts]}]}];
triangle = Graphics[{RGBColor[1,0,0],
Line[{A1, A2, A3, A1}]}];
trajectory = Graphics[{RGBColor[0, 0, 1], Line[pts]}];
image = Show[{ptList, triangle, trajectory},
PlotRange -> All, AspectRatio -> Automatic]; image]
```

```
chaosGame[{0.3, 0.2}, 15];
```

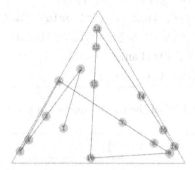

Fig. 20.1. *Sequence of points generated by the chaos game starting from an initial point (labeled 1) inside an equilateral triangle.*

Because the mappings f[1], f[2], and f[3] are contractions, if the initial point init lies outside the triangle, after a few iterations all the successive iterates lie inside as shown below.

```
chaosGame[{1, 1}, 10];
```

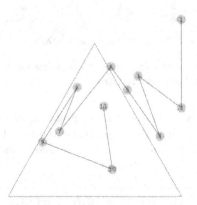

Fig. 20.2. *Sequence of points generated by the chaos game starting from an initial point (labeled 1) outside the triangle.*

If we plot a large number of points (reducing their size to obtained a finer figure), the sequence seems to converge to a Sierpiński triangle.

```
bigChaosGame[init_List, n_Integer] :=
Module[{F, pts, ptList, image},
F[x_] := f[Random[Integer, {1, 3}]][x];
pts = Drop[NestList[F, init, n], Floor[n / 1000]];
ptList = Graphics[{PointSize[0.005], Map[Point, pts]}];
image = Show[ptList, PlotRange → All,
AspectRatio → Automatic]; image]
```

Note that a fraction of 0.1% of the total sequence of points have been dropped.

```
bigChaosGame[{0.3, 0.2}, 10000];
```

Fig. 20.3. *The sequence of a large number of points generated by the chaos game seems to converge to a Sierpiński triangle.*

To help understand why the sequence of points apparently converges to the Sierpiński triangle, consider the respective inverses g[1], g[2], and g[3] of the mappings f[1], f[2], and f[3]. These inverses exist because f[1], f[2], and f[3] are linear. They are defined by

```
A1 = {0, 0}; A2 = {1, 0}; A3 = {0.5, Sqrt[3.0] / 2};
g[1][{x_, y_}] := 2 {x, y} + A1;
g[2][{x_, y_}] := 2 {x, y} + A2;
g[3][{x_, y_}] := 2 {x, y} + A3;
```

Using g[1], g[2], and g[3] we can show that the image by any of these mappings of a point that does not belong to the Sierpiński triangle cannot belong to the Sierpiński triangle. Take, for example, the points $(0.5, \sqrt{3}/4)$ and $(0.25, \sqrt{3}/8)$.

```
{g[1][{0.5, Sqrt[3.0] / 4}], g[1][{0.25, Sqrt[3.0] / 8}]}
{g[2][{0.5, Sqrt[3.0] / 4}], g[2][{0.25, Sqrt[3.0] / 8}]}
{g[3][{0.5, Sqrt[3.0] / 4}], g[3][{0.25, Sqrt[3.0] / 8}]}
```

{{1., 0.866025}, {0.5, 0.433013}}

{{2., 0.866025}, {1.5, 0.433013}}

{{1.5, 1.73205}, {1., 1.29904}}

Hence, the sequence of points $(P_1, P_2, P_3, \ldots)$ converges to the Sierpiński triangle if, and only if, the initial point P_1 belongs to the Sierpiński triangle. Because the area of the Sierpiński triangle is of measure zero (i.e., its Euclidean area is zero), starting from a randomly selected initial point P_1 the probability of converging to the exact Sierpiński triangle is zero! But, on a computer screen, the exact and the approximate Sierpiński triangles cannot be distinguished.

The three mappings f[1], f[2], f[3] are affine transformations which can, respectively, be written

$$
\begin{pmatrix} 0.5 & 0 \\ 0 & 0.5 \end{pmatrix} \begin{pmatrix} x \\ y \end{pmatrix}, \quad \begin{pmatrix} 0.5 & 0 \\ 0 & 0.5 \end{pmatrix} \begin{pmatrix} x \\ y \end{pmatrix} + \begin{pmatrix} 1 \\ 0 \end{pmatrix}, \quad \begin{pmatrix} 0.5 & 0 \\ 0 & 0.5 \end{pmatrix} \begin{pmatrix} x \\ y \end{pmatrix} + \begin{pmatrix} 0.5 \\ \sqrt{3}/2 \end{pmatrix}.
$$

20.2 Variations on the Chaos Game

Many variations of the chaos game can be played. Not all of them generate fractal images. Starting from a set $\mathcal{A}$ of n points $A_1, A_2, \ldots, A_n$, select at random a point P_1 and define a sequence of points $(P_1, P_2, P_3, \ldots)$ such that P_2 on the segment $A_i P_1$ is at a distance from A_i equal to a given fraction r of the length of $A_i P_1$ where A_i is a randomly selected point of $\mathcal{A}$, P_3 on the segment $A_j P_2$ is at a distance from A_j equal to the fraction r of the length of $A_j P_2$ where A_j is a randomly selected point of $\mathcal{A}$, and so on.

Here are a few examples.

20.2.1 Example 1

```
Clear[A1, A2, A3, A4, r, f]
A1 = {0.5, 0.5}; A2 = {- 0.5, 0.5}; A3 = {- 0.5, - 0.5};
A4 = {0.5, - 0.5}; r = 0.4;
f[1][{x_, y_}] := r ({x, y} + A1);
f[2][{x_, y_}] := r ({x, y} + A2);
f[3][{x_, y_}] := r ({x, y} + A3);
f[4][{x_, y_}] := r ({x, y} + A4);
```

```
bigGame1[init_List, n_Integer] :=
Module[{F, pts, ptList, image},
F[x_] := f[Random[Integer, {1, 4}]][x];
pts = Drop[NestList[F, init, n], Floor[n /1000]];
ptList = Graphics[{RGBColor[0, 0, 1], PointSize[0.005],
Map[Point, pts]}];
image = Show[ptList, PlotRange → All,
AspectRatio → Automatic]; image]
```

```
bigGame1[{0, 0}, 10000];
```

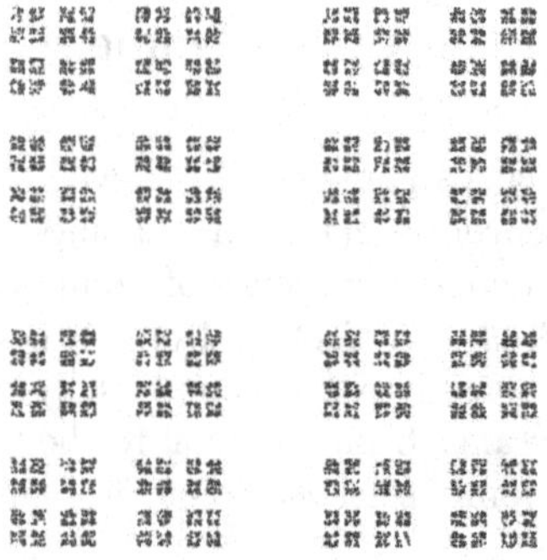

Fig. 20.4. *Sequence of a large number of points generated by the chaos game of Example 1.*

20.2.2 Example 2

```
Clear[A1, A2, A3, A4, A5, r, f]
A1 = {1, 0}; A2 = {Cos[2Pi / 5], Sin[2Pi / 5]};
A3 = {Cos[4Pi / 5], Sin[4Pi / 5]};
A4 = {Cos[6Pi / 5], Sin[6Pi / 5]}; A5 = {Cos[8Pi / 5],
Sin[8Pi / 5]}; r = 0.3;
f[1][{x_, y_}] := r ({x, y} + A1);
f[2][{x_, y_}] := r ({x, y} + A2);
f[3][{x_, y_}] := r ({x, y} + A3);
f[4][{x_, y_}] := r ({x, y} + A4);
f[5][{x_, y_}] := r ({x, y} + A5);
```

```
bigGame2[init_List, n_Integer] :=
Module[{F, pts, ptList, image},
F[x_] := f[Random[Integer, {1, 5}]][x];
pts = Drop[NestList[F, init, n], Floor[n / 1000]];
ptList = Graphics[{RGBColor[0, 0, 1], PointSize[0.003],
Map[Point, pts]}];
image = Show[ptList, PlotRange → All,
AspectRatio → Automatic]; image]
```

```
bigGame2[{0, 0}, 10000];
```

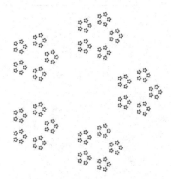

Fig. 20.5. *Sequence of a large number of points generated by the chaos game of Example 2.*

20.2.3 Example 3

```
Clear[A1, A2, A3, A4, A5, r, f]
A1 = {1, 0}; A2 = {Cos[Pi / 3], Sin[Pi / 3]};
A3 = {Cos[2Pi / 3], Sin[2Pi / 3]}; A4 = {- 1, 0};
A5 = {Cos[4Pi / 3], Sin[4Pi / 3]}; A6 = {Cos[5Pi / 3],
Sin[5Pi / 3]}; r = N[1 / 3];
f[1][{x_, y_}] := r ({x, y} + A1);
f[2][{x_, y_}] := r ({x, y} + A2);
f[3][{x_, y_}] := r ({x, y} + A3);
f[4][{x_, y_}] := r ({x, y} + A4);
f[5][{x_, y_}] := r ({x, y} + A5);
f[6][{x_, y_}] := r ({x, y} + A6);
```

```
bigGame3[init_List, n_Integer] :=
Module[{F, pts, ptList, image},
F[x_] := f[Random[Integer, {1, 6}]][x];
pts = Drop[NestList[F, init, n], Floor[n / 1000]];
ptList = Graphics[{RGBColor[0, 0, 1], PointSize[0.005],
Map[Point, pts]}];
image = Show[ptList, PlotRange → All,
AspectRatio → Automatic]; image]
```

```
bigGame3[{0, 0}, 10000];
```

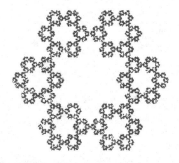

Fig. 20.6. *Sequence of a large number of points generated by the chaos game of Example 3.*

20.3 Barnsley Fern

20.3.1 The Original Barnsley Fern

In the case of the Sierpiński triangle the three mappings were chosen with equal probability. Using nonuniform probabilities, Barnsley [1] found an attractor that bears a startling resemblance to a fern. The IFS consists of the following four affine transformations defined on $\mathbb{R}^2$ and their respective probabilities.

$$f_1(x,y) = \begin{pmatrix} 0 & 0 \\ 0 & 0.16 \end{pmatrix} \begin{pmatrix} x \\ y \end{pmatrix} \quad \text{with probability } p_1 = 0.01,$$

$$f_2(x,y) = \begin{pmatrix} 0.85 & 0.04 \\ -0.04 & 0.85 \end{pmatrix} \begin{pmatrix} x \\ y \end{pmatrix} + \begin{pmatrix} 0 \\ 1.6 \end{pmatrix} \quad \text{with probability } p_2 = 0.85,$$

$$f_3(x,y) = \begin{pmatrix} 0.2 & -0.26 \\ 0.23 & 0.22 \end{pmatrix} \begin{pmatrix} x \\ y \end{pmatrix} + \begin{pmatrix} 0 \\ 1.6 \end{pmatrix} \quad \text{with probability } p_3 = 0.07,$$

$$f_4(x,y) = \begin{pmatrix} -0.15 & 0.28 \\ 0.26 & 0.24 \end{pmatrix} \begin{pmatrix} x \\ y \end{pmatrix} + \begin{pmatrix} 0 \\ 0.44 \end{pmatrix} \quad \text{with probability } p_3 = 0.07.$$

```
f[1][{x_, y_}] := {{0, 0}, {0, 0.16}}. {x, y}
f[2][{x_, y_}] := {{0.85, 0.04}, {- 0.04, 0.85}}. {x, y} +
{0, 1.6}
f[3][{x_, y_}] := {{0.2, - 0.26}, {0.23, 0.22}}. {x, y} +
{0, 1.6}
f[4][{x_, y_}] := {{- 0.15, 0.28}, {0.26, 0.24}}. {x, y} +
{0, 0.44}
```

The following program generates the Barnsley fern

```
BarnsleyFern[n_Integer] :=
Module[{pt, ptList, rnd := Random[Integer, {1,100}]},
pt = {0, 0}; ptList = {pt};
Do[r = rnd; pt = Which[r == 1, f[1][pt], r <= 86, f[2][pt],
r <= 93, f[3][pt], True, f[4][pt]];
ptList = Append[ptList, pt] , {n}];
image = Show[Graphics[{RGBColor[0, 0.4, 0], PointSize[0.003],
Map[Point,ptList]}], PlotRange → All]; image]
```

```
BF = BarnsleyFern[10000];
```

Fig. 20.7. *Barnsley's fern.*

Barnsley's affine transformations are contractions and each of them has a fixed point that is found using the *Mathematica* command FixedPoint.

```
{fxd1, fxd2, fxd3, fxd4} = Table[FixedPoint[f[k],
{0.5, 0.5}], {k, 1, 4}]
```

{{0., 0.}, {2.6556, 9.95851}, {- 0.608365, 1.87189},
{0.153769, 0.631553}}

Figure 20.8 shows their locations on the fern. Point k is the fixed point of f_k ($k = 1, 2, 3, 4$).

```
pos = Graphics[{PointSize[0.04], CMYKColor[0, 0, 1, 0],
Map[Point, {fxd1, fxd2, fxd3, fxd4}]}];
num = Graphics[{Text[1, fxd1], Text[2, fxd2], Text[3, fxd3],
Text[4, fxd4]}];
Show[{BF, pos, num}];
```

Fig. 20.8. *Barnsley's fern with the fixed points of the affine transformations f_1, f_2, f_3, and f_4.*

The following programs help in understanding how the four affine transformations generate the different parts of the Barnsley fern by drawing successive images of an initial triangular shape called init.

```
init = Graphics[Line[{{- 2,1}, {2, 1}, {0, 10}, {- 2,1}}]];
```

Starting from the initial triangular shape, the graphics array below illustrates the action of the different affine transformations. f_1 generates the lower part of the stem; f_2 generates the upper part of the stem, all triangles converging to the fixed point 2 of f_2; starting from the image of the initial shape by f_3, and repeatedly applying f_2 generates the left branches; and similarly, starting from the image of the initial shape by f_4, and repeatedly applying f_2 generates the right branches.

```
f1Action =
Show[Graphics[Map[Line, Transpose[Map[NestList[f[1],#, 50]&,
{{- 2, 1}, {2, 1},{0, 10}, {- 2, 1}}]]]],
PlotRange → {{- 2.6, 2.6}, {1, 11}}, Frame → True,
DisplayFunction → Identity];

f2Action =
Show[Graphics[Map[Line,Transpose[Map[NestList[f[2],#, 50]&,
{{- 2, 1}, {2, 1}, {0, 10}, {- 2, 1}}]]]],
PlotRange → {{- 2.6, 2.6}, {1, 11}},
Frame → True, DisplayFunction → Identity];

f2f3Action =
Show[Graphics[Map[Line,Transpose[Map[NestList[f[2],#, 50]&,
Map[f[3], {{- 2,1}, {2,1}, {0,10}, {- 2, 1}}]]]]],
PlotRange → {{- 2.6, 2.6}, {1, 11}}, Frame →True,
DisplayFunction → Identity];

f2f4Action =
Show[Graphics[Map[Line,Transpose[Map[NestList[f[2],#, 50]&,
Map[f[4], {{- 2, 1}, {2, 1}, {0, 10}, {- 2, 1}}]]]]],
PlotRange → {{-2.6, 2.6}, {1, 11}}, Frame → True,
DisplayFunction → Identity];

fAction = Show[GraphicsArray[
{{f1Action, f2Action}, {f2f3Action, f2f4Action}}]]
```

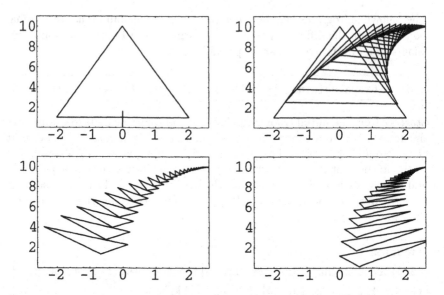

Fig. 20.9. *Action of the four affine transformations on the initial shape. Upper left: f_1 generates the lower part of the stem. Upper right: f_2 generates the upper part of the stem, all triangles converging to the fixed point 2 of f_2. Lower left: starting from the image of the initial shape by f_3, and repeatedly applying f_2 generates the left branches. Lower right: starting from the image of the initial shape by f_4, and repeatedly applying f_2 generates the right branches.*

20.3.2 Modifying the Probabilities

The probabilities p_1, p_2, p_3, and p_4 can actually be modified, but keeping p_1 small, p_2 large, and $p_3 = p_4$ still generates a fernlike image. Here are two examples.

If we choose $p_1 = 0.03$, $p_2 = 0.75$, $p_3 = p_4 = 0.11$, and use the following program,

```
Clear[BarnsleyFern]
BarnsleyFern[n_Integer] :=
Module[{pt, ptList, rnd := Random[Integer, {1, 100}]},
pt = {0, 0}; ptList = {pt};
Do[r = rnd; pt = Which[r == 3, f[1][pt], r <= 78, f[2][pt],
r <= 89, f[3][pt], True, f[4][pt]];
ptList = Append[ptList, pt] , {n}];
image = Show[Graphics[{RGBColor[0, 0.4, 0], PointSize[0.005],
Map[Point,ptList]}], PlotRange → All]; image]
```

we obtain the image below (Figure 20.10).

```
BarnsleyFern[10000];
```

Fig. 20.10. *Barnsley's fern with probabilities $p_1 = 0.03$, $p_2 = 0.75$, $p_3 = p_4 = 0.11$.*

If we choose $p_1 = 0.05$, $p_2 = 0.75$, $p_3 = p_4 = 0.10$, and use the following program,

```
Clear[BarnsleyFern]
BarnsleyFern[n_Integer] :=
Module[{pt, ptList, rnd := Random[Integer, {1, 100}]},
pt = {0, 0}; ptList = {pt};
Do[r = rnd; pt = Which[r == 5, f[1][pt], r <= 80, f[2][pt],
r <= 90, f[3][pt], True, f[4][pt]];
ptList = Append[ptList, pt] , {n}];
image = Show[Graphics[{RGBColor[0, 0.4, 0], PointSize[0.005],
Map[Point,ptList]}], PlotRange → All]; image]
```

we obtain an image (not displayed) difficult to distinguish from the previous one.

20.4 The Collage Theorem

Among important mathematical results, Barnsley's book [1] contains many examples of IFS whose attractors have an amazingly realistic appearance. One of the most important results is Barnsley's Collage theorem which, given a fractal compact subset of $\mathbb{R}^2$, shows how to find an IFS whose attractor is close to the compact subset (refer to Barnsley's book for details).

On page 105 0f [1], for instance, Barnsley defines the IFS that consists of the following transformations defined on $\mathbb{C}$: $w_i(z) = s_i z + (1 - s_i) a_i$, for $i = 1, 2, 3, 4$, where the complex numbers s_i and a_i are given in the table below.

	1	2	3	4
s	0.6	0.6	$0, 4 - 0.3i$	$0.4 + 0.3i$
a	$0.45 + 0.9i$	$0.45 + 0.3i$	$0.6 + 0.3i$	$0.3 + 0.3i$

In order to generate the attractor using **bigGame**, we first write the mappings w_i on $\mathbb{C}$ as affine transformations f_i on $\mathbb{R}^2$.

```
s z + (1 - s) a /. {z → x + y I, s → 0.6,
a → 0.45 + 0.9 I} // ComplexExpand
s z + (1 - s) a /.{z → x + y I, s → 0.6,
a → 0.45 + 0.3 I} // ComplexExpand
s z + (1 - s) a /. {z → x + y I, s → 0.4 - 0.3 I,
a → 0.6 + 0.3 I} // ComplexExpand
s z + (1 - s) a /. {z → x + y I, s → 0.4 + 0.3 I,
a → 0.3 + 0.3 I} // ComplexExpand
```

0.18 + 0.6 x + I (0.36 + 0.6 y)

0.18 + 0.6 x + I (0.12 + 0.6 y)

0.27 + 0.4 x + I (0.36 - 0.3 x + 0.4 y) + 0.3 y

0.27 + 0.4 x + I (0.09 + 0.3 x + 0.4 y) - 0.3 y

The affine transformations f_i are, therefore, defined by

```
f[1][{x_, y_}] := {{0.6, 0}, {0, 0.6}}.{x, y} + {0.18, 0.36}
f[2][{x_, y_}] := {{0.6, 0}, {0, 0.6}}.{x, y} + {0.18, 0.12};
f[3][{x_, y_}] := {{0.4, 0.3}, {- 0.3, 0.4}}.{x, y} +
{0.27, 0.37}
f[4][{x_, y_}] := {{0.4, - 0.3}, {0.3, 0.4}}.x,y +
{0.27, 0.09}
```

Running bigGame generates a leaflike attractor.

```
Clear[bigGame]
bigGame[init_List, n_Integer] :=
Module[{F, pts, ptList, image},
F[x_] := f[Random[Integer, {1, 4}]][x];
pts = Drop[NestList[F, init, n], Floor[n / 1000]];
ptList = Graphics[{RGBColor[0, 0.3, 0], PointSize[0.003],
Map[Point, pts]}];
image = Show[ptList, PlotRange -> All,
AspectRatio -> Automatic]; image]
```

Fig. 20.11. *Leaflike fractal generated using Barnsley's collage theorem.*

Julia and Mandelbrot Sets

21.1 Julia Sets

The French mathematician Gaston Julia (1893–1978) published his famous *Mémoire sur l'itération des fonctions rationelles* in 1918 [26]. He was 25. The paper investigates the behavior of the iterates $z_{n+1} = f(z_n)$, where f is a rational function and, in particular, the set of points whose nth iterates remain bounded as n tends to infinity.

Consider the complex function $f : z \mapsto z^2 + c$, where c is a constant. The Julia set associated with the complex number c is the set of points of the complex plane whose sequence of iterates remains bounded. For example, if $c = -\frac{1}{2}$, we have

```
Clear[f]
f[z_Complex] := z^2 - 1 / 2
Nest[f, 0.8 + 0.5 I, 1000]
Nest[f, 0.9 + 0.5 I, 10]
```

$- 0.366025 - 2.32705 \; 10^{-136} \; \text{I}$

$5027.03 + 2999.2 \; \text{I}$

Apparently the sequence of iterates of $z = 0.8 + 0.5i$ remains bounded whereas the sequence of iterates of $0.9 + 0.5i$ does not. For $c = 0.5$, we can build up a program to visualize the Julia set.

The function `JuliaTest` determines whether the iterates of z escape from the disk of radius 2 centered at the origin for a number of iterates less than n. If it is the case, `JuliaTest` returns an integer $m < n$, indicating the first iterate escaping the disk, otherwise it returns n.

We use the command Compile[variables, expression] that creates a compiled function to evaluate expression assuming numerical values of variables. In compiled functions, the variables being numbers, the evaluation process is faster. Many built-in *Mathematica* commands compile their arguments automatically.

```
Clear[JuliaTest]
JuliaTest = Compile[{x, y, {n, _Integer}, {c, _Complex}},
Module{z, num = 0}, z = x + y I;
While[Abs[z] < 2.0 && num < n, z = z^2 + c; num++]; num]];
```

```
JuliaTest[0.8, 0.5 , 50, - 0.5]
JuliaTest[0.9, 0.5, 50, - 0.5]
```

50

7

Using the command DensityPlot we can visualize the Julia set.

```
DensityPlot[JuliaTest[x, y, 50, - 0.5],
{x, - 2.0, 2.0}, {y, - 2.0, 2.0}, PlotPoints → 500,
Mesh → False];
```

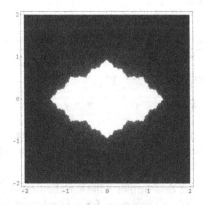

Fig. 21.1. *Julia set of the function* $z \mapsto z^2 - 0.5$.

As shown in Figure 21.1 the Julia set is represented by the white region. The darker the region the faster the iterates of a point in that region escape to infinity. Because the constant is real, if z belongs to the Julia set so does its complex conjugate.

Julia sets often have a complex fractal structure as illustrated below.

```
DensityPlot[JuliaTest[x, y, 50, - 0.75 + 0.5 I],
{x, - 2.0, 2.0}, {y, - 2.0, 2.0}, PlotPoints → 500,
Mesh → False, ColorFunction → Hue];
```

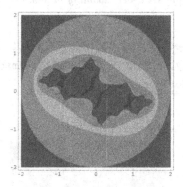

Fig. 21.2. *Julia set of the function* $z \mapsto z^2 - 0.75 + 0.5i$.

Zooming in exhibits more details.

```
DensityPlot[JuliaTest[x, y, 50, - 0.75 + 0.5 I],
{x, 0.9, 1.6}, {y, - 0.7, - 0.1}, PlotPoints → 500,
Mesh → False, ColorFunction → Hue];
```

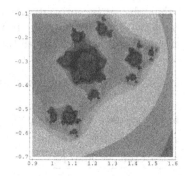

Fig. 21.3. *Julia set above: zooming in* $[0.9, 1.6] \times [-0.7, -0.1]$.

```
DensityPlot[JuliaTest[x, y, 50, - 0.75 + 0.5 I],
{x, 1.21, 1.28}, {y, - 0.2, - 0.1}, PlotPoints → 500,
Mesh → False, ColorFunction → Hue];
```

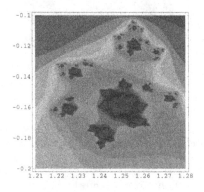

Fig. 21.4. *Julia set above: zooming in* $[1.26, 1.28] \times [-0.2, -0.1]$.

```
DensityPlot[JuliaTest[x, y, 50, - 0.75 + 0.5 I],
{x, 1.24, 1.27}, {y, - 0.13, - 0.1}, PlotPoints → 500,
Mesh → False, ColorFunction → Hue];
```

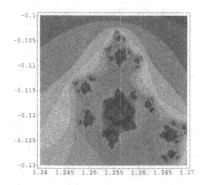

Fig. 21.5. *Julia set above: zooming in* $[1.24, 1.27] \times [-0.13, -0.1]$.

The two examples we presented illustrate the property of Julia sets of being either connected or totally disconnected.

21.2 Julia Sets of Different Functions

21.2.1 $z \mapsto z^3 + c$

```
Clear[JuliaTest]
JuliaTest = Compile[{x, y, {n,_Integer}, {c, _Complex}},
Module[{z, num = 0}, z = x + y I;
While[Abs[z] < 2.0 && num < n, z = z^3 + c;
num++]; num]];
```

```
DensityPlot[JuliaTest[x, y, 50, - 0.5],
{x, - 2.0, 2.0}, {y, - 2.0, 2.0}, PlotPoints → 500,
Mesh → False, ColorFunction → Hue];
```

```
DensityPlot[JuliaTest[x, y, 50, -0.75 + 0.5 I],
{x, - 2.0, 2.0}, {y, - 2.0, 2.0}, PlotPoints → 500,
Mesh → False, ColorFunction → Hue];
```

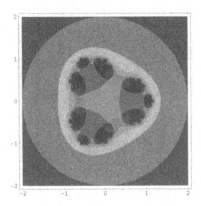

Fig. 21.6. *Julia set of the function $z \mapsto z^3 - 0.5$.*

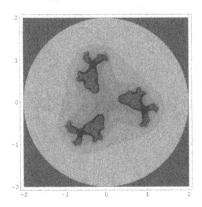

Fig. 21.7. *Julia set of the function $z \mapsto z^3 - 0.75 + 0.5i$.*

Zooming in:

```
DensityPlot[JuliaTest[x, y, 50, - 0.75 + 0.5 I],
{x, - 0.9, 0.1}, {y, 0.1, 1.3}, PlotPoints → 500,
Mesh → False, ColorFunction → Hue];
```

```
DensityPlot[JuliaTest[x, y, 50, - 0.75 + 0.5 I],
{x, - 0.57, - 0.38}, {y, 0.9, 1.25}, PlotPoints → 500,
Mesh → False, ColorFunction → Hue];
```

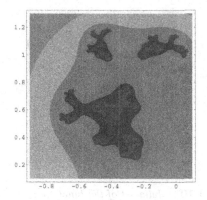

Fig. 21.8. *Julia set above: zooming in* $[-0.9, 0.1] \times [0.1, 1.3]$.

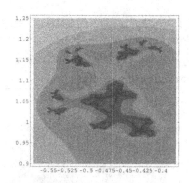

Fig. 21.9. *Julia set above: zooming in* $[-0.57, -0.38] \times [0.9, 1.25]$.

21.2.2 $z \mapsto z^4 + c$

```
Clear[JuliaTest]
JuliaTest = Compile[{x, y, {n, _Integer}, {c, _Complex}},
Module[{z, num = 0}, z = x + y I;
While[Abs[z] < 2.0 && num < n, z = z^4 + c;
num++]; num]];
```

```
DensityPlot[JuliaTest[x, y, 50, - 0.5],
{x, - 2.0, 2.0}, {y, - 2.0, 2.0}, PlotPoints → 500,
Mesh → False, ColorFunction → Hue];
```

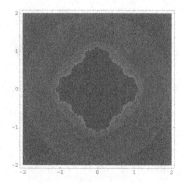

Fig. 21.10.   *Julia set of the function* $z \mapsto z^4 - 0.5$.

And zooming in:

```
DensityPlot[JuliaTest[x, y, 50, - 0.5],
{x, - 0.1, 0.1}, {y, 1.02, 1.22}, PlotPoints → 500,
Mesh → False, ColorFunction → Hue];
```

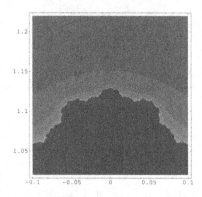

Fig. 21.11.   *Julia set above: zooming in* $[-0.1, 0.1] \times [1.02, 1.22]$.

21.3 Mandelbrot Sets

The Mandelbrot set [34, 35] is the set of points c of the complex plane such that the sequence (z_n) defined by $z_0 = 0$, $z_{n+1} = z_n^2 + c$ does not tend to infinity. Such a set had been already considered by the French mathematician Pierre Fatou (1878–1929) who proved that once an iterate moved to a distance greater than two from the origin, then the orbit would escape to infinity. Points in the Mandelbrot set have connected Julia sets. It can be shown that the Mandelbrot set is simply connected.

The function **MandelbrotTest** is used to test if a point $x + iy$ belongs to the Mandelbrot set. It determines whether the iterates of z escape from the disk of radius two centered at the origin for a number of iterates less than n. Its definition is similar to the function **JuliaTest** defined above.

```
$RecursionLimit = Infinity;
```

```
MandelbrotTest = Compile[{{x, _Real}, {y, _Real},
{n, _Integer}},
Module[{z, num = 0, c = x + y I},
z = x + y I; While[Abs[z] < 2.0 && num < n,
z = z^2 + c; ++num]; num]];
```

For the points that diverge to infinity, and are therefore not in the set, the color reflects the number of iterations it takes to reach a certain distance from the origin.

```
DensityPlot[MandelbrotTest[x, y, 50],
{x,- 2.0, 0.75}, {y,- 1.25, 1.25}, PlotPoints → 500,
Mesh → False, ColorFunction → Hue];
```

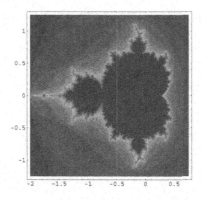

Fig. 21.12. *Mandelbrot set of the function $z \mapsto z^2 + c$.*

Zooming in reveals the very complex structure of the Mandelbrot set.

```
DensityPlot[MandelbrotTest[x, y, 50],
{x, - 1.0, - 0.4}, {y, - 0.3, 0.3}, PlotPoints → 500,
Mesh → False, ColorFunction → Hue];
```

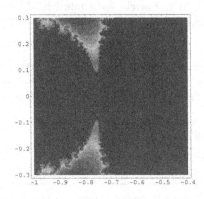

Fig. 21.13. *Mandelbrot set: zooming in* $[-1.0, -0.4] \times [-0.3, 0.3]$.

```
DensityPlot[MandelbrotTest[x, y, 50],
{x, - 0.85, - 0.65}, {y, - 0.2, 0.0}, PlotPoints → 500,
Mesh → False, ColorFunction → Hue];
```

Fig. 21.14. *Mandelbrot set: zooming in* $[-0.85, -0.65] \times [-0.2, 0]$.

```
DensityPlot[MandelbrotTest[x, y, 50],
{x, - 0.77, - 0.72}, {y, - 0.2, - 0.15}, PlotPoints → 500,
Mesh → False, ColorFunction → Hue];
```

Fig. 21.15. *Mandelbrot set: zooming in* $[-0.77, -0.72] \times [-0.2, -0.15]$.

```
DensityPlot[MandelbrotTest[x, y, 50],
{x, - 0.748, - 0.74}, {y, - 0.186, - 0.178},
PlotPoints → 500, Mesh → False, ColorFunction → Hue];
```

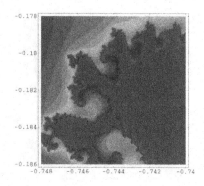

Fig. 21.16.   *Mandelbrot set: zooming in* $[-0.748, -0.74] \times [-0.186, -0.178]$.

21.4 Mandelbrot Sets for Different Functions

21.4.1 $z \mapsto z^3 + c$

```
Mandelbrot3 = Compile[{{x,_Real}, {y,_Real}, {n,_Integer}},
Module[{z, num = 0, c = x + y I},
z = x + y I;
While[Abs[z] < 2.0 && num < n, z = z^3 + c;
++num]; num]];

DensityPlot[Mandelbrot3[x, y, 50],
{x, -1.5, 1.5}, {y, - 1.5, 1.5}, PlotPoints → 500,
Mesh → False, ColorFunction → Hue];
```

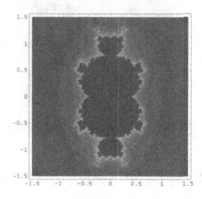

Fig. 21.17. *Mandelbrot set for the function $z \mapsto z^3 + c$.*

21.4.2 $z \mapsto z^4 + c$

```
Mandelbrot4 = Compile[{{x,_Real}, {y,_Real}, {n,_Integer}},
Module[{z, num = 0, c = x + y I},
z = x + y I;
While[Abs[z] < 2.0 && num < n, z = z^4 + c;
++num]; num]];

DensityPlot[Mandelbrot4[x, y, 50],
{x, - 1.5, 1.5}, {y, - 1.5, 1.5}, PlotPoints → 500,
Mesh → False, ColorFunction → Hue];
```

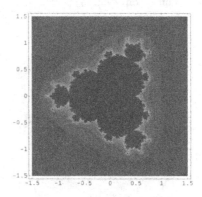

Fig. 21.18. *Mandelbrot set for the function $z \mapsto z^4 + c$.*

Kepler's Laws

Johannes Kepler (1571–1630) came from a modest family but, thanks to a scholarship, he could attend the Lutheran Seminary at the University of Tübingen. At Tübingen, Kepler was taught astronomy by one of the leading astronomers of the day, Michael Maestlin (1550–1631). Maestlin lent Kepler his own annotated copy of Copernicus' book. Kepler quickly grasped the essential ideas of the Copernican system and built up a cosmological theory based on the five regular polyhedra. He collaborated with the Danish astronomer Tycho Brahe (1546–1601) who was one of the most prominent observational astronomer of the time, and succeeded him, when he died in 1601, as Imperial Mathematician (they were both living in the Holy Roman Empire). Among his many achievements, Kepler was probably the first to explicitly use the concept of observational error. He is chiefly known for the three laws bearing his name which Isaac Newton (1643–1727) was able to derive from his gravitational law. The first two laws were published in *Astronomia Nova* (A New Astronomy) in 1609, and the third law in *Harmonices Mundi* (The Harmony of the World) in 1619.

The gravitational force between two mass points m_1 and m_2 derives from the potential

$$U(r) = -G\,\frac{m_1 m_2}{r},$$

where r is the distance between the two mass points, and G is a universal constant whose value, in SI units, can be found loading the *Mathematica* package

```
<<Miscellaneous'PhysicalConstants'
```

```
G = GravitationalConstant
```

$$\frac{6.673 \ 10^{-11} \ \text{Meter}^2 \ \text{Newton}}{\text{Kilogram}^2}$$

Because the force is central, the torque is zero and the angular momentum **L** is conserved.

In order to study the motion of a planet in the gravitational field of the sun, we choose the origin at the center of the sun and denote, respectively, by m, **r**, and **v** the mass, the position vector, and the velocity of the planet. The relations

$$\mathbf{r} \times \mathbf{L} = 0 \text{ and } \mathbf{v} \times \mathbf{L} = 0$$

imply that the motion takes place in the plane passing through the origin and perpendicular to **L**.

Using polar coordinates, we have

$$\mathbf{v} = r'\mathbf{e_r} + r\theta'\mathbf{e_\theta},$$

where the local unit vectors $\mathbf{e_r}$ and $\mathbf{e_\theta}$ are defined in terms of Cartesian unit vectors $\mathbf{e_1}$ and $\mathbf{e_2}$ by

$$\mathbf{e_r} = \mathbf{e_1} \cos\theta + \mathbf{e_2} \sin\theta$$
$$\mathbf{e_\theta} = -\mathbf{e_1} \sin\theta + \mathbf{e_2} \cos\theta.$$

These relations show that $\mathbf{e_r}$ and $\mathbf{e_\theta}$ are orthogonal. Hence,

```
positionVector = {r Cos[theta], r Sin[theta]};
velocityVector = {r' Cos[theta] - r theta' Sin[theta],
r' Sin[theta] + r Cos[theta]};
```

Loading the package `Calculus'VectorAnalysis'` we can use *Mathematica* to derive a few simple results.

```
<<Calculus'VectorAnalysis'
```

In cylindrical coordinates, the components of the position vector **r** and the velocity vector **v** in the local reference frame ($\mathbf{e_r}$, $\mathbf{e_\theta}$, $\mathbf{e_z}$) are

```
positionVector = {r, 0, 0};
velocityVector = {r', r theta' ,0};
```

The angular momentum is therefore given by

```
CrossProduct[positionVector, m velocityVector] // Simplify
```

$$\{0, \ 0, \ m \ r^2 \ theta'\}$$

That is,

$$L = \mid m \ r^2 \ \theta' \mid = \text{constant}.$$

It is now a simple matter to derive Kepler's laws.

Kepler's second law (1609). *A line drawn between the sun and the planet sweeps out equal areas in equal times as the planet orbits the sun.*

The area dA swept by the radius vector during the time interval dt is

$$dA = \frac{1}{2} \parallel \mathbf{r} \times \mathbf{v} \, dt \parallel = \frac{L}{2m} \, dt \implies A' = \text{constant}.$$

Kepler's first law (1609). *The orbit of each planet is an ellipse, with the sun located at one of its foci.*

The equation of motion in polar coordinates is

$$m\mathbf{r}'' = -G \, \frac{Mm}{r^2} \, \mathbf{e_r},$$

where M is the sun mass and m the mass of the planet.

From the definition of the local unit vectors $\mathbf{e_r}$ and $\mathbf{e}_\theta$ given above it follows that

$$\frac{d\mathbf{e_r}}{dt} = \theta', \ \text{and} \ \frac{d\mathbf{e}_\theta}{dt} = -\theta' \mathbf{e_r},$$

so, from differentiating one more time the expression of the velocity

$$\mathbf{v} = \mathbf{r}' = r' \, \mathbf{e_r} + r \, \theta' \, \mathbf{e}_\theta,$$

we obtain the expression of the acceleration

$$\mathbf{r}'' = \left(r'' - (r\theta')^2 \right) \, \mathbf{e_r} + \left(2r'\theta' + r\theta'' \right) \, \mathbf{e}_\theta.$$

Hence, the equation of motion can be written

$$m \left(r'' - (r\theta')^2 \right) = -G \, \frac{Mm}{r^2} \ \text{and} \ m \left(2r' \, \theta' + r \, \theta'' \right) = 0.$$

The second equation implies $r^2\theta' = \text{constant}$, that is, conservation of angular momentum. To determine the orbit of the planet, we use the variable $u = 1/r$. Thus

$$r' = -\frac{u'}{u^2} = -\frac{1}{u^2} \, \theta' \, \frac{du}{d\theta} = -\frac{L}{m} \frac{du}{d\theta},$$

and differentiating once more

$$r'' = -\frac{L}{m}\frac{d}{dt}\left(\frac{du}{d\theta}\right) = -\frac{L}{m}\theta'\frac{d^2u}{dt^2}.$$

The differential equation of the orbit is therefore the very simple equation

$$\frac{d^2u}{d\theta^2} + u = G\frac{Mm^2}{L^2},$$

whose solution is obtained using the command

```
DSolve[u''[theta] + u[theta] == K, u[theta], theta]
```

$$\{\{u[theta] \rightarrow K + C[1] Cos[theta] + C[2] Sin[theta]\}\}$$

where K is the constant GMm^2/L^2; C[1] and C[2] are two constants of integration. Because we are only interested in the shape of the orbit, we can choose C[2] = 0. In polar coordinates the equation of the planar orbit is

$$r = \frac{1}{C\cos\theta + K},$$

where we put C[1] = C. To plot the orbit, instead of the two constants C and K we introduce two new constants r_0 and e and write the polar equation

$$r = r_0\frac{1+e}{1+e\cos\theta}.$$

Changing the constant e, called the *eccentricity*, can modify the shape of the orbit, Note that, in terms of the previous constants

$$r_0 = \frac{1}{C+K} = \frac{GMm^2}{L^2+CGMm^2} e = \frac{C}{K} = \frac{CL^2}{GMm^2}.$$

1. If $0 < e < 1$, the orbit is an ellipse (for $e = 0$ the orbit is a circle). The following commands draw a few elliptical orbits.

```
t1 = Table[PolarPlot[(1 + e) / (1 + e Cos[theta]),
{theta, 0, 2 Pi},
PlotStyle → {Thickness[0.007],
RGBColor[1 - e, 0.2, 3 e / 2]},
TextStyle → {FontSize → 12},
DisplayFunction → Identity], {e, 0.3, 0.6, 0.1}];
pt = Graphics[{PointSize[0.04], RGBColor[1, 0, 0],
Point[{0, 0}]}];
Show[{t1, pt}, AspectRatio → Automatic,
DisplayFunction → $DisplayFunction];
```

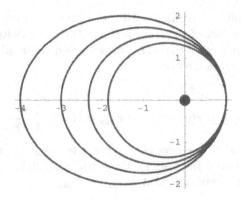

Fig. 22.1. *Elliptical orbits. The big dot represents the sun.*

2. If $e > 1$ the orbit is a hyperbola.

```
e =1.4; pl = PolarPlot[(1 + e) / (1 + e Cos[theta]),
{theta, 0, 2 Pi},
Ticks → {{- 8, - 4, 0, 4, 8, 12, 16}, {- 6,- 3, 3, 6}},
PlotStyle → {Thickness[0.007], RGBColor[0, 0.2, 1]},
TextStyle → {FontSize → 12},
DisplayFunction → Identity];
pt = Graphics[{PointSize[0.05], RGBColor[1, 0, 0],
Point[{0,0}]}];
Show[{pl, pt}, AspectRatio → Automatic,
DisplayFunction → $DisplayFunction];
```

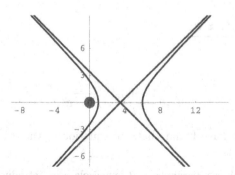

Fig. 22.2. *Hyperbolic orbit. The big dot represents the sun.*

For $e = 1$, the orbit is a parabola. All these orbits are *conic sections*.[1] The origin is one of the foci of the conic section. The eccentricity of the earth's orbit about the sun is equal to 0.017. This value is quite small and the earth's orbit is, therefore, quasi-circular.

The distances

$$r_{\min} = r(0) = r_0 \text{ and } r_{\max} = r(\pi) = \frac{1+e}{1-e}$$

are, respectively, called the *perihelion* (closest to the sun) and the *aphelion* (farthest from the sun). Because, for the earth, $e = 0.017$, $(1+e)/(1-e) = 1.035$, which is very close to 1, that is, $r_{\min} \approx r_{\max}$.

This simple theory neglects the influence of the other planets and the motion of the sun.

Kepler's third law (1618). *The square of the period of revolution of a planet around the sun is proportional to the cube of the semi-major axis of the planet orbit.*

From Kepler's second law, we have

$$A = \frac{L}{2m}\, T,$$

where A is the area swept by the planet during one complete revolution, that is, during a time interval T. Moreover, for an ellipse,

$$A = \pi ab \text{ and } \frac{b}{a} = \sqrt{1 - e^2},$$

where a and b are, respectively, the semi-minor and semi-major axes. Hence,

```
T = 2 A m / L //. {A → Pi a b, b → a Sqrt[1 - e^2]}
```

$$\frac{2\, a^2\ \mathrm{Sqrt}[1 - e^2]\ m\ \mathrm{Pi}}{L}$$

That is,

$$T = \frac{2\pi\, m\, a^2}{L}\, \sqrt{1 - e^2}.$$

In as much as we have found that the equation of the orbit of a planet in polar coordinates is

---

[1]The terms ellipse, parabola, and hyperbola were introduced by the famous Greek geometer Appolonius of Perga (262 BC–190 BC) who is the first to have rigorously studied the conic sections discovered earlier by Menaechmus (380 BC–320 BC).

$$r = \frac{e/C}{1 + e\cos\theta},$$

we have

$$2a = r(0) + r(\pi) = \frac{a}{C}\left(\frac{1}{1+e} + \frac{1}{1-e}\right).$$

that is,

$$1 - e^2 = \frac{e}{aC} = \frac{L^2}{aGMm^2}$$

and replacing in the expression of T, we finally obtain

$$T^2 = \frac{4\pi^2}{GM}\,a^3.$$

Note that the coefficient $4\pi^2/GM$ is the same for all planets in orbit around the sun.

Remark. The fact that a characteristic time (the period of revolution of a planet) behaves as the power $3/2$ of a characteristic length (the semi-major axis of the elliptical trajectory of a planet) can be proved using a scaling argument. The equation of motion of a particle of mass m in a central field derived from a potential $U(r)$ is $mr'' = -\nabla U(r)$. If we change r in λr, and t in μt, the equation of motion becomes $m(\lambda\mu^{-2})\mathbf{r}'' = -\lambda^{k-1}\nabla U(r)$, if $U(r)$ behaves as r^k. The equation of motion is unchanged if, and only if, $\lambda\mu^{-2} = \lambda^{k-1}$, that is, if μ scales as $\lambda^{1-k/2}$. In the case of the gravitational potential, $k = 1$, hence, μ scales as $\lambda^{3/2}$, which is exactly Kepler's third law.

Lindenmayer Systems

Lindenmayer systems (usually referred to as L-systems) are rewriting techniques developed by the biologist Aristid Lindenmayer (1925–1989) in 1968. They are actually symbolic dynamical systems whose evolution can be represented graphically. Thanks to this feature, they have found several applications in computer graphics such as the generation of fractals and realistic models of plants.

23.1 String Rewriting

A string rewriting system is a triple (V, ω, R) where V, called the *alphabet*, is a finite set of symbols called *letters*; ω, called the *axiom* or *initiator*, is a string of symbols belonging to the alphabet V; and R is a set of *replacement rules*. Replacement rules are mappings such as $a \rightarrow w$, where a is a letter and w a *word*, that is, a sequence of letters. Some letters that are mapped to themselves are called *constants*. Conventionally a letter with no explicit production (i.e., image) is a constant.

For example, the L-system with the alphabet $V = \{F, +, -\}$, the axiom $\omega = F$, and the replacement rule $R = \{F \rightarrow F + F - -F + F\}$ can be shown to generate the von Koch curve. The symbols $+$ and $-$ having no explicit production are constants.

Replacement rules can be written using the function StringReplace[''string '', {''s1'' $\rightarrow$ ''ns1'', ''s2'' $\rightarrow$ ''ns2'',...}] which replaces in ''string'' each occurrence of the string ''si'' by the string ''nsi'' for all i.

Here are the first iterations of the replacement rule applied to the axiom F obtained using the function StringReplace.

```
NestList[StringReplace[#, ''F'' → ''F+F--F+F''] &,
''F'' , 2] // TableForm
```

F

F+F--F+F

F+F--F+F+F+F--F+F--F+F--F+F+F+F--F+F

This particular L-system is simple. It is deterministic, in the sense that there is only one production for each symbol, and it is context-free in the sense that the replacement rule refers only to one individual symbol and does not depend upon the neighboring symbols. Deterministic context-free L-systems are called DOL-systems.

23.2 von Koch Curve and Triangle

To show that the example above generates a von Koch curve, the symbols ''F'', ''+'', and ''-'' have to be, respectively, interpreted as "move forward", "turn 60 degrees to the left", and "turn 60 degrees to the right".

In order to build up a program visualizing the dynamics of this L-system, that is, a program generating the line representing the sequence of symbols (i.e., the word) obtained after n iterations of the replacement rule, we represent this line as the path followed by a turtle moving in a plane. The turtle is a robotic creature familiar to the Logo programming language. Moving around the plane, the state of the turtle at a given time is defined by its position and the direction it is facing. With just the two commands moveForward and turnLeft, the turtle can be moved in any path. On iterative graphics by simulating a turtle see [64].

We define the command turtle[symbol] where the argument symbol will take three possible values: f, l, and r, which, respectively, instruct the turtle to move forward one step in the direction it is facing, change direction by turning 60 degrees to the left, or 60 degrees to the right. We make the function turtle listable.

Using first the command Characters, we transform a word in the alphabet {F, +, -} into a list of letters.

```
Characters[''F+F--F+F'']
```

{F, +, F, -, -, F, +, F}

Then we define the symbols f, l, and r, associated with these letters.

```
symbols = {''F'' → f, ''+'' → l, ''-'' → r}                    ;
```

The listable function turtle is defined by

```
Attributes[turtle] = Listable;
turtle[f] := ptList += unit dir;
turtle[l] := dir = turnLeft . dir;
turtle[r] := dir = turnRight . dir;
```

where ptList is the list of points generating the line representing the turtle path, unit is the length of the elementary step, dir the direction faced by the turtle, and turnLeft and turnRight the matrices

$$\begin{pmatrix} \cos\frac{\pi}{3} & -\sin\frac{\pi}{3} \\ \sin\frac{\pi}{3} & \cos\frac{\pi}{3} \end{pmatrix} \quad \text{and} \quad \begin{pmatrix} \cos\frac{\pi}{3} & \sin\frac{\pi}{3} \\ -\sin\frac{\pi}{3} & \cos\frac{\pi}{3} \end{pmatrix}.$$

We assume that the turtle is initially located at the point $(0,0)$ and facing the point $(1,0)$. The path along which the turtle is moving is a line, that is, the list of points ptList. At time $t = 0$, the path has only one point: the initial point.

Putting together all these commands we can write a program generating the von Koch curve.

```
KochCurve[n_Integer] :=
Module[{Lsystem = {''F'' → ''F+F--F+F''}, axiom = ''F'',
symbols, unit = 3.0^(- n), X = {0., 0.}, dir = {1., 0.},
iter = 1, c = Cos[Pi / 3] // N, s = Sin[Pi / 3] // N, turtle,
rewritingrules, ptList, f, l, r, turnLeft, turnRight},
symbols = {''F'' → f, ''+'' → l, ''-'' → r};
ptList = Table[Null, {1 + 4^n}]; ptList[[1]] = X;
turnLeft = {{c, - s}, {s, c}};
turnRight = {{c, s}, {- s, c}};
```

```
Attributes[turtle] = Listable;
turtle[f] := ptList[[++iter]] = (X += unit dir);
turtle[l] := dir = turnLeft . dir;
turtle[r] := dir = turnRight . dir;
rewritingrules = Map[First[Characters[#[[1]]]] →
Characters[#[[2]]] &, Lsystem] /. symbols;
turtle[Nest[# /. rewritingrules &,
Characters[axiom] /. symbols, n]];
image = Show[Graphics[Line[ptList]],
AspectRatio → Automatic,
Axes → None, PlotRange → All]; image]
```

Note that, following Wagon [64], we predefine a path to be a list of Nulls, where Null is a symbol used to indicate the absence of a result.

```
KochCurve[4];
```

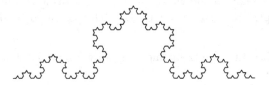

Fig. 23.1. *Fourth stage of the construction of the von Koch curve.*

As shown below, it is not difficult to modify the program above to generate the von Koch triangle. We just have to change the axiom and the length of the predefined path.

```
KochTriangle[n_Integer] :=
Module[{Lsystem = {''F''→ ''F-F++F-F''},
axiom = ''-F++F++F'',
symbols, unit = 3.0^(-n), X = {0., 0.}, dir = {1., 0.},
iter = 1, c = Cos[Pi / 3] // N, s = Sin[Pi / 3] // N, turtle,
rewritingrules, ptList, f, l, r, turnLeft, turnRight},
symbols = {''F'' → f, ''+'' → l, ''-'' → r};
1mm] ptList = Table[Null, {1 + 3*4^n}]; ptList[[1]] = X;
turnLeft = {{c, - s}, {s, c}};
turnRight = {{c, s}, {- s, c}};
Attributes[turtle] = Listable;
turtle[f] := ptList[[++iter]] = (X += unit dir);
turtle[l] := dir = turnLeft . dir;
turtle[r] := dir = turnRight . dir;
rewritingrules = Map[First[Characters[#[[1]]]] →
Characters[#[[2]]] &, Lsystem] /. symbols;
turtle[Nest[# /. rewritingrules &,
Characters[axiom] /. symbols, n]];
image = Show[Graphics[Line[ptList]],
AspectRatio → Automatic, Axes → None,
PlotRange → All]; image]
```

```
KochTriangle[4];
```

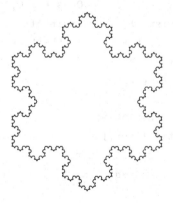

Fig. 23.2. *Fourth stage of the construction of the von Koch triangle.*

23.3 Hilbert Curve

We mentioned earlier (see Chapter 19 "Fractals") that Cantor discovered that
the sets $[0, 1]$ and $[0, 1] \times [0, 1]$ have the same cardinality, which implies the
existence of a bijection between these two sets. In 1879 Eugen Netto (1848–
1919) proved that such a bijection cannot be continuous, but it was shown
that continuous surjective mappings from $[0, 1]$ to $[0, 1] \times [0, 1]$—which are, by
definition, curves—do exist. In 1890, Giuseppe Peano (1858–1932) constructed
the first space-filling curve (on space-filling curves, refer to [49]. More precisely,
Peano gave the parametric representation of a curve defined recursively by a
sequence of functions $\varphi_n : [0, 1] \mapsto [0, 1] \times [0, 1]$ which, in the limit $n \to$
∞, goes through all points of the unit square $[0, 1] \times [0, 1]$. The parametric
representation represents a continuous curve whose Hausdorff dimension is,
however, equal to 2 (see pp 34–36 of [6]).

David Hilbert (1862–1943), who, in particular, is famous for the speech he
delivered to the Second International Congress of Mathematicians in Paris in
which he challenged mathematicians to solve fundamental problems, gave a
simple example of a Peano curve filling the unit square. The Hilbert curve is
defined by the L-system with a three-letter alphabet: {F, +, -}, respectively,
interpreted as "move forward", "turn 90 degrees to the left", and "turn 90
degrees to the right", and the rewriting rules: {''L'' → ''-RF+LFL+FR-'',
''R'' → ''+LF-RFR-FL+''}, where L and R are not letters but represent,
respectively, the following two words: +F-F-F+ and -F+F+F-.

```
HilbertCurve[n_Integer] :=
Module[{Lsystem = {''L'' → ''-RF+LFL+FR-'', ''R'' →
''+LF-RFR-FL+''} ,
axiom = ''R'', symbols, unit = 2.0^(-n), X = {0., 0.},
dir = {1., 0.}, iter = 1, c = 0.0, s = 1.0, turtle,
rewritingrules, ptList, f, l, r, turnLeft, turnRight},
symbols = {''F'' → f, ''+'' → l, ''-'' → r}; ptList =
Table[Null, {4^n}]; ptList[[1]] = X;
turnLeft = {{c, - s}, {s, c}};
turnRight = {{c, s}, {- s, c}};
Attributes[turtle] = Listable;
turtle[f] := ptList[[++iter]] = (X += unit dir);
turtle[l] := dir = turnLeft . dir;
turtle[r] := dir = turnRight . dir;
```

```
rewritingrules = Map[First[Characters[#[[1]]]] →
Characters[#[[2]]] &, Lsystem] /. symbols;
turtle[Nest[# /. rewritingrules &,
Characters[axiom] /. symbols, n]];
ptList = ptList /. {L → lfrfrfl, R → rflflfr}; image =
Show[Graphics[Line[ptList]],
AspectRatio → Automatic, Axes → None,
PlotRange → All]; image]
```

```
HilbertCurve[6];
```

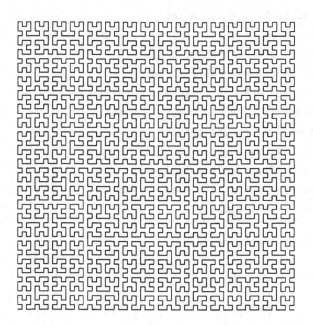

Fig. 23.3. *Sixth stage of the construction of the Hilbert curve.*

23.4 Peano Curve

The Peano curve can be generated following the same method. We should
consider the L-system with the three-letter alphabet: {F, +, -}, respectively,
interpreted as "move forward", "turn 90 degrees to the left", and "turn 90 de-

grees to the right", and the rewriting rules: ''F'' $\rightarrow$ ''F+F-F-F-F+F+F+F-F''.
Here is the program.

```
PeanoCurve[n_Integer] :=
Module[{Lsystem = {''F'' → ''F+F-F-F-F+F+F+F-F''},
axiom = ''F'', symbols, unit = 2.0^(-n), X = {0., 0.},
dir = {1., 0.}, iter = 1, c = 0.0, s = 1.0, turtle,
rewritingrules, ptList, f, l, r, turnLeft, turnRight},
symbols = {''F'' → f, ''+'' → l, ''-'' → r}; ptList =
Table[Null, {1+ 9^n}]; ptList[[1]] = X;
turnLeft = {{c, - s}, {s, c}};
turnRight = {{c, s}, {- s, c}};
Attributes[turtle] = Listable;
turtle[f] := ptList[[++iter]] = (X += unit dir);
turtle[l] := dir = turnLeft . dir;
turtle[r] := dir = turnRight . dir;
rewritingrules = Map[First[Characters[#[[1]]]] →
Characters[#[[2]]] &, Lsystem] /. symbols;
turtle[Nest[# /. rewritingrules &,
Characters[axiom] /. symbols, n]];
image = Show[Graphics[Line[ptList]],
AspectRatio → Automatic, Axes → None,
PlotRange → All]; image]
```

The first iteration generates the line

$$\{\{0, 0\}, \{\tfrac{1}{3}, 0\}, \{\tfrac{1}{3}, \tfrac{1}{3}\}, \{\tfrac{2}{3}, \tfrac{1}{3},\}, \{\tfrac{2}{3}, 0\},$$

$$\{\tfrac{1}{3}, 0\}, \{\tfrac{1}{3}, -\tfrac{1}{3}\}, \{\tfrac{2}{3}, -\tfrac{1}{3}\}, \{\tfrac{2}{3}, 0\}, \{1, 0\}\}$$

As shown below, the numbers from 0 (initial point) to 9 (last point) show the
itinerary along the line.

```
Peano1 = Graphics[Line[{{0.0, 0.0}, {0.333, 0.0},
{0.333, 0.333}, {0.666, 0.333}, {0.666, 0.0}, {0.333, 0.0},
{0.333, - 0.333}, {0.666, - 0.333}, {0.666, 0.0},
{1.0, 0.0}}]];
pts = {{0, 0.03}, {0.303, 0.03}, {0.363, 0.303},
{0.636, 0.303}, {0.636, 0.03}, {0.363, - 0.03},
{0.363, - 0.303},{0.636, - 0.303},{0.696, - 0.03},
{1.0, - 0.03}};
numPts = Graphics[{{PointSize[0.04], CMYKColor[0, 0, 1, 0],
Map[Point, pts]},
Table[Text[i-1, Part[pts,i]], {i, 1, Length[pts]}]}];
Show[{peano1, numPts}, AspectRatio → Automatic];
```

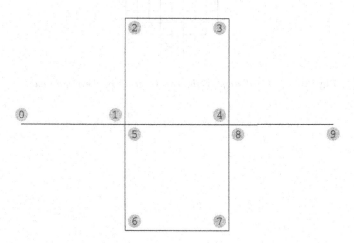

Fig. 23.4. *First stage of the construction of the Peano curve.*

```
PeanoCurve[3];
```

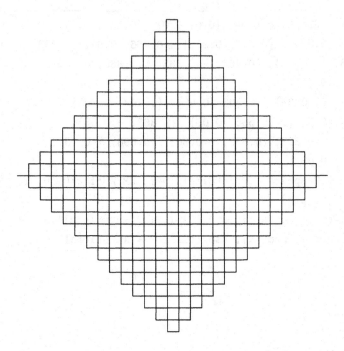

Fig. 23.5. *Third stage of the construction of the Peano curve.*

Logistic Map

The discrete logistic map is the simplest nonlinear map. Its properties are, however, far from being trivial. In 1962, Myrberg [39] already mentioned the existence of considerable difficulties:

<table>
<tr>
<td>En nous limitant dans notre travail au cas le plus simple non linéaire, c'est-à-dire aux polynômes réels du second degré, nous observons que même dans ce cas spécial on rencontre des difficultés considérables, dont l'explication exigera des recherches ultérieures.</td>
<td>Limiting ourselves to the simplest nonlinear case, that is, to quadratic real polynomials, it is observed that even in this special case considerable difficulties are encountered, whose explanation will require more work in the future.</td>
</tr>
</table>

The logistic map, described by the recurrence equation

$$n_{t+1} = f(n_t, r) = rn_t(1 - n_t),$$

can be viewed as the time-discrete evolution of a scaled population n. The word logistic was coined by Pierre Franois Verhulst [62] (1804–1849) who used the differential equation

$$\frac{dN}{dt} = rN\left(1 - \frac{N}{K}\right)$$

for the first time in 1838 to study population growth. The constant r is referred to as the *intrinsic rate of increase* and K is called the *carrying capacity* because it represents the population size that the resources of the environment can just maintain without a tendency to either increase or decrease. Although

Verhulst's paper did not draw much attention when it was first published, it was rediscovered 80 years later by Raymond Pearl (1879–1940) and Lowel J. Reed (1886–1966). After the publication of their paper [42], the logistic model began to be widely used.

Although the solution to the differential equation is trivial, we show that the recurrence equation is, on the contrary, of considerable interest.

24.1 Bifurcation Diagram

If $0 < r < 4$, all the iterates of n belong to the interval $[0, 1]$.

The recurrence equation has two fixed points given by

```
f[n_, r_] := r n (1 - n)
fxdPts = Solve[n == f[n,r], n]
```

$$\{\{n \to 0\}, \{n \to \frac{-1+r}{r}\}\}$$

The stability of these fixed points is determined by the absolute value of the derivative of $f(n, r)$ at these points. Because

```
Clear[r, n]
Abs[Simplify[D[f[n,r], n] /. n → fxdPts[[1,1,2]]]]
Abs[Simplify[D[f[n,r], n] /. n → fxdPts[[2,1,2]]]]
```

Abs[r]

Abs[2 - r]

if $r < 1$, the fixed point $n = 0$ is asymptotically stable and $n = (r - 1)/r$, which does not belong to the interval $[0, 1]$, is unstable; and for $1 < r < 3$, $n = 0$ is unstable whereas $n = (r - 1)/r$ is asymptotically stable.

In the case of one-dimensional maps, such as the logistic map, there exists a simple graphical method to follow the successive iterates of an initial point n_0. First plot the graphs of $n \mapsto f(n, r)$ and $n \mapsto n$. Because the sequence of iterates is generated by the equation $n_{t+1} = f(n_t, r)$, the iterate of the initial value n_0 is on the graph of f at $(n_0, f(n_0, r))$, that is, (n_0, n_1). The horizontal line from this point intersects the diagonal at (n_1, n_1). The vertical line from this point intersects the graph of f at $(n_1, f(n_1, r))$, that is, (n_1, n_2). Repeating this process generates the sequence $((n_0, n_1), (n_1, n_1), (n_1, n_2), (n_2, n_2), \dots)$. The equilibrium point n_* is located at the intersection of the graphs of the two functions. The diagram below that consists of the graphs of the functions $n \mapsto f(n, r)$ and $n \mapsto n$ and the line joining the points of the sequence above is called a *cobweb*. It clearly shows whether the sequence of iterates of the initial point n_0 converges to the equilibrium point.

In order to draw the cobweb given a parameter value r, an initial value n_0, and a maximun number of iterations, we first determine the sequence $((n_0, n_1), (n_1, n_1), (n_1, n_2), (n_2, n_2), \dots)$, then plot the function f for the specific r value, and draw the line (in green) going through all the points of the sequence, and the diagonal joining the origin to the point $(1, 1)$. The initial and final points are respectively colored in blue and red.

```
logisticCobweb[r_, init_, numIter_] :=
Module[{F, seq, pl, l, image},
F[n_] := r n (1 - n);
n = init; seq = {{n, 0}, {n, F[n]}};
For[t = 1, t <= numIter, t++,
n = F[n];
seq = Append[seq, {n, n}];
seq = Append[seq, {n, F[n]}]];
seq = Append[seq, {F[n], F[n]}];
pl = Plot[F[n], {n, 0, 1}, DisplayFunction -> Identity];
l = Graphics[{{RGBColor[0, 1, 0], Line[seq]},
Line[{{0, 0}, {1, 1}}]},
{RGBColor[0, 0, 1], PointSize[0. 03], Point[First[seq]]},
{RGBColor[1, 0, 0], PointSize[0.03], Point[Last[seq]]}}];
image = Show[pl,l, Frame -> True, PlotRange -> All,
AspectRatio -> Automatic,
TextStyle -> {FontSlant -> ''Italic'', FontSize -> 12},
DisplayFunction -> $DisplayFunction]; image]
```

```
logisticCobweb[2.6, 0.9, 15];
```

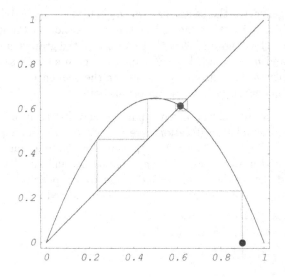

Fig. 24.1. *Logistic map cobweb for $r = 2.6$, $n_0 = 0.9$, and a number of iterations equal to 15.*

Because

```
Clear[r]
{Nest[2.6 # (1 - #) &, 0.9, 15], (r - 1) / r /. r → 2.6}
```

$\{0.615458, 0.615385\}$

after a few iterations, we are quite close to the fixed point.

For $r = 3$, the derivative of f at $n = (r - 1)/r = \frac{2}{3}$ is equal to -1. Because

```
Clear[r, n]
f[f[n, r], r]
```

$(1 - n) \, n \, r^2 \, (1 - (1 - n) \, n \, r)$

we verify that

```
f[f[2 / 3, 3], 3]
D[f[f[n, r], r], n] /. {n → 2/3, r → 3}
```

$$\frac{2}{3}$$

1

and

```
D[f[f[n, r], r], r] /. {n → 2 / 3, r → 3}
D[f[f[n, r], r], {n, 2}] /. {n → 2 / 3, r → 3}
D[D[f[f[n, r], r], n], r] /. {n → 2 / 3, r → 3}
D[f[f[n, r], r], {n, 3}] /. {n → 2 / 3, r → 3}
```

0

0

2

− 108

These results show that the nonhyperbolic fixed point $(\frac{2}{3}, 3)$ is a period-doubling bifurcation point. On bifurcation theory see [9].

The map $f^2(n, r) = f(f(n, r), r)$ has four fixed points that are the solutions of the equation $f^2(n, r) = n$.

```
Clear[n, r]
fxdPtsf2 = Solve[f[f[n, r], r] == n, n] // Simplify
```

$$\{\{n \to 0\}, \{n \to \frac{-1+r}{r}\}, \{n \to \frac{1+r-\mathrm{Sqrt}[-3-2\ r+r^2]}{2\ r}\},$$
$$\{n \to \frac{1+r+\mathrm{Sqrt}[-3-2\ r+r^2]}{2\ r}\}\}$$

Two solutions are already known, namely the two unstable fixed points, $n = 0$ and $n = (r - 1)/r$ of f. The remaining two solutions, denoted $n_{1,1}$ and $n_{1,2}$, are such that

```
Simplify[f[fxdPtsf2[[3, 1, 2]], r]] == fxdPtsf2[[4, 1, 2]]
Simplify[f[fxdPtsf2[[4, 1, 2]], r]] == fxdPtsf2[[3, 1, 2]]
```

True

True

Thus, $f(n_{1,1}) = n_{1,2}$ and $f(n_{1,2}) = n_{1,1}$, which shows that these two points are the components of a two-point cycle. They are defined only for $r \geq 3$. The domain of stability of this cycle is determined by the condition

$$\left| \frac{\partial f}{\partial n}\left(\frac{1}{2r}(1+r+\sqrt{r^2-2r-3}), r \right) \frac{\partial f}{\partial n}\left(\frac{1}{2r}(1+r-\sqrt{r^2-2r-3}), r \right) \right| < 1;$$

that is,

```
(Simplify[D[f[n, r], n] /. n → fxdPtsf2[[3, 1, 2]]] *
Simplify[D[f[n, r], n] /. n → fxdPtsf2[[4, 1, 2]]])
// Simplify
```

$$4 + 2\ r - r^2$$

```
Reduce[Abs[4 + 2 r - r^2] < 1, r, Reals]
```

$$1 - \text{Sqrt}[6] < r < -1 \;||\; 3 < r < 1 + \text{Sqrt}[6]$$

The parameter r being positive, the two-point cycle is stable for $3 < r < 1 + \sqrt{6}$.

Remark. Because the two fixed points n_1 and n_2 of the map $f(n, r)$ are the solutions of the quadratic equation $f(n, r) - n = 0$, and the four fixed points n_1, n_2, $n_{1,1}$ and $n_{1,2}$ of the map $f^2(n, r) = f(f(n, r), r)$ are the solutions of the quartic equation $f^2(n, r) = n$, the fixed points $n_{1,1}$ and $n_{1,2}$ should be the two solutions of the quadratic equation

$$\frac{f^2(n, r) - n}{f(n, r) - n} = 0,$$

which is readily verified using *Mathematica*.

```
eqn = Simplify[(f[f[n, r], r] - n) / (f[n, r] - n)] == 0
```

$$1 + r - n\ r + (-1 + n)\ n\ r^2 == 0$$

```
Solve[eqn, n] // Simplify
```

$$\{\{n \;\rightarrow\; \frac{1 + r - \text{Sqrt}[-3 - 2\ r + r^2]}{2\ r}\},$$

$$\{n \rightarrow \frac{1 + r + \text{Sqrt}[-3 - 2\ r + r^2]}{2\ r}\}\}$$

The bifurcation point $(n, r) = (\frac{2}{3}, 3)$ is therefore determined by equating to zero the discriminant of this quadratic equation. We have

```
Solve[(r + r^2)^2 - 4 (1+r) r^2 == 0, r]
```

$$\{\{r \rightarrow -1\}, \{r \rightarrow 0\}, \{r \rightarrow 0\}, \{r \rightarrow 3\}\}$$

Only $r = 3$ is an acceptable solution, and reporting this value in the expression of the fixed point $n = (r - 1)/r$ of the map $f(n, r)$ we verify that $n = \frac{2}{3}$.

For $r > 1 + \sqrt{6}$, the system undergoes an infinite sequence of period-doubling bifurcations. After 100 iterates have been discarded, 16 iterations of the logistic map for increasing values of r are shown below.

```
Clear[F, r]
r = 2.6;
initialPoint = 0.8;
F[n_] := f[n, r];
seq1 = NestList[F, Nest[F, initialPoint, 100], 16];
pl1 = Show[Graphics[{{RGBColor[1, 0, 0], PointSize[0.02],
Map[Point, Table[{k, seq1[[k]]}, {k,1,16}]]},
Line[Table[{k, seq1[[k]]}, {k, 1, 16}]]}], Frame → True,
FrameTicks → {{0, 4, 8, 12, 16}, Automatic, {}, {}},
DisplayFunction → Identity];

Clear[F,r]
r = 3.23;
initialPoint = 0.8;
F[n_] := f[n, r];
seq2 = NestList[F, Nest[F, initialPoint, 100], 16];
pl2 = Show[Graphics[{{RGBColor[1, 0, 0], PointSize[0.02],
Map[Point,Table[{k, seq2[[k]]}, {k, 1, 16}]]},
Line[Table[{k, seq2[[k]]}, {k,1,16}]]}], Frame → True,
FrameTicks → {{0, 4, 8, 12, 16}, Automatic, {},{}},
DisplayFunction → Identity];
```

```
Clear[F,r]
r = 3.49;
initialPoint = 0.8;
F[n_] := f[n, r];
seq3 = NestList[F, Nest[F,initialPoint, 100], 16];
pl3 = Show[Graphics[{{RGBColor[1, 0, 0], PointSize[0.02],
Map[Point,Table[{k, seq3[[k]]}, {k, 1, 16}]]},
Line[Table[{k, seq3[[k]]}, {k, 1,16}]]}], Frame → True,
FrameTicks → {{0, 4, 8, 12, 16}, Automatic, {}, {}},
DisplayFunction → Identity];

Clear[F,r]
r = 3.554;
initialPoint = 0.8;
F[n_] := f[n, r];
seq4 = NestList[F, Nest[F,initialPoint, 100], 16];
pl4 = Show[Graphics[{{RGBColor[1, 0, 0], PointSize[0.02],
Map[Point,Table[{k, seq4[[k]]}, {k, 1, 16}]]},
Line[Table[{k,seq4[[k]]}, {k, 1, 16}]]}], Frame → True,
FrameTicks → {{0, 4, 8, 12, 16}, Automatic, {}, {}},
DisplayFunction → Identity];
```

Automatic represents an option value that is to be chosen automatically by a built-in function, here FrameTicks.

```
Show[GraphicsArray[{{pl1, pl2}, {pl3, pl4}}],
DisplayFunction → $DisplayFunction];
```

See output in Figure 24.2.

Let $(r_k)_{k \in \mathbb{N}}$ be the sequence of parameter values at which a period-doubling bifurcation occurs. This sequence is such that the 2^k-point cycle is stable for $r_k < r < r_{k+1}$. Then, if $\{n_{k,1}, n_{k,2}, \ldots, n_{k,2^k}\}$ denotes the 2^k-point cycle, for $i = 1, 2, \ldots, 2^k$ and $r_k < r < r_{k+1}$, we have

$$f^{2^k}(n_{k,i}, r) = n_{k,i}.$$

Moreover, for $i = 1, 2, \ldots, 2^k$,

$$\frac{\partial f^{2^k}}{\partial n}(n_{k,i}, r_k) = +1, \quad \frac{\partial f^{2^k}}{\partial n}(n_{k,i}, r_{k+1}) = -1,$$

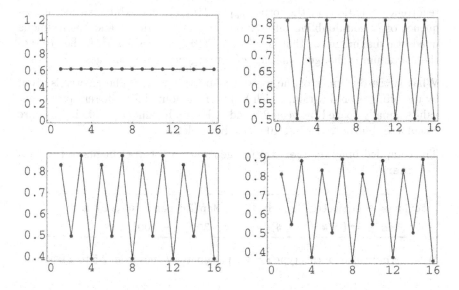

Fig. 24.2. *Sixteen iterations of the logistic map for $r = 2.3$ (fixed point), $r = 3.23$ (period 2), $r = 3.49$ (period 4), and $r = 3.554$ (period 8).*

where numerical values of the $n_{k,i}$ $(i = 1, 2, \ldots, 2^k)$ depend on r.

The first period-doubling bifurcations occur for the following parameter values:

$$r_1 = 3.0 \qquad r_2 = 3.449499 \ldots \quad r_3 = 3.544090 \ldots \quad r_4 = 3.564407 \ldots$$
$$r_5 = 3.568759 \ldots \quad r_6 = 3.569692 \ldots \quad r_7 = 3.569891 \ldots \quad r_8 = 3.569934 \ldots$$

The sequence (r_k) is an increasing bounded sequence of positive numbers. It has, therefore, a limit r_∞ that is found equal to $3.5699456\ldots$. Feigenbaum [14, 15] discovered that the asymptotic behavior of r_k is of the form

$$r_k \sim r_\infty - \frac{a}{\delta^k},$$

where a and δ are two positive numbers. This behavior implies that δ, known as the *Feigenbaum number*, is such that

$$\lim_{k \to \infty} \frac{r_k - r_{k-1}}{r_{k+1} - r_k} = \delta.$$

The interesting fact, found by Feigenbaum, is that the rate of convergence δ of the sequence (r_k) is universal in the sense that it is the same for all recurrence equations of the form $n_{t+1} = f(n_t, r)$ that exhibit an infinite sequence of period-doubling bifurcations, if f is continuous and has a unique quadratic maximum n_c; that is, $f(n, r) - f(n_c, r) \sim (n - n_c)^2$. If the order of the maximum is changed, the rate of convergence δ also changes.

What is the dynamics of the logistic map for $r > r_\infty$? The answer is given by the bifurcation diagram, which has been computed for different parameter values equally spaced between 2.5 and 4. For each value of r, 300 iterates are calculated, but only the last 100 have been plotted.

The compiled function `iterates[r]` generates a list of elements of the form {r, iterate}.

```
iterates = Compile[{{r, _Real}}, Map[Prepend[{#}, r] &,
Take[NestList[r # (1 - #) &, 0.4, 300], - 100]]];
```

We determine lists of this type for 301 parameter values from 2.5 to 4.0 equally spaced.

```
ptsList = Flatten[Table[iterates[r], {r, 2.5, 4.0, 0.005}],
1];
```

Using `ListPlot` we finally obtain the bifurcation diagram.

```
ListPlot[ptsList, PlotStyle → {RGBColor[0, 0, 1],
PointSize[0.002]}, Frame →True];
```

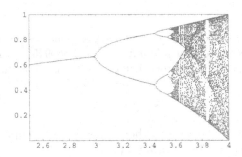

Fig. 24.3. *Bifurcation diagram of the logistic map* $(n, r) \mapsto rn(1-n)$. *The parameter r, plotted on the horizontal axis, varies from 2.5 to 4, and the reduced population n, plotted on the vertical axis, varies between 0 and 1.*

For $r = 4$, the iterates seem to wander in the interval $[0, 1]$ as also shown by the following cobweb in which the initial point is defined with 200 significant digits.

```
logisticCobweb[4, N[Sqrt[3] -1, 200], 300];
```

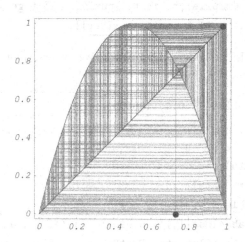

Fig. 24.4. *Logistic map cobweb for $r = 4$, $n_0 = \sqrt{3} - 1$, and a number of iterations equal to 300. The initial point is defined with 200 significant digits.*

For $r = 4$, the trajectory appears to be dense in the interval $[0, 1]$. If I is an interval of $\mathbb{R}$, a subset J of I is *dense* in I if the closure of J coincides with I. In other words, any neighborhood of any point in I contains points in J. For example, the set of rational numbers $\mathbb{Q}$ is dense in $\mathbb{R}$.

It is clear that a numerical experiment cannot, of course, determine whether the trajectory for $r = 4$ converges to an asymptotically stable periodic orbit of very high period or is dense in an interval. A measure μ on an interval I of $\mathbb{R}$ is emphinvariant for the map f defined on I if, for any measurable subset $E \subset I$, $\mu(E) = \mu(f^{-1}(E))$.

If the trajectory is dense, say in $[0, 1]$, we should be able to determine an approximate cumulative distribution function F from a list of iterates. Given a random variable X, its *cumulative distribution function* is a nondecreasing function F_X such that, for any x, $F_X(x)$ is the probability for X to be less than or equal to x.

In order to determine numerically the cumulative distribution function of a random sample: data $= \{x_1, x_2, \ldots\}$, we plot ListPlot(abscissa, ordinate), where abscissa = Sort[data] and ordinate $= \{1/n, 2/n, \cdots, n/n\}$. Note that the cumulative distribution function determined in this way gives the maximum information contained in the numerical **data**. In particular, it gives more information than traditional histograms.

In order to determine a list of iterates, we first define the following compiled function.

```
iterateList = Compile[{r, init, {numIter, _Integer}},
NestList[r # (1 - #) &, init, numIter]];
```

And, following the procedure described above, we can plot an approximate cumulative distribution function for $r = 4$.

```
data = iterateList[4, 0.4, 100000];
```

```
Clear[n]
n = Length[data];
abscissa = Sort[data];
ordinate = Table[k/n, {k, 1, n}];
pts = Table[{abscissa[[j]], ordinate[[j]]}, {j, 1, n}];
logisticCDF = ListPlot[pts, PlotStyle → {RGBColor[1, 0, 0],
PointSize[0.005]}, TextStyle → {FontSlant → ''Italic'',
FontSize → 12}, Frame → True];
```

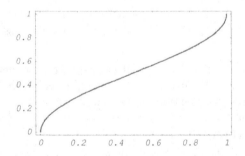

Fig. 24.5. *Approximate cumulative distribution function for the logistic map* $n \mapsto 4n(1-n)$.

As expected for a cumulative distribution function F, this plot shows that $F(0) = 0$ and $F(1) = 1$. Moreover, the derivative of F, if it exists, seems to be infinite at 0 and 1 (see below).

24.2 Exact Dynamics for $r = 4$

24.2.1 Conjugacy and Periodic Orbits

Two maps f and g are said to be *conjugate* if there exists a homeomorphism h such that $h \circ f = g \circ h$, or $h \circ f \circ h^{-1} = g$. This relation implies that, for all $t \in \mathbb{N}$, $h \circ f^t \circ h^{-1} = g^t$. A *homeomorphism* is a continuous bijection having a continuous inverse. The homeomorphism h takes the trajectories of f into the trajectories of g. The logistic map f_4, that is, $n \mapsto f(n, 4)$, and the tent map T_2 defined by

$$T_2(x) = \begin{cases} 2x, & \text{if } 0 \le x < \frac{1}{2}, \\ 2 - 2x, & \text{if } \frac{1}{2} \le x \le 1, \end{cases}$$

are conjugate. To prove these results, it suffices to show that the function $h : x \mapsto \sin^2\left(\frac{\pi}{2}x\right)$ is a homeomorphism such that $h \circ T_2 = f_4 \circ T_2$. We verify that:

1. $h : [0,1] \mapsto [0,1]$ is continuous;

2. $h(x_1) = h(x_2) \Longrightarrow x_1 = x_2$;

3. $h([0,1]) = [0,1]$;

4. h' exists and is continuous, so $h^{-1} : [0,1] \mapsto [0,1]$ exists and is continuous;

5. $(f_4 \circ h)(x) = 4\sin^2\left(\frac{\pi}{2}x\right)\left(1 - \sin^2\left(\frac{\pi}{2}x\right)\right) = \sin^2(\pi x) = (h \circ T_2)(x)$.

This property greatly simplifies the study of the dynamics of f_4.

Let $0.x_1 x_2 x_3 \ldots$ be the binary representation of $x \in [0,1]$, that is,

$$x = \sum_{i=1}^{\infty} \frac{x_i}{2^i},$$

where, for all $i \in \mathbb{N}$, $x_i \in \{0,1\}$. The binary representation of $T_2(x)$ is then given by

$$T_2(x) = \begin{cases} 0.x_2 x_3 x_4 \ldots, & \text{if } 0 \le x \le \frac{1}{2}, \\ 0.(1-x_2)(1-x_3)(1-x_4)\ldots, & \text{if } \frac{1}{2} \le x \le 1. \end{cases}$$

These formulas are both correct for $x = \frac{1}{2}$. The binary representation of $\frac{1}{2}$ being either $0.1000\ldots$ or $0.01111\ldots$, the binary representation of $T_2(x)$ is, in both cases, $0.1111\ldots$, which is equal to 1.

The description of the iterates of x in terms of their binary representation leads to some remarkable results due to James Whittaker [65]

1. If the binary representation of x is finite (i.e., if there exists a positive integer n such that $x_i = 0$ for all $i > n$), then after, at most $n + 1$ iterations, the orbit of x will reach 0 and stay there. Hence, *there exists a dense set of points whose orbit reaches the origin and stays there.*

```
T2[x_] := Which[0   x < 1/2, 2 x, 1/2   x   1, 2 - 2 x]
NestList[T2, 1 / 32, 10]
```

$$\{\frac{1}{32}, \frac{1}{16}, \frac{1}{8}, \frac{1}{4}, \frac{1}{2}, 1, 0, 0, 0, 0, 0\}$$

2. *If the binary representation of x is periodic with period p, then the orbit of x is periodic with a period equal to p or a divisor of p.* For example, if x is equal to $0.001110011100111\ldots = 7/31$,

```
Sum[(1 / 2)^(3 + 5 k) + (1 / 2)^(4 + 5 k) + (1 / 2)^(5 + 5k),
{k, 0, Infinity}]
```

$$\frac{7}{31}$$

we verify that

```
NestList[T2, 7 / 31, 16]
```

$$\frac{7}{31}, \frac{14}{31}, \frac{28}{31}, \frac{6}{31}, \frac{12}{31}, \frac{24}{31}, \frac{14}{31}, \frac{28}{31}, \frac{6}{31}, \frac{12}{31}, \frac{24}{31}, \frac{14}{31},$$
$$\frac{28}{31}, \frac{6}{31}, \frac{12}{31}, \frac{24}{31}, \frac{14}{31}$$

Hence

```
T2[7 / 31] == Nest[T2, 7 / 31, 6] == Nest[T2, 7 / 31, 11]
```

True

Because the binary representation of $7/31$ is periodic with period 5, its trajectory is periodic with period 5.

Using this method, we could show that the binary tent map has periodic orbits of all periods, and the set of all periodic points is dense in $[0.1]$. T_2 and f_4 being conjugate, the logistic map f_4 has the same property.

Regarding the existence of a periodic orbit, an amazing theorem, due to Šarkovskii, indicates which periods imply which other periods. First, define among all positive integers Šarkovskii's order relation by

$$3 \rhd 5 \rhd 7 \rhd \cdots 2 \cdot 3 \rhd 2 \cdot 5 \rhd \cdots \rhd 2^2 \cdot 3 \rhd 2^2 \cdot 5 \rhd \cdots$$
$$\rhd 2^3 \cdot 3 \rhd 2^3 \cdot 5 \rhd \cdots\cdots \rhd 2^3 \rhd 2^2 \rhd 2 \rhd 1.$$

That is, first list all the odd numbers, followed by 2 times the odd numbers, 2^2 times the odd numbers, and so on. This exhausts all the positive integers except the powers of 2 that are listed last in decreasing order. Because $\rhd$ is an order relation, it is transitive (i.e., $n_1 \rhd n_2$ and $n_2 \rhd n_3$ imply $n_1 \rhd n_3$). Šarkovskii's theorem is as follows.

Šarkovskii's Theorem *Let $f : \mathbb{R} \to \mathbb{R}$ be a continuous map. If f has a periodic orbit of period n, then, for all integers k such that $n \rhd k$, f has also a periodic orbit of period k.*

For a proof, see Štefan [54] or Collet and Eckmann [12].

We said above that the binary tent map T_2 and its conjugate, the logistic map f_4, have periodic orbits of all periods. Following Šarkovskii's theorem, to prove this result it suffices to prove that either T_2 or f_4 has a periodic orbit of period 3.

In order to generate a period-3 orbit for f_4 we can start from $x = 0.100100100\ldots$, which generates a period-3 orbit for the binary tent map. Because

```
x0 = Sum[1 / 2^(1+ 3k), {k, 0, Infinity}]
```

$$\frac{4}{7}$$

in order to generate the corresponding period-3 orbit for f_4, we have to start from the point $h(x_0) = \sin^2(2\pi/7)$.

All these periodic orbits are unstable. Hence, periodic orbits computed with a finite precision will always present, after a number of iterations depending upon the precision, an erratic behavior. This feature is illustrated in the plots below using a precision either equal to $MachinePrecision, which is approximately 16, or to 100.

```
initialPoint = (Sin[2 Pi / 7])^2;
f4[n_] := 4 n (1- n)
seq1 = NestList[f4, N[initialPoint], 100];
pl1 = Show[Graphics[{{RGBColor[1, 0, 0], PointSize[0.02],
Map[Point,Table[{k, seq1[[k]]}, {k, 1, Length[seq1]}]]},
Line[Table[{k, seq1[[k]]}, {k, 1, Length[seq1]}]]}],
TextStyle -> {FontSlant -> ''Italic'', FontSize -> 12},
Frame -> True, FrameTicks ->
{{0, 20, 40, 60, 80, 100}, {0.2, 0.4, 0.6, 0.8, 1},{},{}}];
```

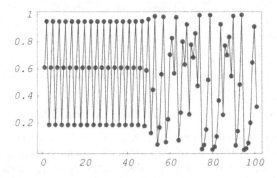

Fig. 24.6. *One hundred iterates of the logistic map* $n \mapsto 4n(1 - n)$ *starting from* $n_0 = sin^2(2\pi/7)$, *defined with* **$MachinePrecision**, *showing the instability of the period-3 point.*

```
initialPoint = (Sin[2 Pi / 7])^2;
f4[n_] := 4 n (1- n)
seq1 = NestList[f4, N[initialPoint, 70], 100];
pl1 = Show[Graphics[{{RGBColor[1, 0, 0], PointSize[0.02],
Map[Point,Table[{k, seq1[[k]]}, {k, 1, Length[seq2]}]]},
Line[Table[{k, seq1[[k]]}, {k, 1, Length[seq2]}]]}],
TextStyle -> {FontSlant -> ''Italic'', FontSize -> 12},
Frame -> True, FrameTicks ->
{{0, 20, 40, 60, 80, 100}, {0.2, 0.4, 0.6, 0.8, 1},{},{}}];
```

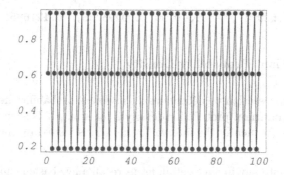

Fig. 24.7. *One hundred iterates of the logistic map $n \mapsto 4n(1-n)$, starting from $n_0 = sin^2(2\pi/7)$ defined with 70 significant digits.*

24.2.2 Exact Solution of the Recurrence Equation

Mathematica can, in some cases, solve recurrence equations using the command RSolve.

```
RSolve[{n[t+1] == 4 n[t] (1-n[t]), n[0] == n0}, n[t], t]
```

Solve: :ifun : Inverse functions are being used by Solve,

so some solutions may not be found; use

Reduce for complete solution

information. More

$$\{\{n[t] \rightarrow \frac{1 - Cos[2^t \, ArcCos[1 - 2 \, n0]]}{2}$$

This exact result shows that the map f_4 has sensitive dependence on initial conditions, which, according to Devaney [13], is one of the necessary conditions for a map to be chaotic [9]. The following results show how a very small increase of the initial value n_0 (from $\sqrt{3} - 1$ to $\sqrt{3} - 1 + 10^{-4}$) modifies the value of n_t after $10, 20, \ldots, 100$ iterations.

```
N[Table[(1- Cos[2^t ArcCos[1 - 2(Sqrt[3] - 1) -
2 / 10^4]]) /2 - (1 - Cos[2^t ArcCos[1 - 2(Sqrt[3] - 1)]]) /
2,
{t, 10, 100, 10}]]
```

$\{- \, 0.102199, \, - \, 0.459289, \, 0.287665, \, 0.906013, \, - \, 0.819842,$

0.154034, 0.520761, $-$ 0.310196, 0.0295783, -0.121685}

24.2.3 Invariant Probability Density

To conclude this section we determine the invariant probability density for the map f_4. The invariant probability density ρ for a map f is such that $\rho(x)\,dx$ measures how frequently the interval $[x, x + dx]$ is visited by the dense orbit of a point x_0.

Before going into any further detail, let us recall a few basic results of ergodic theory (for more details see Shields [51]).

Let $f : S \to S$ be a map. A subset A of S is invariant if $f(A) = A$. A measure μ is invariant for the map f if, for all measurable subsets A of S, $\mu(f^{-1}(A)) = \mu(A)$. The map f is ergodic with respect to the invariant measure μ if any measurable invariant subset A of S is such that either $\mu(A) = 0$ or $\mu(A) = \mu(S)$.

If f is ergodic with respect to the invariant probability measure μ (μ is a probability measure if $\mu(S) = 1$), and $g : S \to \mathbb{R}$, an integrable function with respect to μ, then, for almost all $x_0 \in S$,

$$\lim_{t \to \infty} \frac{1}{t} \sum_{i=1}^{t} g \circ f^i(\mathbf{x_0}) = \int g(\mathbf{x})\,d\mu(\mathbf{x}).$$

That is, the time average is equal to the space average. If there exists a positive real function ρ such that $d\mu(x) = \rho(x)\,dx$, ρ is called an *invariant probability density*.

If the map $f : [0,1] \to [0,1]$ is such that any point x has k preimages $y_1, y_2, \ldots, y_k$ by f (i.e., for all $i = 1, 2, \ldots, k$, $f(y_i) = x$), the probability of finding an iterate of f in the interval $[x, x + dx]$ is then the sum of the probabilities of finding its k preimages in the intervals $[y_i, y_i + dy_i]$. Hence, from the relation

$$\rho(x)\,dx = \sum_{i=1}^{k} \rho(y_i)\,dy_i,$$

we obtain the Perron–Frobenius equation:

$$\rho(x) = \sum_{i=1}^{k} \frac{\rho(y_i)}{|f'(y_i)|}, \tag{24.1}$$

where we have taken into account that

$$\frac{dx}{dy_i} = f'(y_i).$$

In the case of the binary tent map T_2, the Perron–Frobenius equation reads

$$\rho(x) = \tfrac{1}{2} \left(\rho \left(\tfrac{1}{2} x \right) + \rho \left(1 - \tfrac{1}{2} x \right) \right).$$

This equation has the obvious solution $\rho(x) = 1$. That is, the map T_2 preserves the Lebesgue measure. If $I_1 \cup I_2$ is the preimage by T_2 of an open interval I of $[0, 1]$, we verify that

$$m(I_1 \cup I_2) = m(I_1) + m(I_2) = m(I),$$

where m denotes the Lebesgue measure. Using the relation

$$f_4 \circ h = h \circ T_2,$$

where $h : x \mapsto \sin^2(\pi/2\, x)$, the density ρ of the invariant probability measure for the map f is given by

$$\rho(x)\, dx = \frac{dh^{-1}}{dx}\, dx \tag{24.2}$$

$$= \frac{dx}{\pi \sqrt{x(1 - x)}}, \tag{24.3}$$

which can be checked using *Mathematica*

```
hInverse[x_] := 2 ArcSin[Sqrt[x]] / Pi;
D[hInverse[x], x]
```

$$\frac{1}{\text{Pi Sqrt}[1 - x]\ \text{Sqrt}[x]}$$

The graph of the invariant probability density ρ is

```
rho[x_] := (1 / Pi) 1 / Sqrt[x (1 - x)]
Plot[rho[x], {x, 0, 1}, ]
TextStyle → {FontSlant → ''Italic'',
FontSize → 12}, Frame → True];
```

As suggested by the numerical data used to plot logisticCDF, we verify that the invariant probability density function is infinite at $x = 0$ and $x = 1$.

The cumulative distribution function F is the integral from 0 to x; that is,

```
F[x_] := Assuming[0 < x < 1,
Integrate[(1 / Pi) 1 / Sqrt[u (1 - u)], {u, 0, x}]]
F[x]
```

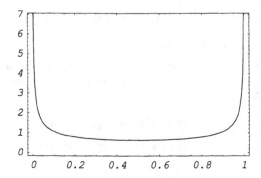

Fig. 24.8. *Invariant probability density of the logistic map* $n \mapsto 4n(1 - n)$.

$$\frac{2\ \text{ArcSin}[\text{Sqrt}[x]]}{\text{Pi}}$$

```
plF = Plot[F[x], {x, 0, 1},PlotStyle → {RGBColor[0, 0,1]},
TextStyle → {FontSlant → ''Italic'',
FontSize → 12}, Frame → True];
```

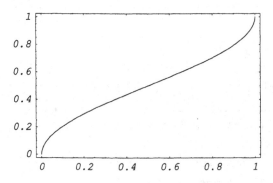

Fig. 24.9. *Invariant cumulative distribution function of the logistic map* $n \mapsto 4n(1-n)$.

This plot is very similar to the `ListPlot` of the numerical data (called `logisticCDF`) obtained above. The following plot, in which the numerical CDF (in red) and the exact one (in blue) are represented shows a very good agreement between the two results.

```
Show[{logisticCDF, plF}];
```

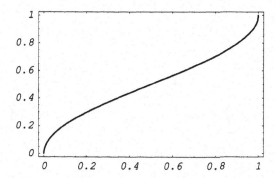

Fig. 24.10. *Comparing the exact invariant cumulative distribution function (in blue) with the approximate one (in red) obtained above. The two curves cannot be distinguished.*

Lorenz Equations

In his historical paper, published in 1963, Lorenz [31] derived, from a model of fluid convection, a three-parameter family of three ordinary differential equations that appeared, when integrated numerically, to have extremely complicated solutions. In particular, he discovered that all nonperiodic solutions of his deterministic model were bounded but showed irregular fluctuations. Thirty years later [32] he described how he

> had come across a phenomenon that later came to be called "chaos"—
> —seemingly random and unpredictable behavior that nevertheless
> proceeds according to precise and often easily expressed rules.

The study of chaos can be traced back to Henri Poincaré (1854–1912). In *Science et Méthode*, first published in 1909, he already indicates the possibility for certain systems to be subject to sensitive dependence on initial conditions [43].

The Lorenz equations are

$$\frac{dx}{dt} = \sigma(y - x),$$
$$\frac{dy}{dt} = rx - y - xz,$$
$$\frac{dz}{dt} = xy - bz,$$

where σ, r, and b are real positive parameters. The system is invariant under the transformation $(x, y, z) \rightarrow (-x, -y, z)$. Although this system is rather complicated, we can use NDSolve to solve it numerically.

NDSolve follows the general procedure of reducing step size until it tracks solutions accurately. When solutions have a complicated structure, however, occasionally larger settings may need to be chosen for MaxSteps. With the setting `MaxSteps → Infinity` there is no upper limit on the number of steps used. Chosing $\sigma = 10$, $r = 28$, and $b = 8/3$ we have

```
sol = NDSolve[ {x'[t] == 10 (y[t] - x[t]),
y'[t] == 28 x[t] - y[t] - x[t] z[t],
z'[t] == x[t] y[t] - (8/3) z[t],
x[0] == z[0] == 0, y[0] == 1},
{x, y, z}, {t, 0, 40}, MaxSteps → Infinity]
```

```
{{x → InterpolatingFunction[{{0., 40.}}, <>],
y → InterpolatingFunction[{{0., 40.}}, <>],
z → InterpolatingFunction[{{0., 40.}}, <>]}}
```

We can view the projection of the three-dimensional trajectory on the planes xOy, yOz, and xOz.

```
LorenzXY = ParametricPlot[Evaluate[{x[t], y[t]} /. sol],
1mm] {t, 0, 40}, Frame → True, AspectRatio → 1,
TextStyle → {FontSlant → ''Italic'', FontSize → 12},
Axes → None, FrameLabel → {x, y}, PlotRange → All,
PlotPoints → 1000];
```

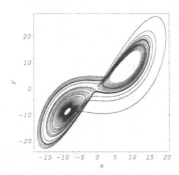

Fig. 25.1. *Projection on the xOy-plane of a numerical solution of the Lorenz equations for $t \in [0, 40]$ and $(x_0, y_0, z_0) = (0, 0, 1)$.*

```
LorenzYZ = ParametricPlot[Evaluate[{y[t], z[t]} /. sol],
1mm] {t, 0, 40}, Frame → True, AspectRatio → 1,
TextStyle → {FontSlant → ''Italic'', FontSize → 12},
Axes → None, FrameLabel → {y, z}, PlotRange → All,
PlotPoints → 1000];
```

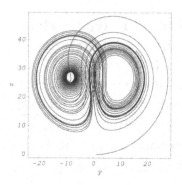

Fig. 25.2. *Projection on the yOz-plane of a numerical solution of the Lorenz equations for* $t \in [0, 40]$ *and* $(x_0, y_0, z_0) = (0, 0, 1)$.

```
LorenzXZ = ParametricPlot[Evaluate[{x[t], z[t]} /. sol],
1mm] {t, 0, 40}, Frame → True, AspectRatio → 1,
TextStyle → {FontSlant → ''Italic'', FontSize → 12},
Axes → None, FrameLabel → {x, z}, PlotRange → All,
PlotPoints → 1000];
```

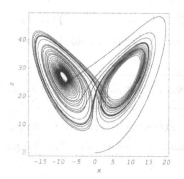

Fig. 25.3. *Projection on the xOz-plane of a numerical solution of the Lorenz equations for* $t \in [0, 40]$ *and* $(x_0, y_0, z_0) = (0, 0, 1)$.

The orbit is obviously not periodic. As t increases, the orbit winds first around the unstable fixed point $(x_1, y_1, z_1) = (-8.45528\ldots, -8.45528\ldots, 27)$ and then around the unstable fixed point $(x_1, y_1, z_1) = (8.45528\ldots, 8.45528\ldots, 27)$ without ever settling down. Its shape does not depend upon a particular choice of the initial conditions.

If the initial point (x_0, y_0, z_0) is either the first or the second unstable point, the orbit turns around this point slowly moving away.

```
sol1 = NDSolve[ {x'[t] == 10 (y[t] - x[t]),
y'[t] == 28 x[t] - y[t] - x[t] z[t],
z'[t] == x[t] y[t] - (8/3) z[t],
x[0] == - 8.45528, y[0] == - 8.45528, z[0] == 27},
{x, y, z}, {t, 0, 40}, MaxSteps → Infinity]
```

```
{{x → InterpolatingFunction[{{0., 40.}}, <>],
y → InterpolatingFunction[{{0., 40.}}, <>],
z → InterpolatingFunction[{{0., 40.}}, <>]}}
```

```
sol2 = NDSolve[ {x'[t] == 10 (y[t] - x[t]),
y'[t] == 28 x[t] - y[t] - x[t] z[t],
z'[t] == x[t] y[t] - (8/3) z[t],
x[0] == 8.45528, y[0] == 8.45528, z[0] == 27},
{x, y, z}, {t, 0, 40}, MaxSteps → Infinity]
```

```
{{x → InterpolatingFunction[{{0., 40.}}, <>],
y → InterpolatingFunction[{{0., 40.}}, <>],
z → InterpolatingFunction[{{0., 40.}}, <>]}}
```

```
LorenzYZ1 = ParametricPlot[Evaluate[y[t], z[t] /. sol1],
{t, 0, 40}, Frame → True, AspectRatio → 1,
TextStyle → {FontSlant → ''Italic'', FontSize → 12},
Axes → None, FrameLabel → {y,z}, PlotRange -> All,
FrameTicks →
{{- 9.5, - 8.5, - 7.5}, {25.5, 26.5, 27.5, 28.5},{},{}},
PlotPoints → 1000, DisplayFunction → Identity];
```

```
LorenzYZ2 = ParametricPlot[Evaluate[y[t], z[t] /. sol2],
{t, 0, 40}, Frame → True, AspectRatio → 1,
TextStyle → {FontSlant → ''Italic'', FontSize → 12},
Axes → None, FrameLabel → {y,z}, PlotRange -> All,
FrameTicks →
{{7.5, 8.5, 9.5}, {25.5, 26.5, 27.5, 28.5},{},{}},
PlotPoints → 1000, DisplayFunction → Identity];
```

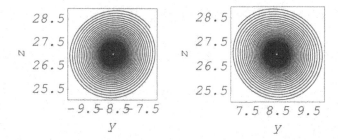

Fig. 25.4. *Projection on the yOz-plane showing the trajectory slowly moving away from the unstable fixed points.*

The divergence of the flow (trace of the Jacobian matrix) is equal to $-(\sigma + b + 1)$. Thus a three-dimensional volume element contracts, as a function of time t, by a factor $e^{-(\sigma+b+1)t}$. It can be shown that there is a bounded ellipsoid $E \subset \mathbb{R}^3$ that all trajectories eventually enter. Taken together, the existence of the bounded ellipsoid and the negative divergence of the flow imply that there exists a bounded set of zero Lebesgue measure inside the ellipsoid E towards which all trajectories tend. For the most complete study of the Lorenz model, consult Sparrow [53].

The Morse Potential

In a paper published in July 1929 Philip Morse [38] used the potential

$$V(r) = De^{-2a(r-r_0)} - 2De^{-a(r-r_0)}$$

to model the vibrational energy of a diatomic molecule, where r is the bond length, r_0 the equilibrium bond length, a a parameter controlling the potential width, and D the dissociation energy of the molecule measured from the potential minimum.

The minimum of $V(r)$ is equal to $-D$ for $r = r_0$, and $V(r)$ tends to zero when r tends to infinity.

Introducing the variable $u = a(r - r_0)$, the potential takes the form $De^{-2u} - 2De^{-u}$. It is represented below by the blue curve.

```
V[r_] := D Exp[- 2 a (r - r0)] - 2 D Exp[- a (r - r0)]
```

```
plV = Plot[V[r] / D /. r → u / a + r0, {u, - 1, 3},
PlotRange → All, AxesLabel → {''a(r-r0)'', ''V(r) / D'},
AxesOrigin → {0, 0}, PlotStyle → {RGBColor[0, 0, 1]},
DisplayFunction → Identity];
plH = Plot[u^2 - 1, {u, - 2, 2},
PlotStyle → {RGBColor[1, 0, 0]},
DisplayFunction → Identity];
Show[{plV, plH}, TextStyle → {FontSlant → ''Italic'',
FontSize → 12}, DisplayFunction → $DisplayFunction];
```

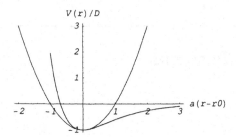

Fig. 26.1. *The Morse potential (in red) and its harmonic part (in blue).*

The red curve is the harmonic part of the Morse potential. The angular frequency ω_0 of the classical harmonic oscillations of a particle of mass m about r_0 is $a\sqrt{2D/m}$ because

```
Series[V[r], {r, r0, 2}]
```

$$- D + a^2 D (r - r0)^2 + O[r - r0]^3$$

Although the Morse potential realistically allows for a diatomic molecule to dissociate at infinite bond length, its behavior at very short bond length is not realistic. The bond length should never be negative, and for $r \rightarrow 0$ the potential should, therefore, tend to infinity. This is not the case for the Morse potential. But, numerically, $V(0)$ is very large and, according to Morse, as far as its effect on the energy levels and wave function goes, it is as good as infinity (see Morse's paper [38]).

The Schrödinger equation for a particle of mass m in the Morse potential is

$$-\frac{h^2}{8\pi^2 m}\frac{d^2\psi}{dr^2} + V(r)\psi = E\psi,$$

where h is the Planck constant.

Note that this equation is actually the radial part of the three-dimensional Schrödinger equation using separation of variables in spherical coordinates.

In terms of the variable u the equation takes the form:

$$\frac{d^2\psi}{du^2} + \frac{8\pi^2 m}{a^2 h^2}(E - V(u))\psi = 0.$$

And if we make a second transformation, letting $x = e^{-u}$, then

$$\frac{d^2\psi}{dx^2} + \frac{1}{x}\frac{d\psi}{dx} + \frac{8\pi^2 m}{a^2 h^2}\left(\frac{E}{x^2} + \frac{2D}{x} - D\right)\psi = 0,$$

where $\psi(x)$ should be finite and continuous for $x \geq 0$. The form of the equation above shows that it only depends upon two parameters $8\pi^2 m/a^2 h^2$ and D.

Following Morse, we, once more, transform the equation. Let

$$\psi(x) = e^{-dx}(2dx)^{b/2} F(x).$$

Then, if $d = 2\pi\sqrt{md}/(ah)$, $E = -a^2 h^2 b^2/(32\pi^2 m)$, and $z = 2dx$, the equation becomes

$$z \frac{d^2 F}{dz^2} + (b+1-z) \frac{dF}{dz} + \left(\frac{8\pi^2 mD}{a^2 dh^2} - \frac{b+1}{2}\right) F = 0.$$

At this point we can try using *Mathematica* to solve this equation. Let A denote the constant $8\pi^2 mD/a^2 dh^2$.

```
DSolve[z F''[z] + (b + 1 - z) F'[z] +
(A - (b + 1) / 2) F[z] == 0, F[z], z]
```

$$\{\{F[z] \rightarrow C[1] \text{ HypergeometricU}[\frac{1 - 2 A + b}{2}, 1 + b, z] +$$

$$C[2] \text{ LaguerreL}[\frac{-1 + 2 A - b}{2}, b, z]\}\}$$

The general solution is a linear combination of the confluent hypergeometric function $U((b+1-2A)/2, b+1, z)$ and the generalized Laguerre polynomial $L((2A-b-1)/2, b+1, z)$ when $(2A-b-1)/2$ is a positive integer n. The wave function should be well behaved on the positive semi-axis. Hence, the only acceptable solution of the equation above is the generalized Laguerre polynomial L_n^b for Re $b > 1$; the confluent hypergeometric function U behaves as z^{1-b} for small z.

Because we made many changes of variables, we use *Mathematica* to find the simplified expressions of A and b.

```
A = (8 Pi^2 m D) / (a^2 d h^2) /.
d → 2 Pi Sqrt[2 m D] / (a h) // Simplify
b = 2A - 2 n - 1 / /Simplify
```

$$\frac{2 \text{ Sqrt}[2] \text{ Sqrt}[D \, m] \text{ Pi}}{a \, h} - 1 - 2 n + \frac{4 \text{ Sqrt}[2] \text{ Sqrt}[D \, m] \text{ Pi}}{a \, h}$$

For $n \geq 0$, the last expession shows that $4\pi\sqrt{2Dm}/ah$ has to be greater than 1. The energy of the ground state is obtained by replacing n by 0 in the general expression of the energy given above. We find

```
groundStateEnergy =
- a^2 h^2 (2 A - 2 n - 1)^2 / (32 Pi^2 m) /. n → 0 // Expand
```

$$- D - \frac{a^2 \ h^2}{32 \ m \ Pi^2} + \frac{a \ h \ Sqrt[D \ m]}{2 \ Sqrt[2] \ m \ Pi}$$

We mentioned above that the expression of the angular frequency ω_0 of the classical harmonic oscillations of a particle of mass m about r_0 is $a\sqrt{2D/m}$.

The ground state energy can, therefore, be written

$$E_0 = -D + \frac{1}{2} \frac{h\omega}{2\pi} - \frac{1}{16D} \frac{h^2\omega^2}{4\pi^2},$$

which is the ground state energy of the harmonic oscillator shifted by D with a correction in $\left(n + \frac{1}{2}\right)^2$ due to the anharmonicity of the asymmetric Morse potential. This result suggests that the general expression of the energy levels may also be written as the energy levels of the harmonic oscillator shifted by D with an extra anharmonic correction. This can be checked using *Mathematica*.

```
energyLevels[n] =
- a^2 h^2 (2 A - 2 n - 1)^2 / (32 Pi^2 m) // Expand
```

$$- D - \frac{a^2 \ h^2}{32 \ m \ Pi^2} - \frac{a^2 \ h^2 \ n}{8 \ m \ Pi^2} - \frac{a^2 \ h^2 \ n^2}{8 \ m \ Pi^2} +$$

$$\frac{a \ h \ Sqrt[D \ m]}{2 \ Sqrt[2] \ m \ Pi} + \frac{a \ h \ Sqrt[D \ m] \ n}{Sqrt[2] \ m \ Pi}$$

That is,

$$E_n = -D + \frac{h\omega}{2\pi}\left(n + \frac{1}{2}\right) - \frac{1}{4D} \frac{h^2\omega^2}{4\pi^2}\left(n + \frac{1}{2}\right)^2.$$

Because $b = 2A - 2n - 1$ must be positive for the wave function to be finite, the number of discrete energy levels is finite. They correspond to the values $n = 0, 1, 2, \ldots, \lfloor A - \frac{1}{2} \rfloor$, where $\lfloor x \rfloor$ denotes the largest integer less than x (i.e., the *Mathematica* function Floor[x]). Historically, it was the first example of a Schrödinger equation giving a finite number of discrete energy levels.

The theoretical results obtained by Morse are in good agreement with the experimental data known at that time (see his paper [38]). For recent results on diatomic molecules obtained using the Morse potential, visit the Web site http://www.opticsexpress.org/abstract.cfm?URI=OPEX-10-8-376 which presents animated wave functions.

Prime Numbers

27.1 Primality

A positive integer p is a *prime (number)* if, and only if, it has only two distinct divisors: 1 and p itself. The only divisors of 227 are 1 and 227 making 227 a prime whereas 237, which has four distinct divisors 1, 3, 79, and 237, is not a prime.

The *Mathematica* function PrimeQ tests primality. Using this function we verify that 227 is a prime and 237 is not.

```
{PrimeQ[227], PrimeQ[237]}
```

{True, False}

The divisors of an integer can be found using the built-in function Divisors.

```
Divisors[237]
```

{1, 3, 79, 237}

shows that 237 is not a prime.

The command Prime[n] gives the nth prime number. Remember that 1 is not a prime, so Prime[1] = 2.

```
Prime[57]
```

The package NumberTheory'PrimeQ' implements primality proving. If ProvablePrimeQ[n] returns True, then the number n can be mathematically proven to be prime. In addition, according to the *Mathematica* book [68], while the built-in primality testing function PrimeQ does not actually give a proof that a number is prime, there are no known examples where PrimeQ fails.

```
<<NumberTheory'PrimeQ'
```

```
PrimeQ[22507410677]
```

True

```
ProvablePrimeQ[22507410677]                  .
```

True

Loading the package

```
<<NumberTheory'NumberTheoryFunctions'
```

we can use many other interesting functions concerning primality. NextPrime[n] and PreviousPrime[n] give, respectively, the smallest prime greater than n and the largest prime less than n.

```
NextPrime[22507410677]
```

22507410691

```
PreviousPrime[22507410677]
```

22507410637

The command Random[Prime, {n1, n2}] gives a prime in the range {n1, n2}.

```
Random[Prime, {100, 200}]
```

191

A Few Simple Exercises

1. *Write a program generating twin primes, that is, pairs of primes that differ by 2.*

Here is a simple program that selects all pairs of primes in a list of pairs of odd integers which differ by 2.

```
Cases[Table[{2k + 1, 2k + 3}, {k,1, 500}],
{x_, y_} /; PrimeQ[{x, y}] == {True, True}]
```

{{3, 5}, {5, 7}, {11, 13}, {17, 19}, {29, 31}, {41, 43},
{59, 61}, {71, 73}, {101, 103}, {107, 109}, {137, 139},
{149, 151},{179, 181}, {191, 193}, {197, 199}, {227, 229},
{239, 241},{269, 271}, {281, 283}, {311, 313}, {347, 349},
{419, 421},{431, 433}, {461, 463}, {521, 523}, {569, 571},
{599, 601}, {617, 619}, {641, 643}, {659, 661}, {809, 811},
{821, 823}, {827, 829}, {857, 859}, {881, 883}}

We can also generate twin primes in a specific range, as done below.

```
Cases[Table[{2k + 1, 2k + 3}, {k, 50000, 50500}],
{x_, y_} /; PrimeQ[{x, y}] == {True, True}]
```

{{100151, 100153}, {100361, 100363}, {100391, 100393},
{100517, 100519}, {100547, 100549}, {100799, 100801}}

2. *Study the primality of the numbers belonging to the sequence (31, 331, 3331, 33331, ...).*

The function

```
f[n_] := (1/3) (10^n - 7)
```

generates the numbers of this sequence:

```
Table[f[n], {n, 2, 10}]
```

{31, 331, 3331, 33331, 333331, 3333331, 33333331,
333333331, 3333333331}

Testing primality, we find

```
lis = ;
For[n = 2, n ≤ 1000, n++,
If[PrimeQ[f[n]] == True, lis = Append[lis, n]]] // Timing
```

{15.625 Second, Null}

```
lis
```

{2, 3, 4, 5, 6, 7, 8, 18, 40, 50, 60, 78, 101, 151, 319,
382, 784}

The first seven elements of the sequence are prime numbers; then there are only 10 more primes less than $f(1000)$. Note that $f(n)$ has n digits; the prime number f[784] has, therefore, 784 digits!

3. *Verify that for all integer values of n from* -39 *to* 40, $n^2 - n + 41$ *takes only prime values.*

This is very simple. First note that if we change n into $1 - n$, the quadratic polynomial is left unchanged. To verify the property it is, therefore, sufficient to verify it for n varying from 1 to 40. The command

```
lis = Cases[Table[n^2 - n + 41, {n, 1, 40}],
p_ /; PrimeQ[p] == True]
```

{41, 43, 47, 53, 61, 71, 83, 97, 113, 131, 151, 173, 197, 223,
251, 281, 313, 347, 383, 421, 461, 503, 547, 593, 641, 691,
743, 797, 853, 911, 971, 1033, 1097, 1163, 1231, 1301, 1373,
1447, 1523, 1601}

```
Length[lis]
```

40

shows that all 40 different integers are primes.

4. *Find all primes that can be written* $m^2 + n^2$ *where* m *and* n *are integers not greater than 500.*

An ordered sequence of integers of the form $m^2 + n^2$, where m and n are integers not greater than 500, is generated using the command

```
Clear[seq]
seq = Intersection[Sort[Flatten[Table[m^2 + n^2, {m, 1, 500},
{n, 1, 500}]]]];
Length[seq]
```

78901

Then, using the program

```
Clear[primesList]
primesList = ;
For[j = 1, j ≤ Length[seq], j++,
If[PrimeQ[seq[[j]]] == True,
primesList = Append[primesList,seq[[j]]]]]
Length[primesList]
```

13724

we find that in the sequence of 78,901 integers that are the sum of two squares, 13,724 are primes. They are listed in **primesList**. **Short [expression]** prints a short form of **expression** less than about one line long.

```
Short[primesList]
```

{2, 5, 13, 17, 29, <<13717>>, 493049, 495017}

Actually it can be shown that a prime p is the sum of two squares if, and only if, $p = 2$ or $p \equiv 1 \mod 4$. For a proof and a *Mathematica* program that, given p, find m and n such that $p = m^2 + n^2$ see [64]. As shown below, we can verify that except the first element of **primesList**, which is 2, all other elements are indeed congruent to 1 modulo 4.

```
For[k = 1, k <= Length[primesList], k++,
If[Mod[primesList[[k]], 4] != 1, Print[k]]] // Timing
```

1

{0.058051 Second, Null}

5. *Find ten primes in arithmetic progression.*

The longest sequence of primes in arithmetic progression is

$$56211383760397 + 44546738095860\, k, \text{ where } k = 0, 1, \ldots, 22.$$

It was discovered in 2004 by Markus Frind, Paul Jobling, and Paul Underwood (see http://primes.plentyoffish.com/).

Mathematica can easily confirm this result.

```
primesAP = Table[56211383760397 + k 44546738095860,
{k, 0, 22}];
```

```
PrimeQ[primesAP]
```

{True, True}

```
Length[%]
```

23

Coming back to our much simpler problem, we are looking for a sequence of primes of the form $p_1 + kr$, where p_1 and r are given and $k = 0, 1, 2, \ldots, 9$.

To solve this problem we use a theorem stating that if n elements of an arithmetic progression are odd, then r can be divided by all primes less than n.

Hence for $n = 10$, r can be divided by 2, 3, 5, and 7, that is, by 210. If we choose $r = 210$, because this number can be divided by 3, 5, and 7, none of these numbers can be the first term of the progression. The first term cannot be 11 because the second one would be 221 which is not a prime. The relation $210 = 1 \bmod 11$ implies that the remainder of each term of the progession divided by 11 should increase by one unit and, consequently, the first term p_1 of the progression should also be equal to 1 mod 11. Because p_1 must be odd we can try a prime of the form $22j + 1$, where j is an integer.

A list of possible first terms is the list np1 obtained below.

```
np1 = {};
For[j = 1, j ≤ 100, j++,
If[PrimeQ[22 j + 1] == True, np1 = Append[np1, 22 j + 1]]]
```

```
Length[np1]
```

30

We then build all 30 arithmetic progressions of length 10 whose first term is an element of np1.

```
Clear[seq]
seq = Table[Table[np1[[i]] + 210 k, {k, 0, 9}],
{i, 1, Length[np1]}];
```

And check the primality of each arithemetic progression.

```
For[i = 1, i ≤ Length[np1], i++,
l = {};
For[k = 1, k ≤ 10, k++,
If[PrimeQ[seq[[i, k]]] == False,
l = Append[l, seq[[i,k]]]]];
If[l == {}, Print[seq[[i]]]]]
```

{199, 409, 619, 829, 1039, 1249, 1459, 1669, 1879, 2089}

Among the 30 arithmetic progressions, only one has prime terms.

Increasing significantly the number of elements of the list np1 does not give any new sequence of primes. For example:

```
Clear[np1]
np1 = {};
For[j=1, j ≤ 10000, j++,
If[PrimeQ[22 j + 1] == True, np1= Append[np1, 22 j+1]]]
```

```
Length[np1]
```

1952

```
Clear[seq]
seq = Table[Table[np1[[i]] + 210 k, {k, 0, 9}],
{i, 1, Length[np1]}];
```

```
For[i = 1, i ≤ Length[np1], i++,
l = {};
For[k = 1, k ≤ 10, k++,
If[PrimeQ[seq[[i, k]]] == False,
l = Append[l, seq[[i,k]]]]];
If[l == {}, Print[seq[[i]]]]]
```

{199, 409, 619, 829, 1039, 1249, 1459, 1669, 1879, 2089}

There is no new arithmetic progression.

27.2 Mersenne Numbers

A number of the form $2^n - 1$ is called a *Mersenne number* after the French theologian and mathematician Marin Mersenne (1588–1648) who observed that if $2^n - 1$ is prime, then n must be prime, but that the converse is not necessarily true. Note that if n is not a prime it can be written as the product of two positive integers $n = ab$, and the relation

$$2^{ab} - 1 = (2^a - 1)(2^{a(b-1)} + 2^{a(b-2)} + \cdots + 1)$$

shows that the Mersenne number $2^n - 1$ cannot be a prime if n is not a prime.

In order to define a function giving a list of the first Mersenne numbers, we first generate a table of Mersenne numbers and the corresponding exponents, asking *Mathematica* to only display those that are prime using the command Select.

```
Select[Table[{n, PrimeQ[2^n - 1]},
{n, 1, 100}], #[[2]] == True &]
```

{{2, True}, {3, True}, {5, True}, {7, True}, {13, True},
{17, True}, {19, True}, {31, True}, {61, True}, {89, True}}

To obtain the list of the exponents generating prime Mersenne numbers, we transpose the list above and extract the first of the two sublists.

```
exponents = Transpose[Select[Table[{n, PrimeQ[2^n - 1]},
{n, 1, 100}], #[[2]] == True &]][[1]]
```

{2, 3, 5, 7, 13, 17, 19, 31, 61, 89}

A list of first prime Mersenne numbers is then obtained entering either the command

```
Table[2^exponents[[k]] -1, {k, 1, Length[exponents]}];
```

or

```
Map[(2^# - 1) &, exponents];
```

which is a bit faster.

These different steps can now be grouped together in the following function.

```
MersenneNumbersList[n_] := Module[{exponents, Mlist},
exponents = Transpose[Select[Table[{j, PrimeQ[2^j-1]},
{j, 1, n}], #[[2]] == True &]][[1]];
Mlist = Map[(2^# - 1) &, exponents]; Mlist]
```

```
MersenneNumbersList[200]//Timing
```

{0.025853 Second, {3, 7, 31, 127, 8191, 131071, 524287,
2147483647, 2305843009213693951, 618970019642690137449562111,

162259276829213363391578010288127, 17014118346046923173164\

87303715884105727}}

Remark. In as much as a Mersenne number is of the form $2^n - 1$, its expression in base 2 contains only 1s.

```
Table[BaseForm[2^n - 1, 2], {n, 1,10}]
```

$\{1_2 , 11_2 , 111_2 , 1111_2 , 11111_2 , 111111_2 , 1111111_2 ,$
$11111111_2 , 111111111_2 , 1111111111_2 \}$

It is not difficult to modify the function `MersenneNumbersList` into a function searching for prime Mersenne numbers in a given range.

27.3 Perfect Numbers

Mersenne numbers are related to perfect numbers. A *perfect number* is an integer that equals the sum of its proper divisors. Pythagoreans (*circa* 525 BC), who believed that all things are numeric, studied perfect numbers for their mystical properties. The first perfect number is 6 which is equal to $1 + 2 + 3$.

The following function gives the list of all perfect numbers less than a given integer n.

```
perfectNumbersList[n_Integer] :=
Module[{pNums = {}},
For[k=1, k <= n, k++,
If[Total[Most[Divisors[k]]] == k,
pNums = Append[pNums, k]]]; pNums]
```

`Most[expression]` removes the last element of `expression`.

```
perfectNumbersList[10000]
```

$\{6, 28, 496, 8128\}$

As shown below, there is no perfect number between 10,000 and 1,000,000.

```
perfectNumbersList[1000000] // Timing
```

$\{26.5845 \text{ Second}, \{6, 28, 496, 8128\}\}$

We can build up a function `perfectNumberQ[n]` which gives True if n is perfect.

```
perfectNumberQ[n_Integer] := If[Total[Most[Divisors[n]]] ==
n, True, False]
```

We can use this test to verify that 33550336, 8589869056, 137438691328, and 2305843008139952128 are perfect numbers. Let us first make `perfectNumberQ` listable.

```
SetAttributes[perfectNumberQ, Listable]
```

```
perfectNumberQ[{33550336, 8589869056, 137438691328,
2305843008139952128}]
```

$\{True, True, True, True\}$

Leonhard Euler (1707–1783) proved that *every even perfect number must be of the form* $2^{n-1}(2^n - 1)$, where the Mersenne number $2^n - 1$ is a prime, a result already suggested by Euclid (*circa* 300 BC). For instance, 28, which is equal to $1 + 2 + 4 + 7 + 14$ is perfect and it can be written as $2^2(2^3 - 1)$. Perfect numbers have many remarkable properties: They are triangular, that is, of the form $k(k + 1)/2$ and the sum of the reciprocals of the divisors of a perfect number (including the reciprocal of the number itself) is always equal to 2. Thus, for 28

$$\frac{1}{1} + \frac{1}{2} + \frac{1}{4} + \frac{1}{7} + \frac{1}{14} + \frac{1}{28} = 2.$$

Using *Mathematica*, we can easily check this property.

```
Total[1 / Divisors[496]]
```

2

```
Total[1 / Divisors[8128]]
```

2

We can also verify that only perfect numbers have this property.

```
sumReciprocalsTwo[n_Integer] :=
Module[{seq = {}},
For[k = 1, k <= n, k++,
If[Total[1 / Divisors[k]] == 2, seq = Append[seq, k]]]; seq]
```

```
sumReciprocalsTwo[10000]
```

{6, 28, 496, 8128}

```
sumReciprocalsTwo[1000000]
```

{6, 28, 496, 8128}

Two questions concerning perfect numbers are still unanswered: the first is whether there are any odd perfect numbers, and the other is whether there are infinitely many perfect numbers.

Public-Key Encryption

Public-key encryption is a cryptographic system that allows users to communicate securely without having prior access to a shared secret key. It uses two keys: a public key known to everyone and a private or secret key known only to the recipient of the message. These two keys are related mathematically.

28.1 The RSA Cryptosystem

The RSA cryptosystem, named after its inventors R. Rivest, A. Shamir, and L. Adleman is the most widely used public-key system. It can be described as follows [47].

1. Choose two large prime numbers p and q.

2. Choose an integer e satisfying the conditions $1 < e < pq$ and such that e and $(p-1)(q-1)$ are relatively prime. Because $(p-1)(q-1)$ is necessarily even, e, called the *public exponent*, has to be odd but does not have to be prime.

3. Find an integer d, called the *secret exponent*, satisfying $de = 1$ modulo $(p-1)(q-1)$; that is, d is the inverse of e modulo $(p-1)(q-1)$.

4. If t is a positive integer representing the plaintext, the ciphertext, c is the positive integer $t^e \bmod pq$. Clearly t must be less than pq.

5. Then $t = c^d \bmod pq$.

The public key is the pair (e, n), where $n = pq$. The private key is d. Because there is no easy method to obtain d, p, and q knowing n and e, the two numbers n and e can be made public.

The security of RSA is based upon the difference in the (short) computer time needed to find a prime and the (huge) computer time needed to factorize a very large prime.

28.1.1 ToCharacterCode and FromCharacterCode

The first task is to transform the plaintext message into an integer. There exist many methods. One method is to use the ASCII code. The printable ASCII characters have ASCII codes ranging from 32 to 126 with 32 being the code of the space key.

```
ToCharacterCode[" "]
```

{32}

The function ToCharacterCode["string"] gives the list of the integer codes of the string characters. For instance,

```
ToCharacterCode["Hello!"]
```

{72, 101, 108, 108, 111, 33}

Its inverse FromCharacterCode[n1, n2,...] gives a sequence of characters with codes n1,n2,....

```
FromCharacterCode[{72, 101, 108, 108, 111, 33}]
```

Hello!

28.1.2 Obtaining the Integer t

To obtain the integer t representing the plaintext string message we use the function encoding that associates a two-digit number to each ASCII character.

FromDigits[digitsList, b] constructs an integer from the digitsList of digits given in base b.

```
encodingString[str_String] :=
FromDigits[ToCharacterCode[str] - 30 /.
{2 → 02, 3 → 03, 4 → 04,5 → 05,6 → 06,
7 → 07, 8 → 08, 9 → 09}, 100]
```

For example, if we want to send the following credit card number generated below,

```
cardNumber = NumberForm[Random[Integer, {10^15, 10^16 - 1}],
DigitBlock → 4, NumberSeparator → " "]
```

8788 1596 0150 8954

we will send the message:

```
t = encodingString["My credit card number is 8788 1596 0150
8954"]
```

47910269847170758602696784700280877968718402758502262526260219\
232724021819231802262772322

The inverse of the function encodingString is the function decoding defined by

```
decoding[num_Integer] :=
FromCharacterCode[IntegerDigits[num, 100] + 30 ]
```

IntegerDigits[num, b] gives a list of the digits in base b of the integer num. For example:

```
IntegerDigits[567891235, 100]
```

{5, 67, 89, 12, 35}

```
decoding[t]
```

My credit card number is 8788 1596 0150 8954

28.1.3 Choosing the Integer $n = pq$

Prime[n] gives the nth prime number. Remember that 1 is not a prime number so the first prime is 2. This function does not accept very large arguments as shown below.

```
p = Prime[10^15]
```

Prime::largp: Argument 1000000000000000 in

Prime[1000000000000000]

is too large for this implementation. More

Prime[1000000000000000]

If we want to find the smallest prime greater than a given integer n, we use the *Mathematica* function NextPrime but first we have to load the package NumberTheory'NumberTheoryFunctions'.

```
<<NumberTheory'NumberTheoryFunctions'
```

```
NextPrime[10^100] // Timing
```

{0.051037 Second, 10000000000000000000000000000000\
00000000000000000000000000000000000000000000000000000000000\
0000000000000000267}

Using this command we select two large prime numbers p and q and determine $n = pq$.

```
p = NextPrime[Random[Integer, {10^51, 10^53}]]
```

11891461639329158254011406810889410238185522367888077

```
q = NextPrime[Random[Integer, {10^51, 10^53}]]
```

217289469371050684534817988919220845459351483895462 69

```
n = p q
```

2583889389656037294623718776432403118356838148793644975408\
11259358522106357481051858327332657241804934713

```
Length[IntegerDigits[n]]
```

105

We verify that $t < n$.

```
t < n
```

True

The basic idea underlying public-key cryptography is that factoring the public number n is impossible.

28.1.4 Choosing the Public Exponent e

The public exponent has to be less than n and have no common divisor with $(p-1)(q-1)$. Let

```
e = 341353751;
```

and we verify that the exponent e has no common divisor with $(p-1)(q-1)$.

```
GCD[341353751, (p-1) (q-1)]
```

1

28.1.5 Coding t

The function PowerMod[a, b, n] gives a^b modulo n. It is much more efficient than directly evaluating Mod[a^b, n].

```
Mod[t^e, n]
```

General::ovfl:

Overflow occurred in

computation. More

Overflow[]

```
PowerMod[t, e, n] // Timing
```

{0.0002 Second,

2235531538206408257315405511509680332538175929461060136500\

900768599354930321445990458249202387578114 9113}

```
c = %[[2]];
```

28.1.6 Choosing the Secret Exponent d

Because c is the transform of t, in order to obtain t from c we have to determine the inverse transform. That is, we have to find the inverse d of the exponent e. This exponent d has therefore, to satisfy the relation $de = 1 \mod (p-1)(q-1)$. It can be kept secret if, and only if, the two factors p and q of n cannot be found.

```
d = PowerMod[e, -1, (p-1) (q-1)]
```

1826323531841738319297396097085759136432278658130885720109497 0\

6838356319152957727482473924887886784336967

28.1.7 Decrypting t

Knowing d, we can define the function **decoding** inverse of the function **encodingString**.

```
decoding[PowerMod[c, d, n]]
```

My credit card number is 8788 1596 0150 8954

28.2 Summing Up

Finally, to encode a message we define the function encrypt[message] knowing the public key {e, n},

```
encrypt[message_String] :=
PowerMod[FromDigits[ToCharacterCode[message] - 30 /.
{2 → 02, 3 → 03, 4 → 04, 5 → 05, 6 → 06,
7 → 07, 8 → 08, 9 → 09}, 100], e, n]
```

and decode the encrypted number using the function decrypt[number] knowing the secret key {d,p,q} (the knowledge of p and q is essential for finding d).

```
decrypt[num_Integer] :=
FromCharacterCode[IntegerDigits[PowerMod[num, d, n], 100] +
30]
```

```
cryptedMessage = encrypt["My credit card number is:
2889 3038 0146 0363"]
```

5390853114482414827877369848455283410953373515802659418 9691251\
90681724907278197466716718962982 9977068233

```
originalMessage = decrypt[cryptedMessage]
```

My credit card number is: 2889 3038 0146 0363

Remark 1. Although in ordinary arithmetic the multiplicative inverse of an integer is a rational, using modulo arithmetic the multiplicative inverse of an integer is also an integer. But, an integer having a common factor with the modulus being not invertible, the exponent e and the modulus $(p-1)(q-1)$ have to be relatively prime.

Remark 2. Let φ denote the Euler function, that is, the mapping which associates with each integer n the number $\varphi(n)$ of positive integers relatively prime to n. If the factorization of n in prime factors is $p_1^{a_1} p_2^{a_2} \ldots$, then $\varphi(n) = n(1 - 1/p_1)(1 - 1/p_2) \ldots$. In the case of the RSA cryptosystem, we used the public number n, the product of two prime numbers p and q, therefore, $\varphi(n) = n(1 - 1/p)(1 - 1/q) = (p-1)(q-1)$.

In practice, the numbers p and q are much larger than those we considered in the example above.

Many more details on crytography can be found in [37]. See also *The Magic of Public Key Encryption*, a 40-page paper that can be downloaded from www.hifn.com/docs/a/The-Magic-of-Public-Key-Encryption.pdf.

Quadratrix of Hippias

In 430 BC, Hippias (460 BC–400 BC) of Elis (in the Peloponnese, Greece), a contemporary of Socrates, discovered the quadratrix, a curve he used for trisecting an angle. As a matter of fact, the quadratrix may be used for dividing an angle into any number of equal parts. In 350 BC Dinostratus (390BC–320BC) used the quadratrix to square the circle.[1]

Trisecting an angle and squaring the circle were classical problems of Greek geometry. The solutions of these two problems given by Hippias and Dinostratus, are not solutions using a ruler and compass. They are based on the possibility of dividing a segment into a given number of equal parts. Hippias and Dinostratus are mentioned in the works of the famous Greek geometer Pappus of Alexandria (*circa* 290–*circa* 350).

29.1 Figure

Let $ABCD$ be the vertices of a unit square and draw the arc of the circle centered at A of radius $AB = 1$ (in orange in Figure 29.1 below). The quadratrix (in red in the figure) is the locus of the point P intersection of the segment $B'C'$ (in blue in the figure) moving down uniformly parallel to itself from BC to AD and the segment AA' (in blue in the figure) rotating clockwise about A uniformly from AB to AD. The thicker blue lines indicate the positions at time t of the line $B'C'$ and AA'.

The parametric equation of the quadratrix is then given by

---

[1]Dinostratus was the brother of Menaechmus (380 BC–320 BC) who is credited for having discovered that the ellipse, parabola, and hyperbola are conic sections that were later rigorously studied by Appolonius of Perga (262 BC–190 BC).

```
x[t_] := (1-t) Cot[Pi (1-t) / 2]
y[t_] := 1 - t
```

Note that the point of the quadratrix on *AD* is not defined and can only be obtained as a limit.

```
Limit[(1 - t) Cot[Pi (1 - t) / 2], t → 1]
```

$$\frac{2}{Pi}$$

To draw the figure we first load the package Graphics`Arrow` in order to be able to draw arrows.

```
<<Graphics`Arrow`
```

```
Clear[txt]
plLocus = ParametricPlot[{x[t], y[t]}, {t, 0, 1},
PlotStyle → {RGBColor[1, 0, 0]},
AspectRatio → Automatic, DisplayFunction → Identity];
plCircle = Graphics[{RGBColor[1, 0.5, 0],
Circle[{0, 0}, 1, {0, Pi / 2}]}];
plAngle = Graphics[{Thickness[0.007], RGBColor[0, 0, 1],
Circle[{0, 0}, 0.3, {51 Degree, 90 Degree}]}];
plTime = Graphics[{Thickness[0.007], RGBColor[0, 0, 1],
Line[{{0.9, 0.567}, {0.9,1}}]}];
square = Graphics[Line[{{0, 0}, {0, 1}, {1, 1}, {1, 0},
{0, 0}}]];
segmentABprime = Graphics[{RGBColor[0, 0, 1],
Line[{{0, 0}, {Cos[51 Degree], Sin[51 Degree]}}]}];
segmentAQ = Graphics[{RGBColor[0, 1, 0],
Line[{{0, 0}, {Cos[17 Degree], Sin[17 Degree]}}]}];
plHorizontal1 = ParametricPlot[{t, 0.567}, {t, 0, 1},
PlotStyle → {RGBColor[0, 0, 1]},
AspectRatio → Automatic, DisplayFunction → Identity];
plHorizontal2 = ParametricPlot[{t, 0.189}, {t, 0, 1},
PlotStyle → {RGBColor[0, 1, 0]},
AspectRatio → Automatic, DisplayFunction → Identity];
```

```
drawArrows = Graphics[{RGBColor[0, 0, 1],
{Arrow[{0.898, 0.7}, {0.898, 0.68}],
Arrow[{0.131, 0.271}, {0.143, 0.265}]}}];
txtFont = {''Helvetica'', 12};
txt = Graphics[{Text[FontForm[''A'', txtFont], {-0.03, 0}],
Text[FontForm[''B'', txtFont], {- 0.03, 1}],
Text[FontForm[''C'', txtFont], {1.03, 1}],
Text[FontForm[''D'', txtFont], {1.03, 0}],
Text[FontForm[''P'',txtFont], {0.46, 0.6}],
Text[FontForm[''A''', txtFont], {0.65, 0.8}],
Text[FontForm[''B''', txtFont], {- 0.03, 0.567}],
Text[FontForm[''C''', txtFont], {1.03, 0.567}],
Text[FontForm[''H'', txtFont], {- 0.03, 0.189}],
Text[FontForm[''Q'', txtFont], {0.64, 0.2}],
Text[FontForm[''L'', txtFont],{0.637, - 0.04}],
Text[FontForm[TraditionalForm[t], txtFont], {0.92, 0.8}],
Text[FontForm[TraditionalForm[π t / 2], txtFont],
{0.1, 0.36}]}];

Show[{plLocus, plCircle, plAngle, plTime, drawArrows,
square, segmentABprime, segmentAQ, plHorizontal1,
plHorizontal2, txt}, Ticks → None, Axes → None,
DisplayFunction → $DisplayFunction];
```

Output represented in Figure 29.1.

29.2 Trisecting an Angle

Let θ be the angle to divide in three equal parts. We have

$$2\,\frac{\theta}{\pi} = \frac{\text{arc } A'D}{\text{arc } BA'D} = \frac{AB'}{AB} = AB' \quad \text{since} \quad AB = 1.$$

So

$$\theta = \text{arc } A'D = \frac{\pi}{2}\, AB'$$

is proportional to AB'.

To trisect the angle θ, we consider the point H on AB such that $AH = AB'/3$ and draw a parallel to AD through H that meets the quadratrix at Q. The angle QAD is exactly the third of the angle $PAD = \theta$.

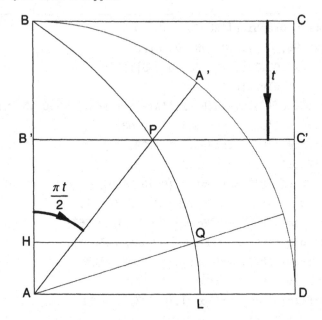

Fig. 29.1. *Construction of the quadratrix).*

Note that this method allows draw an angle equal to any rational fraction of θ.

29.3 Squaring the Circle

Once we admit the existence of the limit point L on the side AD of the square, using only a ruler and a compass we can construct a square with a side length equal to $\sqrt{\pi}$. First from a segment of length $2/\pi$ we construct a segment half the length . There exist various methods . As shown below, you draw two circles of radius $2/\pi$ centered respectively at $O = (0,0)$ and $L = (2/\pi, 0)$ and draw the line going through the points of intersection of these two circles to determine the point M on the line joining the centers of the two circles. We have $OM = 1/\pi$.

```
circle1 = Graphics[Circle[{0, 0}, 2 / Pi]];
circle2 = Graphics[Circle[ {2 / Pi, 0}, 2 / Pi]];
centersSegment = Graphics[Line[{{0, 0}, {2 / Pi, 0}}]];
perpSegment = Graphics[Line[{{1 / Pi, - 1}, {1 / Pi, 1}}]];
txtFont = {''Helvetica'', 12};
txt = Graphics[{Text[FontForm[''O'', txtFont], {- 0.05, 0}],
Text[FontForm[''L'', txtFont], {2 / Pi + 0.05, 0}],
Text[FontForm[''M'', txtFont], {1 / Pi + 0.05, 0.05}]}];
Show[{circle1, circle2, centersSegment, perpSegment, txt},
AspectRatio → Automatic];
```

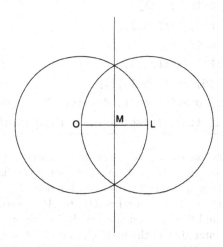

Fig. 29.2. *Construction of a segment of length $1/\pi$.*

```
Clear[txt]
drawCircle = Graphics[Circle[{0, 0}, (1 + Pi) / 2, {0, Pi}]];
drawDiameter = Graphics[Line[{{- (1 + Pi) / 2, 0},
{(1 + Pi) / 2, 0}}]];
draw1overPi = Graphics[Line[{{(1 - Pi) / 2, 0},
{(1 - Pi) / 2, - 1 / Pi}}]];
drawAC = Graphics[Line[{{- (1 + Pi) / 2, 0},
{(1- Pi) / 2, -1 / Pi}}]];
```

```
drawSqPi = Graphics[{RGBColor[1, 0, 0],
Line[{{(1 - Pi) / 2, 0}, {(1 - Pi) / 2, Sqrt[Pi]}}]}];
drawDE = Graphics[Line[ {{(1 - Pi) / 2, 1},
{(1 + Pi) / 2, 0}}]]]; txtFont = {''Helvetica'', 12};
txt = Graphics[{Text[FontForm[''A'', txtFont],
{- (1 + Pi) / 2 - 0.15, 0}],
Text[FontForm[''B'', txtFont],
{(1 - Pi) / 2 + 0.15, 0.15}],
Text[FontForm[''C'', txtFont],
{(1 - Pi) / 2 + 0.15, -1 / Pi}],
Text[FontForm[''D'', txtFont],
{(1 - Pi) / 2 + 0.15, 1. 1}],
Text[FontForm[''E'', txtFont],
{(1 + Pi) / 2 + 0.15, 0}],
Text[FontForm[''F'', txtFont],
{(1 - Pi) / 2+0.15,Sqrt[]-0.15}]}];
Show[{drawCircle, drawDiameter, draw1overPi, drawAC,
drawDE, drawSqPi, txt}, AspectRatio → Automatic];
```

In Figure 29.3 below, we show how to construct a segment of length $\sqrt{\pi}$. We draw a horizontal segment $AB = 1$, and determine the list of points C, D, E, F as follows. Draw a vertical downward segment $BC = 1/\pi$, a vertical upward segment $BD = 1$, the parallel DE to AC where E is on the line AB, and $BE = \pi$, and finally F on the line BD at the intersection with the upper semi-circle centered at O the middle point of AE and radius $(1+\pi)/2$. Because $BF^2 = BA \times BE = \pi$, we have constructed a segment BF of length $\sqrt{\pi}$.

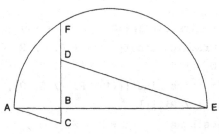

Fig. 29.3. *Construction of a segment of length $\sqrt{\pi}$.*

Quantum Harmonic Oscillator

After having finished his chemistry studies, Erwin Schrödinger (1887–1961) devoted himself to Italian painting for many years and then took up botany and published a series of papers on plant phylogeny. During the years 1906 to 1910, as a student at the University of Vienna, he was greatly influenced by Fritz Hasenöhrl's lectures on theoretical physics. He then acquired a mastery of eigenvalue problems in the physics of continuous media, thus laying the foundation of his future important work. Moving very often, he occupied many academic positions starting as assistant to Max Wien (1866–1938). His most fruitful period took place when he replaced Max von Laue (1879–1960) at the University of Zürich, where he enjoyed contacts, in particular, with Hermann Weyl (1885–1955) who was to provide the deep mathematical knowledge that would prove so helpful to Schrödinger. Having never been very satisfied by the quantum condition on orbits in Niels Bohr's (1885–1962) atomic model, he believed that atomic spectra should be determined by some kind of eigenvalue problem. In 1926, he discovered the wave equation that bears his name. In 1933, "for the discovery of new productive forms of atomic theory," he shared with Paul Adrien Dirac (1902–1984) the Nobel Prize in Physics.

The quantum harmonic oscillator is among the most important model systems in quantum mechanics because the dynamics of many systems near an equilibrium configuration can often be modeled by one or more harmonic oscillators.

30.1 Schrödinger Equation

The Schrödinger equation for the harmonic oscillator is

$$-\frac{\hbar^2}{2m}\frac{d^2\psi}{dx^2} + \frac{1}{2}m\,\omega^2\,x^2\psi = E\,\psi.$$

If we define the reduced units of length ξ and energy e such that

$$x = \sqrt{\frac{\hbar}{m\omega}}\, \xi \quad \text{and} \quad E = \frac{\hbar\,\omega}{2}\, e,$$

the reduced Schrödinger equation reads

$$-\frac{d^2\psi}{d\xi^2} + \xi^2\,\psi = e\,\psi.$$

For very large ξ, the term $\xi^2\psi$ becomes preponderant and we can neglect the term $e\psi$. This suggests defining a new function φ by $\psi(\xi) = \varphi(\xi)\exp(-\xi^2/2)$. The differential equation satified by φ is then

$$\frac{d^2\varphi}{d\xi^2} - 2\,\xi\,\frac{d\varphi}{d\xi} + (2e-1)\varphi = 0.$$

We can solve this equation in many different ways. If we put $e = 2n+1$ where n is an integer, we obtain the Hermite equation:

$$\frac{d^2\varphi}{d\xi^2} - 2\,\xi\,\frac{d\varphi}{d\xi} + 2\,n\,\varphi = 0.$$

It can be solved using the command

```
DSolve[phi''[xi] - 2 xi phi'[xi] + 2 n phi[xi] == 0,
phi[xi], xi]
```

$$\left\{\left\{\text{phi[xi]} \rightarrow \text{C[1] HermiteH[n, xi]} + \text{C[2] Hypergeometric1F1}\left[-\frac{n}{2}, \frac{1}{2}, \text{xi}^2\right]\right\}\right\}$$

Because a wave function has to be square integrable, the only acceptable solution is the Hermite polynomial $H_n(\xi)$. Plotting the first Hermite polynomials we obtain

```
Table[HermiteH[n, xi], {n, 0, 4}]
```

$$\{1,\ 2\ \text{xi},\ -2 + 4\ \text{xi}^2,\ -12\ \text{xi} + 8\ \text{xi}^3,\ 12 - 48\ \text{xi}^2 + 16\ \text{xi}^4\}$$

```
Plot[{2 xi, - 2 + 4 xi^2, , - 12 + 8 xi^2,
12 - 48 xi^2 + 16 xi^4}, {xi, - 2, 2}]
PlotStyle → {RGBColor[0, 0, 1], RGBColor[0, 1, 0],
RGBColor[0, 1, 1], RGBColor[1, 0, 0]}]
```

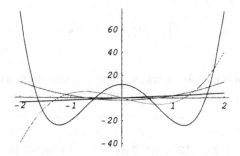

Fig. 30.1. *Hermite polynomials H_1, H_2, H_3, and H_4.*

The ground state corresponds to $n = 0$; that is,

$$\psi_0(\xi) = \exp(-\xi^2/2),$$

and the first excited states are

$$\psi_1(\xi) = 2e^{-\xi^2/2},$$
$$\psi_2(\xi) = \left(-2 + 4\xi^2\right) e^{-\xi^2/2},$$
$$\psi_3(\xi) = \left(-12\xi + 8\xi^2\right) e^{-\xi^2/2},$$
$$\psi_4(\xi) = \left(12 - 48\xi^2 + 16\xi^4\right) e^{-\xi^2/2}.$$

The eigenstates should have a norm equal to 1. Using *Mathematica* we find the norms of the first wave functions.

```
normPsi = Table[Integrate[(HermiteH[n,xi] Exp[-(1/2) xi^2],
{xi ,- Infinity, Infinity}], {n,0, 5}]
```

{Sqrt[Pi], 2 Sqrt[Pi], 8 Sqrt[Pi], 48 Sqrt[Pi], 384 Sqrt[Pi], 3840 Sqrt[Pi]}

The first normed eigenstates are, therefore, given by

$$\psi_0(\xi) = \frac{e^{-\xi^2/2}}{\pi^{1/4}}, \qquad \psi_1(\xi) = \frac{\sqrt{2}\xi e^{-\xi^2/2}}{\pi^{1/4}},$$
$$\psi_2(\xi) = \frac{(-1+2\xi^2)e^{-\xi^2/2}}{\sqrt{2}\pi^{1/4}}, \qquad \psi_3(\xi) = \frac{\xi(-3+2\xi^2)e^{-\xi^2/2}}{\sqrt{3}\pi^{1/4}},$$
$$\psi_4(\xi) = \frac{(3-12\xi^2+4\xi^4)e^{-\xi^2/2}}{2\sqrt{6}\pi^{1/4}}, \quad \psi_5(\xi) = \frac{\xi(15-20\xi^2+4\xi^4)e^{-\xi^2/2}}{2\sqrt{15}\pi^{1/4}}.$$

More generally, the normed eigenstates are given by

$$\psi_n(\xi) = \frac{e^{-\xi^2/2}}{\sqrt{2^n n!}\,\pi^{1/4}}\,H_n(\xi).$$

Because the Hamiltonian is invariant under the transformation $x \to -x$, we verify that the eigenstates are either even or odd functions.

```
psi[n_, xi_] :=
(Exp[- xi^2/2] / Sqrt[2^n n! Sqrt[Pi]] HermiteH[n, xi]
Table[Plot[ psi[n, xi], {xi, - 6, 6},
PlotStyle → {RGBColor[0, 0, 1]}], {n, 0, 5}];
```

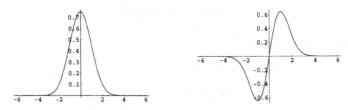

Fig. 30.2. *Normed wave functions ψ_0 and ψ_1.*

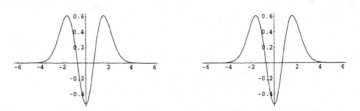

Fig. 30.3. *Normed wave functions ψ_2 and ψ_3.*

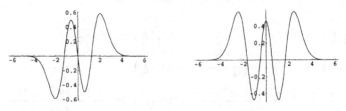

Fig. 30.4. *Normed wave functions ψ_4 and ψ_5.*

30.2 Creation and Annihilation Operators

Let us now define the creation a^+ and annihilation a^- operators by

$$a^+ = \frac{1}{\sqrt{2}}\left(x - \frac{d}{dx}\right) \quad \text{and} \quad a^- = \frac{1}{\sqrt{2}}\left(x + \frac{d}{dx}\right).$$

```
creationOperator[f_] := (- D[f, x] + x f) / Sqrt[2]
annihilationOperator[f_] := (D[f, x] + x f) / Sqrt[2]
```

The commutation relation is

```
annihilationOperator[creationOperator[f[x]]] -
creationOperator[annihilationOperator[f[x]]] // Simplify
```

f[x]

that is

$$[a^-, a^+] = a^- a^+ - a^+ a^- = 1.$$

We can express the Hamiltonian of the harmonic oscillator in terms of the creation and annihilation operators. Because

```
annihilationOperator[creationOperator[psi[x]]] +
creationOperator[annihilationOperator[psi[x]]] // Simplify
```

x^2 psi[x] - psi''[x]

we can write

$$H\psi = \frac{1}{2}(a^- a^+ + a^+ a^-)\psi.$$

If we now consider the action of the product $a^+ a^-$ on the different normed wave functions ψ_n defined above, we find

```
creationOperator[annihilationOperator[psi[0,x]]]
```

0

```
creationOperator[annihilationOperator[psi[1,x]]] ==
psi[1,x]
```

True

```
creationOperator[annihilationOperator[psi[2,x]]] ==
2 psi[2,x]) // Simplify
```

True

and more generally,

```
creationOperator[annihilationOperator[psi[n,x]]] ==
n psi[n,x]) // Simplify
```

True

that is,

$$a^+ a^- \psi_n = n\psi_n.$$

Hence, taking into account the expression of the Hamiltonian and the commutation relation, we have

$$
\begin{aligned}
H\psi_n &= \frac{1}{2}(a^- a^+ + a^+ a^-)\psi_n \\
&= \frac{1}{2}(a^+ a^- + 1 + a^+ a^-)\psi_n \\
&= \left(n + \frac{1}{2}\right)\psi_n.
\end{aligned}
$$

Quantum Square Potential

The problem is simple to solve analytically. Here we use *Mathematica* to obtain numerical results.

31.1 The Problem and Its Analytical Solution

The Schrödinger equation for the square-well potential is

$$-\frac{\hbar^2}{2m}\frac{d^2\psi}{dx^2} + V\psi = E\psi,$$

where

$$V(x) = \begin{cases} 0, & \text{if } |x| > a; \\ V_0, & \text{if } |x| \le a. \end{cases}$$

Let

$$k = \frac{\sqrt{2m|E|}}{\hbar}, \qquad q = \frac{\sqrt{2m(V_0 - |E|}}{\hbar}$$

and

$$\lambda = \frac{2mV_0a^2}{\hbar^2}, \qquad y = qa.$$

Then (because E is negative)

$$E = -\left(1 - \frac{y^2}{\lambda}\right)V_0, \qquad k = \frac{\sqrt{\lambda - y^2}}{a}.$$

The symmetry of the Schrödinger equation implies that the eigenfunctions are either even or odd.

The even solutions are

$$\psi(x) = \begin{cases} A\cos qx, & \text{if } |x| \le a; \\ Be^{kx}, & \text{if } x < -a; \\ Be^{-kx}, & \text{if } x > +a; \end{cases}$$

and the continuity of the eigenfunctions and their derivatives at $\pm a$ imposes

$$A\cos qa = Be^{-ka}$$

$$k = q\tan qa \Rightarrow \frac{\sqrt{\lambda - y^2}}{y} = \tan y.$$

The odd solutions are

$$\psi(x) = \begin{cases} C\sin qx, & \text{if } |x| \le a; \\ De^{kx}, & \text{if } x < -a; \\ -De^{-kx}, & \text{if } x > +a; \end{cases}$$

with

$$-C\sin qa = De^{-ka}$$

$$k = q\cot qa \Rightarrow \frac{\sqrt{\lambda - y^2}}{y} = -\cot y.$$

31.2 Numerical Solution

Choosing a as the unit of length (i.e., replacing a by 1), the Schrödinger equation becomes

$$\frac{d^2\psi}{dx^2} = U\psi,$$

where

$$U(x) = \begin{cases} -\lambda e, & \text{if } |x| > 1; \\ \lambda(e+1), & \text{if } |x| \le 1; \end{cases}$$

with $e = E/V_0$. Bound states correspond to $-1 < e < 0$.

In order to determine the numerical values of y in the case of even eigenfunctions, we draw the graphs of the two functions

$$y \mapsto \frac{\sqrt{\lambda - y^2}}{y} \quad \text{and} \quad y \mapsto \tan y.$$

to estimate starting points for FindRoot.

31.2.1 Energy Levels for $\lambda = 16$

```
lambda = 16;
Plot[{Sqrt[lambda - y^2] / y, Tan[y]},
{y, 0, 4}, PlotRange -> {0, 4},
TextStyle -> {FontSlant -> ''Italic'', FontSize -> 12}];
```

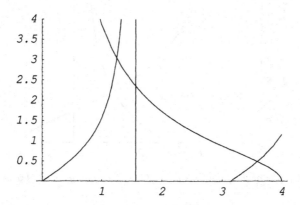

Fig. 31.1. *Graphs of the functions* $y \mapsto \sqrt{\lambda - y^2}/y$ *and* $y \mapsto \tan y$ *in the interval* $[0, 4]$.

There are two solutions. We can find their approximate values by selecting the graphics and pointing the intersection points while pressing the command key. We obtain $y_1 = 1.24$ and $y_3 = 3.6$. Using **FindRoot** we get more precise values.

```
y1 = y /. FindRoot[Sqrt[lambda-y^2] / y == Tan[y], {y, 1.24}]
```

1.25235

```
y1 = y /. FindRoot[Sqrt[lambda-y^2] / y == Tan[y], {y, 3.6}]
```

3.5953

Repeating the same procedure for odd eigenfunctions, we first draw the graphs

```
lambda = 16;
Plot[{Sqrt[lambda - y^2] / y, - Cot[y]},
{y, 0, 4}, PlotRange → {0, 4},
TextStyle → {FontSlant → ''Italic'', FontSize → 12}];
```

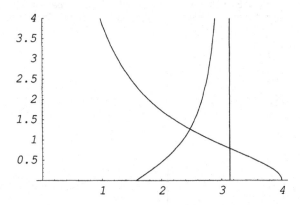

Fig. 31.2. *Graphs of the functions* $y \mapsto \sqrt{\lambda - y^2}/y$ *and* $y \mapsto -\cot y$ *in the interval* $[0, 4]$.

Pointing the intersection points while pressing the command key, we obtain $y_2 = 2.47$. Using FindRoot we get the more precise value:

```
y2 = y /. FindRoot[Sqrt[lambda - y^2] / y == - Cot[y],
{y, 2.47}]
```

2.47458

The corresponding values of the energy levels (in unit V_0) are:

```
energy1 = (- 1 + y1^2 / lambda) V0
energy2 = ( - 1 + y2^2 / lambda) V0
energy3 = ( - 1 + y3^2 / lambda) V0
```

− 0.192111 V0

− 0.617279 V0

− 0.901976 V0

31.2.2 Figure Representing the Potential and the Energy Levels

The following commands generate the figure showing the potential and the energy levels.

```
V[x_ /; Abs[x] > 1] := 0
V[x_ /; Abs[x] < 1] := - 1
Plot[V[x], {x, - 2, 2},
PlotStyle → Thickness[0.015],
PlotRange → {0.05, - 1.05}, Axes → False,
Ticks → {None, Automatic},
DefaultFont → {''Times'', 12},
Epilog → ({Line[{{1, energy1}, {- 1, energy1}}],
Text[''E1'', { 1.25, energy1}],
Text[ToString[energy1], {0, energy1 + 0.05}],
Line[{{1, energy2}, {- 1, energy2}}],
Text[''E2'', {1.25, energy2}],
Text[ToString[energy2], {0, energy2 + 0.05}],
Line[{{1, energy3}, {- 1, energy3}}],
Text[''E3'', {1.25, energy3}],
Text[ToString[energy3], {0, energy3 + 0.05}],
Text[''- V0'', {- 1.25, - 1}],
Text[''0'', {- 0.8, 0}]} /. V0 → 1)];
```

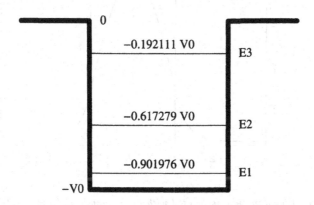

Fig. 31.3. *Square potential well and energy levels.*

31.2.3 Plotting the Eigenfunctions

We can plot the corresponding eigenfunctions. The operator StringJoin concatenates strings.

```
Clear[q, k, PsiEven1]
q = y1; k = Sqrt[lambda - y1^2];
PsiEven1[x_ /; x < - 1] := (Cos[q] Exp[k]) Exp[k x]
PsiEven1[x_ /; Abs[x] ≤ 1] := Cos[q x]
PsiEven1[x_ /; x > 1] := (Cos[q] Exp[k]) Exp[-k x]
PsiEven1norm = Sqrt[NIntegrate[(PsiEven1[x])^2, {x, -
Infinity, Infinity}]];
normedPsiEven1[x_] = (1 / PsiEven1norm) PsiEven1[x];
Plot[normedPsiEven1[x], {x, - 3, 3},
PlotStyle → {RGBColor[0, 0, 1]},
Epilog → {Line[{{1, - 1}, {1, 1}}],
Line[{{- 1, - 1}, {- 1, 1}}]},
TextStyle → {FontSlant → ''Italic'', FontSize → 12},
Frame → True, FrameLabel → {''x'', ''ψ''},
PlotLabel → StringJoin[''E = '', ToString[energy1]]];
```

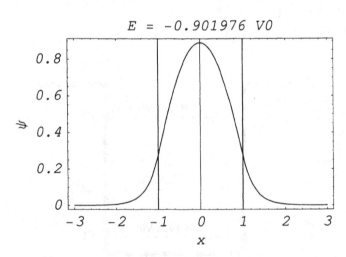

Fig. 31.4. *Eigenfunction associated with the energy level $E_1 = -0.901976\ V_0$.*

```
Clear[q, k, PsiOdd2]
q = y2; k = Sqrt[lambda - y2^2];
PsiOdd2[x_ /; x < -1] := (- Sin[q] Exp[k]) Exp[k x]
PsiOdd2[x_ /; Abs[x] ≤ 1] := Sin[q x]
PsiOdd2[x_ /; x > 1] := (Sin[q] Exp[k]) Exp[- k x]
PsiOdd2norm = Sqrt[NIntegrate[(PsiOdd2[x])^2, {x, - Infinity,
Infinity}]];
normedPsiOdd2[x_] = (1 / PsiOdd2norm) PsiOdd2[x];
eigenF2 = Plot[normedPsiOdd2[x], {x, - 3, 3},
PlotStyle → {RGBColor[0, 0, 1]},
Epilog → {Line[{{1, - 1}, {1, 1}}],
Line[{{- 1, - 1}, {- 1, 1}}]},
TextStyle → {FontSlant → ''Italic'', FontSize → 12},
Frame → True, FrameLabel → {''x'', ''ψ''},
PlotLabel → StringJoin[''E = '', ToString[energy2]]];
```

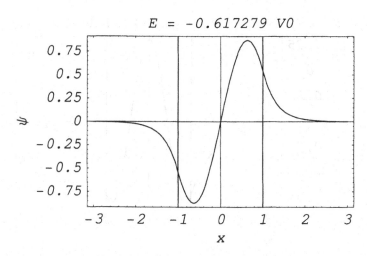

Fig. 31.5. *Eigenfunction associated to the energy level $E_2 = -0.617279\,V_0$.*

```
Clear[q, k, PsiEven1]
q = y3; k = Sqrt[lambda - y3^2];
PsiEven3[x_ /; x < - 1] := (Cos[q] Exp[k]) Exp[k x]
PsiEven3[x_ /; Abs[x] ≤ 1] := Cos[q x]
PsiEven3[x_ /; x > 1] := (Cos[q] Exp[k]) Exp[- k x]
PsiEven3norm = Sqrt[NIntegrate[(PsiEven3[x])^2, {x, -
Infinity, Infinity}]];
normedPsiEven3[x_] = (1 / PsiEven3norm) PsiEven3[x];
Plot[normedPsiEven3[x], {x, - 3, 3},
PlotStyle → {RGBColor[0, 0, 1]},
Epilog → {Line[{{1, - 1}, {1, 1}}],
Line[{{- 1, - 1}, {- 1, 1}}]},
TextStyle → {FontSlant → ''Italic'', FontSize → 12},
Frame → True, FrameLabel → {''x'', ''ψ''},
PlotLabel → StringJoin[''E = '', ToString[energy3]]];
```

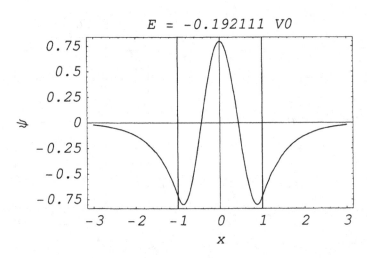

Fig. 31.6. *Eigenfunction associated with the energy level $E_3 = -0.192111\,V_0$.*

Our results agree with the general property of one-dimensional bound states:

If bound states are listed in the order of increasing energies, the nth eigenfunction has $n-1$ nodes.

Skydiving

We study various physical properties of a diver who jumps and falls freely from an airplane at a moderate altitude before pulling the ripcord of a parachute.

Upon leaving the airplane, the diver accelerates downwards due to the force of gravity. As her velocity increases, the air resistance (proportional to the square of her velocity) exerts a greater retarding force, and eventually balances the pull of gravity. From this time onward, the diver descends at a uniform velocity.

32.1 Terminal Velocity

We first determine the free-fall diver's velocity as a function of time and find the value of the terminal uniform velocity. g is the acceleration due to gravity and kdiver the coefficient of air resistance of the diver. Note that the terminal velocity in m/s is given by

```
terminalVelocity = Sqrt[ g / kdiver]
```

49.4975

```
g = 9.8; kdiver = 0.004;
sol1 = NDSolve[{v1'[t] == - g + kdiver *( v1[t])^2,
v1[0] == 0}, v1, {t, 0, 100}];
plv1 = Plot[Evaluate[v1[t] /. sol1], {t, 0, 40},
PlotRange → {0, - 60},
PlotStyle → {RGBColor[0, 0, 1]}, Axes → False,
Frame → True, FrameLabel → {''time'', ''velocity''},
RotateLabel → False, DefaultFont → {''Helvetica'', 14}];
```

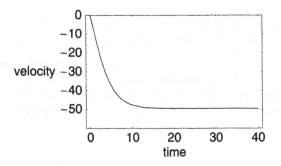

Fig. 32.1. *Free-fall diver's velocity as a function of time.*

32.2 Delaying Parachute Opening

We determine the diver's velocity as a function of time assuming that the diver opens the parachute after 30 seconds. kparachute is the coefficient of air resistance of the parachute.

```
g = 9.8; kdiver = 0.004; kparachute = 0.4;
sol2 = NDSolve[{v2'[t] == - g +
If[t < 30, kdiver, kparachute] *( v2[t])^2, v2[0] == 0},
v2, {t, 0, 100}];
plv2 = Plot[Evaluate[v2[t] /. sol2], {t, 0, 40},
PlotRange → {0, - 55}, PlotStyle → {RGBColor[0, 0, 1]},
Axes → False, Frame → True,
FrameLabel → {''time'', ''velocity''},
RotateLabel → False, DefaultFont → {''Helvetica'', 14}];
```

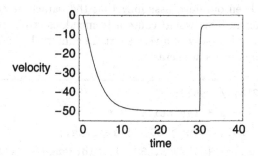

Fig. 32.2. *Diver's velocity as a function of time when parachute opening is delayed.*

In this case, the terminal velocity is

```
terminalVelocity = Sqrt[g / kparachute]
```

4.94975

The diver's velocity is found to decrease from approximately 50 m/s to 5 m/s in less than one second as confirmed by the plot below.

```
Plot[Evaluate[v2[t] /. sol2], {t, 29.9, 31},
PlotRange → {0, - 55}, PlotStyle → {RGBColor[0, 0, 1]},
Axes → False, Frame → True,
FrameLabel → {''time'', ''velocity''},
RotateLabel → False, DefaultFont → {''Helvetica'', 14}];
```

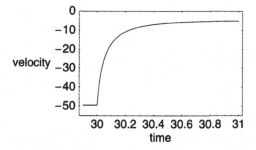

Fig. 32.3. *Rapid change of the diver's velocity when the parachute takes less than one second to fully open.*

This result has been obtained assuming that the parachute went from fully closed to fully open. This would cause a tremendous strain on the body as the velocity changed rapidly in a very short time interval as illustrated below on the plot of the diver's acceleration.

```
velocity = v2[t] /. sol2;
acceleration = - g + (velocity)^2 *
Which[t < 30, kdiver, t ≥ 30, kparachute];
Plot[acceleration, {t, 29.9, 30.3}, PlotRange → All,
PlotStyle → {RGBColor[0, 0, 1]}, Axes → False,
Frame → True, FrameLabel → {''time'', ''acceleration''},
RotateLabel → False, DefaultFont → {''Helvetica'', 14}];
```

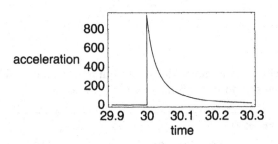

Fig. 32.4. *Diver's acceleration when the parachute is opened in a very short time.*

32.3 Taking into Account Time for Parachute to Open

To improve the model we should allow, say three seconds, for the parachute to open.

```
g = 9.8; kdiver = 0.004; kparachute = 0.4; dt = 3;
sol3 = NDSolve[{v3'[t] == - g + (v3[t])^2 *
Which[t < 30, kdiver,
t < 30 + dt, kdiver + (kparachute - kdiver) * (t - 30) / dt,
t ≥ 30 + dt, kparachute], v3[0] == 0}, v3, {t, 0, 100}];
plv3 = Plot[Evaluate[v3[t] /. sol3], {t, 0, 40},
PlotRange → {0, - 55},
PlotStyle → {RGBColor[0, 0, 1]}, Axes → False,
Frame → True, FrameLabel → {''time'', ''velocity''},
RotateLabel → False, DefaultFont → {''Helvetica'', 14}];
```

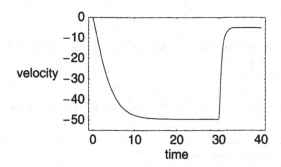

Fig. 32.5. *Diver's velocity when the parachute takes three seconds to fully open.*

Here is a more detailed plot showing the evolution of the diver's velocity and acceleration when she opens the parachute.

```
Plot[Evaluate[v3[t] /. sol3], {t, 29.9, 32},
PlotRange → {0, - 55}, PlotStyle → {RGBColor[0, 0, 1]},
Axes → False, Frame → True,
FrameLabel → {''time'', ''velocity''},
RotateLabel → False, DefaultFont → {''Helvetica'', 14}];
```

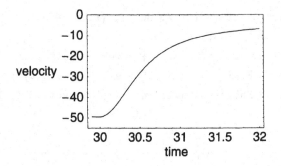

Fig. 32.6. *More detailed plot of the diver's velocity when the parachute takes three seconds to fully open.*

Plotting the acceleration, we can observe that the strain on the diver's body is less.

```
Clear[velocity, acceleration]
velocity = v3[t] /. sol3;
acceleration = - g + (velocity)^2 *
Which[t < 30, kdiver,
t < 30 + dt, kdiver + (kparachute - kdiver) * (t - 30) / dt ,
t ≥ 30 + dt, kparachute];
Plot[acceleration, {t, 29.5, 30.6},
PlotRange → {0, 60}, PlotStyle → {RGBColor[0, 0, 1]},
Axes → False, Frame → True,
FrameLabel → {''time'', ''acceleration''},
RotateLabel → False, DefaultFont → {''Helvetica'', 12}];
```

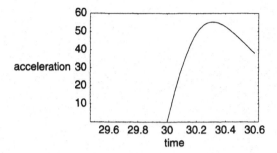

Fig. 32.7. *Diver's acceleration when the parachute takes three seconds to fully open.*

Tautochrone

The constancy of the period of a pendulum, when the amplitude of the oscillations is small, is said to have been discovered by Galileo Galilei (1564–1642) *circa* 1583 by comparing the period of the oscillations of a swinging lamp in a Pisa cathedral with his pulse rate. This property led Galileo and the Dutch mathematician, astronomer, and physicist Christiaan Huygens (1629–1695) to use a pendulum as a clock regulator.

As an astronomer, Huygens' interest in the accurate measurement of time led him to the discovery of a pendulum whose period is truly constant. In 1673, while living in Paris, he published his *Horologium Oscillatorium* that contained complete solutions of many problems of dynamics. In particular, he showed that, if a pendulum's bob swings along an arc of an inverted cycloid rather than that of a circle, the period of the oscillations is constant whatever the amplitude of these oscillations. The inverted cycloid is, therefore, a *tautochrone* curve (from the Greek, *tauto* meaning "the same" and *chronos* meaning "time"); that is, a curve such that the time required for a particle subjected to specific forces (here gravity) to slide down without friction to its lowest point is independent of its initial position on the curve.

33.1 Involute and Evolute

The *involute* of a planar curve γ is the locus Γ in the plane of γ of the endpoint of a thread kept taut as it is unwound from γ. The curve γ, which is then the locus of the centers of curvature of the curve Γ, is called the *evolute* (or *envelope* of the normals) of Γ. Because the involute of a cycloid is a cycloid (see below), if a pendulum is swung between two arches of inverted cycloids, the pendulum's bob traces out the involute of a cycloid, that is, a cycloid. If the parametric representation of a curve Γ is $(x(u), y(u))$, the parametric representation of its evolute is $(x(u) - R \sin \varphi, y(u) + R \cos \varphi)$, where R is the radius of curvature of γ, and φ the angle between the tangent vector and the Ox-axis. Because

$$R = \frac{(x^2 + y^2)^{3/2}}{x'y'' - x''y'},$$

$$\cos \varphi = \frac{x'}{(x'^2 + y'^2)^{1/2}},$$

$$\sin \varphi = \frac{y'}{(x'^2 + y'^2)^{1/2}},$$

the parametric representation of the evolute of a curve Γ is

$$\left(x(u) - \frac{(x'(u)^2 + y'(u)^2)\, y'(u)}{x'(u)y''(u) - x''(u)y'(u)}, \quad y(u) + \frac{(x'(u)^2 + y'(u)^2)\, x'(u)}{x'(u)y''(u) - x''(u)y'(u)} \right).$$

If the curve γ is the cycloid (on the cycloid see section 6.4.1) represented by

$$x(u) = u - \sin u \quad \text{and} \quad y(u) = 1 - \cos u = 2\sin^2(u/2),$$

its evolute is represented by

```
evolute = {x[u] - ((x'[u]^2 + y'[u]^2) y'[u]) /
(x'[u] y''[u] - x''[u] y'[u]),
y[u] + ((x'[u]^2 + y'[u]^2) x'[u]) /
(x'[u] y''[u] - x''[u] y'[u])} // Simplify
ParametricPlot[evolute, {u, 0, 2 Pi},
AspectRatio → Automatic];
```

$$\{u + \sin[u], \, - 2 \sin[\tfrac{u}{2}]^2\}$$

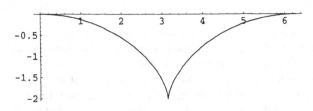

Fig. 33.1. *The evolute of a cycloid is a cycloid.*

The evolute of a cycloid is also a cycloid.

Figure 33.2 illustrates the property of a cycloid mentioned above, that is, *the involute of a cycloid is a cycloid,* and how to build a Huygens pendulum using this property.

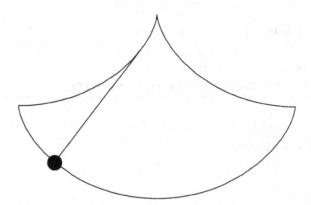

Fig. 33.2. *Huygens pendulum.*

33.2 The Cycloid

In this section we verify that the cycloid is a tautochrone curve.

The motion of the mass point is taking place in a vertical plane. In this plane, we choose the Ox-axis vertical pointing to the right and the Oy-axis vertical pointing upwards. The kinetic and potential energies of a particle of mass m are, respectively, $\frac{1}{2}mv^2$ and mgy, where v is the particle velocity and g the acceleration due to gravity.

The parametric equations of the inverted cycloid are

```
x[u_] := u - Sin[u];
y[u_] := - 1 + Cos[u];
```

Conservation of energy implies $\frac{1}{2}mv^2 + mgy = mgy_0$ assuming that when at $t = 0$, the particle located at (x_0, y_0), has a velocity equal to zero. Taking into account that

$$v^2 = \left(\frac{dx}{dt}\right)^2 + \left(\frac{dy}{dt}\right)^2,$$

we have

$$\frac{\sqrt{dx^2 + dy^2}}{dt} = \sqrt{2g(y_0 - y)}.$$

Hence,

$$dt = \frac{\sqrt{dx^2 + dy^2}}{\sqrt{2g(y_0 - y)}},$$

and replacing y_0 by $-1 + \cos u_0$, we obtain

```
Sqrt[(D[x[u], u])^2 + (D[y[u], u])^2] /
Sqrt[2 g (- (1 - Cos[u0]) -y[u])] // Simplify
```

$$\frac{\text{Sqrt}[1 - \text{Cos}[u]}{\text{Sqrt}[g(-\text{Cos}[u] + \text{Cos}[u0])]}$$

that is,

$$dt = \sqrt{\frac{1 - \cos u}{g(\cos u_0 - \cos u)}}\, du.$$

Note that for $y_0 = 0$ (i.e., $u_0 = 0$ and $\cos u_0 = 1$), which corresponds to the case of the swinging bob moving back and forth along the whole curve, $dt = dt/\sqrt{g}du$. The period T, equal to 4 times the integral from 0 to π, is equal to $4\pi\sqrt{1/g}$. If $0 < u_0 < \pi/2$, the period of the cycloidal pendulum is given by

$$T = 4 \int_{u_0}^{\pi} \sqrt{\frac{1 - \cos u}{g(\cos u_0 - \cos u)}}\, du.$$

Substituting $\cos u$ for $2\cos^2(u/2) - 1$, the expression of the period becomes

$$T = 4\sqrt{\frac{1}{g}} \int_{u_0}^{\pi} \frac{\sin\frac{u}{2}\, du}{\sqrt{\cos^2\frac{u_0}{2} - \cos^2\frac{u}{2}}}.$$

This expression suggests the further change of variable: $\xi = \cos(u/2)/(u_0/2)$, which finally gives

$$T = 8\sqrt{\frac{1}{g}} \int_0^1 \frac{d\xi}{\sqrt{1 - \xi^2}}.$$

This result shows that the period does not depend upon the initial position of the bob.

The integral is elementary. We find

$$T = 8\sqrt{\frac{1}{g}} \arcsin(1) = 4\pi\sqrt{\frac{1}{g}}.$$

Remark 1. Because the change of variable shows that the period T does not depend upon the initial position of the bob, we could have used the expression of the period with $u_0 = 0$, which gives $T = 4\sqrt{1/g} \int_0^\pi du = 4\pi\sqrt{1/g}$.

Remark 2. The period T should have the dimension of time, which is apparently not the case. Actually, the parametric equation of the cycloid should be written $\big(a(u - \sin u), -a(1 - \cos u)\big)$, where a is a characteristic length of the cycloid. In the above derivation of T we took $a = 1$.

33.3 Fractional Calculus

In order to be able to determine tautochrone curves for a particle in a potential U, we briefly describe the basic principles of fractional calculus [41, 37, 40].

The fractional derivative of a function f is an extension of the usual derivative $d^n f/dx^n$[1] to nonintegral values of n. The concept of the fractional derivative was mentioned for the first time in a letter from Guillaume de l' Hôpital (1661–1704) to Gottfried von Leibniz (1646–1716) dated September 30, 1695.

If n is a positive integer greater than 1, we have

$$\int_0^x \int_0^{u_{n-1}} \cdots \int_0^{u_1} f(u)\, du_1 \ldots du_{n-1} du = \frac{1}{(n-1)!} \int_0^x f(u)(x-u)^{n-1}\, du,$$

and if denoting by F_n the function

$$x \mapsto \begin{cases} 0, & \text{if } x \le 0; \\ \dfrac{x^{n-1}}{\Gamma(n)}, & \text{if } x > 0; \end{cases}$$

where Γ is the Euler Gamma function, we can write the result of integrating n times the function f as the convolution of f and F_n.

If α and β are two complex numbers such that Re α and Re β are both positive, assuming $x > 0$, then the convolution of F_α and F_β, denoted $F_\alpha * F_\beta$, is given by

$$\int_0^x \frac{\xi^{\alpha-1}}{\Gamma(\alpha)} \frac{(x-u)^{\beta-1}}{\Gamma(\beta)}\, d\xi = x^{\alpha+\beta-1} \int_0^1 \frac{u^{\alpha-1}(1-u)^{\beta-1}}{\Gamma(\alpha)\Gamma(\beta)}\, du = \frac{x^{\alpha+\beta-1}}{\Gamma(\alpha+\beta)},$$

where we took into account the relation

$$B(\alpha, \beta) = \int_0^1 u^{\alpha-1}(1-u)^{\beta-1}\, du = \frac{\Gamma(\alpha)\Gamma(\beta)}{\Gamma(\alpha+\beta)}.$$

---

[1] Leibniz is the inventor of the notation $d^n f/dx^n$ for the derivative of order n of a function f.

B is the Beta function. The result obtained above can be simply written $F_\alpha * F_\beta = F_{\alpha+\beta}$. By analytic continuation, the result remains valid for all values of α and β different from $0, -1, -2, \ldots$. Considering F_α and F_β as distributions belonging to $\mathcal{D}'_+$, which are distributions with support in $\mathbb{R}_+$ (set of nonnegative reals), we can further extend the relation $F_\alpha * F_\beta = F_{\alpha+\beta}$ for $\alpha = 0, -1, -2, \ldots$ and $\beta = 0, -1, -2, \ldots$, as a consequence of the relation $\delta^{(n)} * F_n = \delta$, where n is a positive integer and δ the Dirac distribution. Hence $\lim_{\alpha \to -n} F_\alpha = \delta^{(n)}$.

Note that $I(z) = \int_0^\infty x^{\alpha-1} e^{-zx} \, dx$ can be written $\Gamma(\alpha)/z^\alpha$ replacing in the integral zx by u and using the Cauchy theorem where the function $z \mapsto z^\alpha$ is such that it takes the value 1 for $z = 1$. Hence, the Laplace transform of the distribution F_α is $1/z^\alpha$, that is valid for any value of α by analytic continuation. This result gives another proof of the relation $\lim_{\alpha \to -n} F_\alpha = \delta^{(n)}$. For more details on distribution theory see [8].

The considerations above suggest defining the fractional derivative of order $\alpha > 0$ with respect to x of $f(x)$ as the convolution $D_x^\alpha * f$, where D_x^α is the distribution $x^{-\alpha+1}/\Gamma(-\alpha)$. Hence, the derivative of order $\alpha > 0$ with respect to x of the function $x \mapsto x^\lambda$ is $x^{\lambda-\alpha}/\Gamma(\lambda + 1 - \alpha)$. This last result shows that the fractional derivative of a constant c is not equal to zero! In fact

$$D_x^\alpha * c = c \lim_{\lambda \to 0} \frac{x^{\lambda-\alpha}}{\Gamma(\lambda + 1 - \alpha)} = \frac{cx^{-\alpha}}{\Gamma(1 - \alpha)}.$$

33.4 Other Tautochrone Curves

In the previous section, we contented ourselves with verifying that the cycloid was a tautochrone curve for the gravitational potential. In this section we are interested in the more general problem of finding tautochrone curves for a particle in a general potential.

If the particle is in a potential $U(y)$, assuming that for $t = 0$ the particle located at (x_0, y_0) has a velocity equal to zero, the conservation of energy implies $\frac{1}{2}mv^2 + U(y) = U(y_0)$, and reasoning as above, we obtain

$$dt = \frac{ds}{\sqrt{2(U(y_0) - U(y))}},$$

where we replaced $\sqrt{dx^2 + dy^2}$ by ds, the element of the arc. Integrating over a quarter of a period yields

$$T = 4 \int_0^{y_0} \frac{ds}{\sqrt{2(U(y_0) - U(y))}},$$

where the period T is a constant not depending upon y_0. The lower limit of integration assumes that when the velocity of the bob is maximum the potential is minimum at $y = 0$. We also assume that $U(0) = 0$, which can always be achieved by adding an appropriate constant.

Writing the integral above under the form

$$T = \frac{4\pi}{\Gamma(1/2)} \int_0^{U_0} \frac{\frac{ds}{dU}}{\sqrt{2(U_0 - U(y))}} \, dU,$$

shows that the period T is equal to 4 times π the derivative of order $-\frac{1}{2}$ with respect to $U_0 = U(y_0)$ of the function ds/dU. Note that $\Gamma(1/2) = \pi$. Taking the derivative of s with respect to $U(y)$ has the same form as the derivative of s with respect to U_0, thus $T = 4\pi D_U^{-1/2} D_U^1 s = 4\pi D_U^{1/2} s$, or, equivalently $s = D_U^{-1/2}(T/4\pi)$.

From the expression of the fractional derivative of a constant we derived in the previous section we finally obtain

$$s = \frac{T}{4\pi} \frac{\sqrt{U(u)}}{\Gamma(3/2)} = \frac{T}{2\pi^{3/2}} \sqrt{U(y)}$$

or

$$U(y) = \frac{4\pi^3}{T^2} s^2.$$

With these two relations we can either determine the tautochrone curve for a specific potential or determine the potential for which a given curve is tautochrone.

The cycloid whose parametric representation is $(x(u), y(u)) = (u + \sin u, \cos u - 1)$ is tautochrone for the potential $U(y)$ proportional to

```
(Integrate[Sqrt[(1 + Cos[u])^2 + (Sin[u])^2], u])^2 //
Simplify
```

$$16 \; \mathrm{Sin}[\frac{u}{2}]^2$$

Because $2\sin^2(u/2) = 1 - \cos u$, the potential $U(y)$ is a linear function of y corresponding to a gravitational potential.

For an arbitrary potential $U(y)$ we have

$$\frac{ds}{dU} = \frac{T}{4\pi^{3/2}} \frac{U'(y)}{\sqrt{U(y)}}$$

or

$$1 + (x'(y))^2 = \left(\frac{T}{4\pi^{3/2}}\right)^2 \frac{(U'(y))^2}{U(y)}.$$

Solving for $x'(y)$ and integrating, we obtain

$$x(y) = \int_0^y \sqrt{\left(\frac{T}{4\pi^{3/2}}\right)^2 \frac{(U'(y))^2}{U(y)} - 1}\, du.$$

If, for instance, we consider the quadratic potential $U(y) = ky^2$, the expression under the radical is constant and the corresponding tautochrone curve is a straight line passing through the origin.

Many other examples can be found in [16].

van der Pol Oscillator

The differential equation

$$\frac{d^2x}{dt^2} + \lambda \left(x^2 - 1\right) \frac{dx}{dt} + x = 0,$$

describes the dynamics of the first relaxation oscillator named after the Dutch electrical engineer Balthasar van der Pol (1889–1959) [59, 60]. It is a harmonic oscillator that includes a nonlinear friction term $\lambda \left(x^2 - 1\right) \dot{x}$. If the amplitude of the oscillations is large, the amplitude-dependent "coefficient" of friction $\lambda \left(x^2 - 1\right)$, is positive, and the oscillations are damped. As a result, the amplitude of the oscillations decreases, and the amplitude-dependent "coefficient" of friction eventually becomes negative, corresponding to a sort of antidamping. If we put

$$x_1 = x \quad \text{and} \quad x_2 = \dot{x},$$

the van der Pol equation takes the form

$$\frac{dx_1}{dt} = x_2, \ \frac{dx_2}{dt} = -x_1 - \lambda \left(x_1^2 - 1\right) x_2.$$

The solution of this system of two differential equations gives the trajectory in the phase space, that is, the (x_1, x_2)-plane.

It can be proved that, according to the sign of the parameter λ, the van der Pol oscillator exhibits two different behaviors when the time t goes to infinity.

Let us solve numerically the system of two differential equations for the initial condition $x_1(0) = x_2(0) = 0.2$ and plot the trajectories for $\lambda = 0.5$ and $\lambda = -0.5$. In all phase space plots, the initial point is represented by a blue dot.

```
lambda = - 0.5;
solution2 = NDSolve[{x1'[t] == x2[t],
x2'[t] == - x1[t] - lambda (x1[t]^2 - 1) x2[t],
x1[0] == x2[0] == 0.2}, {x1,x2}, {t, 0, 30}]
```

{{x1 → InterpolatingFunction[{{0., 30.}}, <>],

x2 → InterpolatingFunction[{{0., 30.}}, <>]}}

```
plNeg = ParametricPlot[Evaluate[{x1[t], x2[t]} /. solution2],
{t, 0, 30}, DisplayFunction → Identity];
init = Graphics[{PointSize[0.04], RGBColor[0, 0, 1],
Point[{0.2, 0.2}]}];
stEq = Graphics[{PointSize[0.04], RGBColor[1, 0, 0],
Point[{0., 0.}]}];
Show[{init, stEq, plNeg}, Axes → False, Frame → True,
AspectRatio → Automatic, PlotRange → All,
TextStyle → {FontSlant → ''Italic'', FontSize → 12},
DisplayFunction → $DisplayFunction];
```

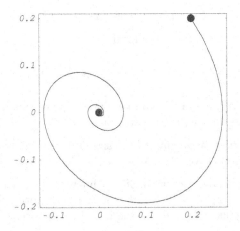

Fig. 34.1. *Trajectory in the (x_1, x_2)-phase space of the van der Pol oscillator for $\lambda = -0.5$ and $t \in [0, 30]$.*

The trajectory converges to a fixed point (red dot).

```
lambda = 0.5;
solution3 = NDSolve[{x1'[t] == x2[t],
x2'[t] == - x1[t] - lambda (x1[t]^2 - 1) x2[t],
x1[0] == 0, x2[0] == 0.2}, {x1, x2}, {t, 0, 50}]
```

{{x1 → InterpolatingFunction[{{0., 50.}}, <>],
x2 → InterpolatingFunction[{{0., 50.}}, <>]}}

```
plPos = ParametricPlot[Evaluate[{x1[t], x2[t]} /. solution3],
{t, 0, 20}, DisplayFunction → Identity];
init = Graphics[{PointSize[0.04], RGBColor[0, 0, 1],
Point[{0.2, 0.2}]}];
stCycle = ParametricPlot[Evaluate[{x1[t], x2[t]} /.
solution3],
{t, 20, 50}, PlotStyle → {RGBColor[1, 0, 0]},
DisplayFunction → Identity];
Show[{init, stCycle, plPos}, Axes → False, Frame → True,
AspectRatio → Automatic, PlotRange → All,
TextStyle → {FontSlant → ''Italic'', FontSize → 12},
DisplayFunction → $DisplayFunction];
```

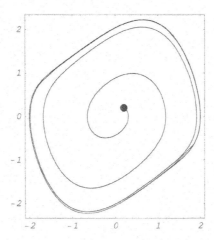

Fig. 34.2. *Trajectory in the* (x_1, x_2)-*phase space of the van der Pol oscillator for* $\lambda = 0.5$ *and* $t \in [0, 50]$.

The trajectory converges to a limit cycle (red curve).

Note that the asymptotic behaviors are already clearly visible after a time t of the order of 30.

van der Waals Equation

In 1873 the Dutch physicist Johannes Diederik van der Waals (1837–1923) obtained his doctoral degree for a thesis on the continuity of the gas and liquid state in which he put forward his famous equation of state that included both the gaseous and liquid states. He showed that these two states could merge in a continuous manner and are in fact the same, their only difference being of a quantitative and not of a qualitative nature. These results were considered very important,[1] and he was awarded the Nobel Prize in Physics in 1910 "for his work on the equation of state for gases and liquids."

35.1 Equation of State

If we denote by P the pressure, V the volume, N the number of moles, T the absolute temperature, and R the molar gas constant equal to 8.31441 joules per mole-kelvin, the ideal-gas equation of state is $PV = NRT$. This equation of state, which describes approximately the behavior of a gas at low pressures, is a simple consequence of the following laws.

1. *Charles' law* (1787). At constant volume V, the pressure P is a linear function of the temperature T.

2. *Gay–Lussac's law* (1802). At constant pressure P, the volume V is a linear function of the temperature T.

3. *Boyle–Mariotte's law* (Boyle 1662, Mariotte 1676). At constant temperature T, the product PV is constant.

4. *Avogadro's law* (1811). At given temperature T and pressure P, the volume of one mole of gas is the same for all gases.

---

[1] Maxwell regarded van de Waals as one of the foremost physicists in molecular science.

The ideal-gas equation of state can also be derived from the kinetic theory of gases assuming that gas molecules have zero volume and do not interact. These two assumptions are obviously not true, and in 1881 van der Waals introduced into the ideal-gas equation of state two parameters a and b that, respectively, take into account the nonzero volume of the molecules and the existence of interactions between the molecules. He thus obtained the following more realistic equation, which, for a mole of gas, is written

$$\left(P + \frac{a}{V^2}\right)(V - b) = RT \ (a > 0, b > 0),$$

where here V denotes the molar volume.

For what follows, it is convenient to express P as a function of V and T. Using *Mathematica* we obtain

```
vdWeqn = (P + a/V^2) (V - b) == R T
solP = Solve[vdWeqn, P
```

$$\{\{P \rightarrow \frac{-(a \ b) + a \ V - R \ T \ V^2}{(b - V) \ V^2}\}\}$$

The function $(T, V) \mapsto P(T, V)$ can therefore be defined by

```
P[T_, V_] := solP[[1, 1, 2]]
```

Because

$$\left(\frac{\partial P}{\partial V}\right)_T = -\frac{RT}{(V - b)^2} + \frac{2a}{V^3}$$

it follows that, for not too high values of T, there exist values of the molar volume V for which the isothermal compressibility is negative. This violates the stability condition, and for these values of V the van der Waals equation cannot be valid. We show below how van der Waals used this feature to describe the gas-liquid first-order phase transition.

35.2 Critical Parameters

There exists an isotherm $P = P(T_c, V)$ that has an inflection point. The coordinates (V_c, T_c, P_c) of that point satisfy the equations

$$\left(\frac{\partial P}{\partial V}\right)_{T=T_c} = \left(\frac{\partial^2 P}{\partial V^2}\right)_{T=T_c} = 0.$$

That is,

```
eqn1 = D[P[T, V], {V, 1}] == 0;
eqn2 = D[P[T, V], {V, 2}] == 0;
Solve[{eqn1, eqn2}, {V, T}]
```

$$\left\{\left\{T \to \frac{8\,a}{27\,b\,R}, \ V \to 3\,b\right\}\right\}$$

Replacing T and V in the expression of $P(T, V)$ by these values yields

```
P[T, V] /. {T → 8 a / (27 b R), V → 3 b}
```

$$\frac{a}{27\,b^2}$$

Hence, the coordinates (V_c, T_c, P_c) of the critical point are

$$P_c = \frac{a}{27\,b^2}, \ T_c = \frac{8\,a}{27\,b\,R}, \ V_c = 3\,b.$$

Combining these three coordinates in a dimensionless ratio, we obtain

```
Pc = a (27 b^2); , Tc = (8 a) / (27 b R); Vc = 3 b;
Pc Vc / (R Tc)
```

$$\frac{8}{3}$$

The dimensionless ratio does not depend upon a, b, and R, and should, therefore, be universal. This is not the case. The following table gives the value of this ratio for some gases.

Hydrogen	Oxygen	Water
3.29	3.42	4.35

35.3 Law of Corresponding States

Expressing the pressure, the volume, and the temperature as a fraction of their critical values we obtain the following dimensionless expression of the van der Waals equation.

```
dimensionlessEqn = Simplify[
vdWeqn /. {V → V Vc, T → T Tc, P → P Pc}
```

$$\frac{a \ (3 - 9 \ V + (P + 8 \ T) \ V^2 - 3 \ P \ V^3)}{b \ V} \ == \ 0$$

Using **Reduce**, this result can be further simplified.

```
reducedP = Solve[dimensionlessEqn, P]
```

$$\{\{P \ \rightarrow \ \frac{3 - 9 \ V + 8 \ T \ V^2}{V^2 \ (-1 + 3 \ V)}\}\}$$

Because the reduced van der Waals equation, defined by

```
P[T_, V_] := reducedP[[1,1,2]]
```

does not contain any adustable parameter, it is universal. It should, therefore, be valid for all gases. This result, discovered by van der Waals, is known as the *law of corresponding states*. This law is very approximately verified.

In this case, the critical coordinates are $(1, 1, 1)$. Thus, if we want to draw the isotherms in the vicinity of the critical temperature, we have to consider the critical isotherm obtained for $t = 1$, and a few other isotherms for t close to 1.

```
functionsTable = Table[P[V,T], {T, 0.9, 1.1, 0.05}];
Plot[{functionsTable[[1]], functionsTable[[2]],
functionsTable[[3]], functionsTable[[4]],
functionsTable[[5]]}, {V, 0.5, 2.5},
PlotStyle → {RGBColor[0, 0, 1], RGBColor[0, 1, 0],
RGBColor[1, 0, 0], RGBColor[0, 1, 0], RGBColor[0, 0, 1]},
TextStyle → {FontFamily → ''Helvetica'', FontSize → 10},
Frame → True];
```

We used the options **PlotStyle** to specify the colors of the different isotherms and **TextStyle** to specify the font and its size.

Remark. Note that from any equation of state that, in as much as the van der Waals equation contains exactly three independent parameters, we can derive

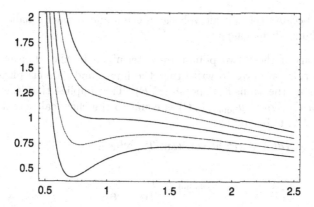

Fig. 35.1. *Dimensionless van der Waals isotherms.*

a universal form of the equation of state which implies a law of corresponding states. Here are two other historical examples:

Berthelot's equation

$$\left(P + \frac{a}{TV^2}\right)(V - b) = RT \ (a > 0, b > 0)$$

Dieterici's equation

$$P(V - b) \exp\left(\frac{a}{RTV}\right) = RT \ (a > 0, b > 0)$$

35.4 Liquid–Gas Phase Transition

For $T < T_c$, the two points on the van der Waals isotherm where

$$\frac{\partial P}{\partial V} = 0$$

represent states that have reached the limit of stability.

The point corresponding to the largest volume is interpreted as the limit of stability of the gaseous phase while the point corresponding to the smallest volume is the limit of stability of the liquid phase.

According to the Clapeyron equation, at a given temperature, below T_c, the liquid $\rightarrow$ gas first-order phase transition occurs at fixed values of P and T and is characterized by a discontinuity ΔV of the specific volume. Therefore, if, at temperature T, vaporization occurs at pressure P, the system should not follow the isotherm but follow a horizontal line joining the point (V_{liquid}, P) to the point (V_{gas}, P), the discontinuity being equal to $\Delta V = V_{\text{liquid}} - V_{\text{gas}}$ (ΔV

is negative because the specific volume of the liquid is much smaller than the specific volume of the gas).

The positions of these two points are determined by the Maxwell rule. To derive this rule, we have to write that the liquid and gaseous phases are in equilibrium at the transition point (T, P). This implies that the chemical potential of the liquid $\mu_{\text{liquid}}(T, P)$ is equal to the chemical potential of the gas $\mu_{\text{gas}}(T, P)$ at that point.

Because $d\mu - S\,dt + V\,dP$, at constant T we have

$$\mu_{\text{gas}} - \mu_{\text{liquid}} = \int_{(V_{\text{liquid}}, P)}^{(V_{\text{gas}}, P)} V\,dP.$$

Integrating by parts yields

$$\mu_{\text{gas}} - \mu_{\text{liquid}} = P\left(V_{\text{gas}} - V_{\text{liquid}}\right) - \int_{(V_{\text{liquid}}, P)}^{(V_{\text{gas}}, P)} P\,dV.$$

That is, the horizontal line cuts the isotherm so that the area between the isotherm above the straight line and the straight line is equal to the area between the isotherm below the straight line and the straight line.

Using the reduced van der Waals equation of state we define a function `transitionPoint` that given a `temperature` below the critical point temperature (in reduced coordinates the critical temperature and critical pressure are both equal to one) determine the pressure and the reduced volumes of the gas and liquid phases.

The idea is to solve the following system of two equations

```
(vGas - vLiquid) P[temperature, vLiquid] == Integrate[P[T,
V], {V, vLiquid, vGas}]
```

and

```
P [temperature, vGas] == P[temperature,vLiquid]
```

where `temperature` is given and vGas and vLiquid are the two unknowns; the transition pressure is then obtained once vGas and vLiquid are found. Because we use FindRoot to solve this system, we need to find starting values for vGas and vLiquid.

vGas and vLiquid should both be between the two volume values' solution of the equation $\partial P / \partial V = 0$, therefore let us first determine these values.

```
Clear[temperature]
temperature = 0.85;
volumeBoundaries = Solve[D[P[temperature,V], V] == 0, V] ==
0, V]
```

{{V → 0.254445}, {V → 0.67168},
{V → 1.72093}}

From the expression of the reduced van der Waals equation, we see that any volume value less than $\frac{1}{3}$ is not physical. Looking at the graph of the function $P[T, V]$, we see that the straight line representing the Maxwell construction should be closer to the maximum than to the minimum of p[t,v]. We therefore define startingPressure by

```
startingPressure = 0.2 P[temperature,V /.
volumeBoundaries[[2]]] +
0.8 P[temperature, V /. volumeBoundaries[[3]]]
```

0.506369

and we choose the starting volumes values' from the solutions of

```
startingVolumes = Solve[P[temperature, V] ==
startingPressure, V]
```

{{V → 0.553196}, {V → 1.1487},
{V → 3.10775}}

When solving the system of two equations mentioned above, in order to find vLiquid we start from startingVolumes[[1,1, 2]] and look for a solution lying between 0 and startingVolumes[[2,1, 2]], and to find vGas, we start from startingVolumes[[3,1, 2]] and look for a solution lying between startingVolumes[[2,1, 2]] and Infinity.

Taking all these considerations into account the function transitionPoint can be defined as follows:.

```
transitionPoint[temperature_] :=
Module[{volumeBoundaries, startingPressure, startingVolumes,
volumes},
volumeBoundaries = Solve[D[P[temperature, V], V] == 0, V];
startingPressure = 0.2 P[temperature, V /.
volumeBoundaries[[2]]] + 0.8 P[temperature, V /.
volumeBoundaries[[3]]];
startingVolumes = Solve[P[temperature, V] ==
startingPressure, V];
area1 = (vGas - vLiquid) P[temperature, V] /. V → vLiquid;
F[v_] := Integrate[P[T, V], V] /. T → temperature;
area2 = F[vGas] - F[vLiquid];
volumes = FindRoot[Evaluate[{area1 == area2 /.
T → temperature,
P[temperature,vLiquid] == P[temperature,vGas]}],
{vLiquid, startingVolumes[[1,1, 2]], 0 ,
startingVolumes[[2,1, 2]]}, {vGas, startingVolumes[[3,1,2]],
startingVolumes[[2,1, 2]], Infinity}];
pTransition = P[temperature, vLiquid] /.volumes[[1]];
NumberForm[{temperature, pTransition,
volumes[[1, 2]], volumes[[2, 2]]}, 6]]
```

```
transitionPoint[0.85]
```

{0.85, 0.504492, 0.55336, 3.12764}

Figure 35.2 shows the Maxwell construction.

```
vdW = Plot[P[0.85, V], {v, 0.5, 4},
PlotStyle → {RGBColor[1, 0, 0]}, Frame → True,
DisplayFunction → Identity];
maxwell = Graphics[{RGBColor[0, 0, 1],
Line[{{0.553, 0.504}, {3.128, 0.504}}]}];
Show[{vdW, maxwell}, DisplayFunction → $DisplayFunction];
```

This function is valid for an interval of temperature values between 0.8438—
value below which there exist points on the isotherm with a negative pressure—

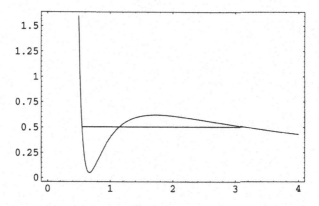

Fig. 35.2. *Maxwell construction.*

and 0.98—the value above which machine precision in not sufficient. Here is a table from which we could find the coexistence boundary in the (V,P) diagram.

```
Table[transitionPoint[k], {k,0.85, 0.98, 0.01}]
```

{{0.85, 0.504492, 0.55336, 3.12764},
{0.86, 0.531249, 0.561955, 2.9545},
{0.87, 0.55887, 0.571159, 2.79091},
{0.88, 0.587363, 0.581059, 2.63597},
{0.89, 0.616737, 0.591763, 2.48888},
{0.9, 0.646998, 0.603402, 2.34884},
{0.91, 0.678155, 0.616148, 2.2151},
{0.92, 0.710215, 0.630225, 2.0869},
{0.93, 0.743184, 0.645932, 1.96343},
{0.94, 0.77707, 0.663692, 1.84383},
{0.95, 0.811879, 0.684122, 1.72707},
{0.96, 0.847619, 0.708189, 1.61181},
{0.97, 0.884294, 0.737556, 1.49603},
{0.98, 0.921912, 0.775539, 1.3761}}

Bidirectional Pedestrian Traffic

36.1 Self-Organized Behavior

Animal groups display a variety of remarkable coordinated behaviors. For example, all the members in a school of fish change direction simultaneously without any obvious cue; in the same way, while foraging, birds in a flock alternate feeding and scanning. Self-organized motion in schools of fish or flocks of birds is not specific to animal groups. Pedestrian crowds display self-organized spatiotemporal patterns that are not imposed by any controller: on a crowded sidewalk, pedestrians walking in opposite directions tend to form lanes along which walkers move in the same direction.

All these behaviors have in common the following characteristics.

1. They consist of a large number of interacting agents.

2. They exhibit emergence; that is, a self-organizing collective behavior difficult to anticipate from knowledge of the agents' behavior.

3. Their emergent behavior does not result from the existence of a central controller.

The appearance of emergent properties is the single most distinguishing feature of the so-called complex systems [9].

In what follows we build up a *Mathematica* program of two groups of pedestrians moving in opposite directions in a passageway.

36.2 Initial Configuration of Pedestrians Moving in Opposite Directions in a Passageway

The passageway is represented by a square lattice of length L and width $W < L$. Each cell is either empty or occupied by a pedestrian. Pedestrians are divided in two groups moving in opposite directions. Group 1 moves to the west (i.e., the right) and group 2 to the east (i.e., the left). The following function generates a configuration of pedestrians in a multilane passageway with exact numbers of westbound and eastbound pedestrians.

```
initialMultiLaneConfig[numberOneParticles_Integer,
numberTwoParticles_Integer, latticeWidth_Integer,
latticeLength_Integer] :=
Module[{sum1, sum2, s},
sum1 = 0; sum2 = 0;
s = Table[0, {i, 1, latticeWidth}, {j, 1, latticeLength}];
While[sum1 < numberOneParticles,
i = Random[Integer, {1, latticeWidth}];
j = Random[Integer, {1, latticeLength}];
If[s[[i, j]] == 0, (s[[i, j]] = 1; sum1 = sum1 + 1), sum1 =
sum1]];
While[sum2 < numberTwoParticles,
i = Random[Integer, {1, latticeWidth}];
j = Random[Integer, {1, latticeLength}];
If[s[[i, j]] == 0, (s[[i, j]] = 2; sum2 = sum2 + 1), sum2 =
sum2]]; s]
```

The While loops are used to add pedestrians belonging to either group 1 or group 2 until the number of pedestrians in each group reaches its exact value (numberOneParticles or numberTwoParticles). The total number of cells is latticeLength * latticeWidth. The cell states are either 0 (empty), 1 (occupied by a pedestrian of group 1), or 2 (occupied by a pedestrian of group 2).

```
(config = initialMultiLaneConfig[7,7, 3, 13])//TableForm
```

```
1 0 0 1 0 2 0 0 2 0 0 2 1
2 0 0 0 0 0 0 0 0 0 0 0 0
2 1 0 2 0 1 0 1 2 1 0 0 0
```

The next step is to write down a sequence of functions representing the moving rules. As for most lattice models we assume that the motion takes place on a two-dimensional torus; that is, we assume so-called periodic boundary conditions. This trick is frequently used, in particular, in solid-state physics to remove surface effects. It makes a simulation dealing with a finite number of atoms as if this number were infinite. For example, if we consider a finite two-dimensional lattice of length L and width W, that is, a set of cells $\{(i,j) \mid 1 \le i \le W, 1 \le j \le L\}$, then the cells to the east of cells (i, L) $(1 < i < W)$ are, respectively, the cells $(i, 1)$ $(1 \le i \le W)$ which are to the west of cells (i, L), and the cells to the south of cells (W, j) $(1 \le j \le L)$ are, respectively, the cells $(1, j)$ $(1 \le j \le L)$ which are to the north of cells (W, j). Taking $L = 3$ and $W = 2$, instead of the finite lattice

$c_{1,1}$ $c_{1,2}$ $c_{1,3}$
$c_{2,1}$ $c_{2,2}$ $c_{2,3}$
$c_{3,1}$ $c_{3,2}$ $c_{3,3}$

we consider the following infinite lattice.

$$
\begin{array}{cccccccccccc}
\bullet \\
\bullet\bullet\bullet & c_{3,3} & c_{3,1} & c_{3,2} & c_{3,3} & c_{3,1} & c_{3,2} & c_{3,3} & c_{3,1} & \bullet\bullet\bullet \\
\bullet\bullet\bullet & c_{1,3} & c_{1,1} & c_{1,2} & c_{1,3} & c_{1,1} & c_{1,2} & c_{1,3} & c_{1,1} & \bullet\bullet\bullet \\
\bullet\bullet\bullet & c_{2,3} & c_{2,1} & c_{2,2} & c_{2,3} & c_{2,1} & c_{2,2} & c_{2,3} & c_{2,1} & \bullet\bullet\bullet \\
\bullet\bullet\bullet & c_{3,3} & c_{3,1} & c_{3,2} & c_{3,3} & c_{3,1} & c_{3,2} & c_{3,3} & c_{3,1} & \bullet\bullet\bullet \\
\bullet\bullet\bullet & c_{1,3} & c_{1,1} & c_{1,2} & c_{1,3} & c_{1,1} & c_{1,2} & c_{1,3} & c_{1,1} & \bullet\bullet\bullet \\
\bullet\bullet\bullet & c_{2,3} & c_{2,1} & c_{2,2} & c_{2,3} & c_{2,1} & c_{2,2} & c_{2,3} & c_{2,1} & \bullet\bullet\bullet \\
\bullet\bullet\bullet & c_{3,3} & c_{3,1} & c_{3,2} & c_{3,3} & c_{3,1} & c_{3,2} & c_{3,3} & c_{3,1} & \bullet\bullet\bullet \\
\bullet\bullet\bullet & c_{1,3} & c_{1,1} & c_{1,2} & c_{1,3} & c_{1,1} & c_{1,2} & c_{1,3} & c_{1,1} & \bullet\bullet\bullet \\
\bullet
\end{array}
$$

One of the first pedestrian traffic models is due to Fukui and Ishibashi [17, 18]. Here, adopting a slightly different point of view, we build up a much simpler deterministic bidirectional pedestrian traffic model. As in most pedestrian traffic models, a pedestrian moves forward to the cell in front of him if it is empty. If this cell is occupied by another pedestrian moving in the same direction, the pedestrian does not move, but, if it is occupied by a pedestrian moving in the opposite direction, the pedestrian moves to the cell in front of his right adjacent cell (with respect to the moving direction), and if this cell is also occupied, he moves to his right adjacent cell. If both cells are occupied, the pedestrian does not move. This deterministic behavior simplifies the moving rule. In all cases, pedestrians who can move forward have the right of way. Eastbound (resp., westbound) pedestrians move at odd (resp., even) time steps. As a result of the lane-changing rule, to avoid collisions, the local walking rule, which gives the state $s(i, j, t + 1)$ at time $t + 1$ of cell (i, j) as a function of the states of the 9 cells $\{(p, q) \mid p \in \{i - 1, i, i + 1\}, q \in \{j - 1, j, j + 1\}\}$ at time t, is of the form

$$s(i, j, t + 1) = F\big(s(i - 1, j - 1, t), s(i - 1, j, t), s(i - 1, j + 1, t),$$
$$s(i, j - 1, t), s(i, j, t), s(i, j + 1, t),$$
$$s(i + 1, j - 1, t), s(i + 1, j, t), s(i + 1, j + 1, t)\big)$$

where $i \in \{1, W\}$ and $j \in \{1, L\}$, depends on a lesser number of variables and is not probabilistic. At each time step the motion of a given pedestrian is determined by his state and the states of his 8 neighbors. Such a 9-cell neighborhood is known in cellular automata theory as a *Moore neighborhood*. Because each cell can be in 3 different states, the Moore neighborhood can be in $3^9 = 19,683$ different configurations. To apply the motion rule F to a given lattice, taking into account the interaction of each pedestrian with his Moore neighborhood, following Gaylord and D'Andria [19] (this reference contains many interesting *Mathematica* programs of agent-based cellular automata models), we use a function **neighborhood** defined by

```
neighborhood[F__, lattice_] :=
MapThread[F, Map[RotateRight[lattice, #] &,
{{1, 1}, {1, 0}, {1, -1}, {0, 1}, {0, 0},
{0, -1}, {-1, 1}, {-1, 0}, {-1, -1}}], 2]
```

Remember that the **Blanksequence** __ (two underscores) is a pattern object that can stand for any sequence of one or more *Mathematica* expressions.

The function **neighborhood** takes correctly into account the cyclic boundary conditions as shown below.

```
(lattice = {{1, 2, 3, 4, 5}, {6, 7, 8, 9, 10},
{11,12,13,14,15}, {16,17,18,19, 20}}) // TableForm
```

```
1    2    3    4    5
6    7    8    9    10
11   12   13   14   15
16   17   18   19   20
```

```
{neighborhood[F, #] & [lattice] [[1,1]],
neighborhood[F, #] & [lattice] [[2,3]]}
```

```
{F[20, 16, 17, 5, 1, 2, 10, 6, 7],
F[2, 3, 4, 7, 8, 9, 12, 13, 14]}
```

The command RotateRight[list, n] cycles the elements of list n positions to the right. For example:

```
RotateRight[{a, b, c, d, e, f}, 2]
```

```
{e, f, a, b, c, d}
```

The command RotateLeft[list, n] cycles the elements of list n positions to the left, and we obviously have

```
RotateRight[{a, b, c, d, e, f}, 2] == RotateLeft[{a, b, c, d,
e, f}, - 2]
```

True

36.3 Moving Rules for Type 1 Pedestrians

Here is the list of the moving rules for type 1 pedestrians according to the structure of the Moore neighborhood. A Blank _ (single underscore) stands for any site value 0, 1, or 2, and x | y stands for x OR y.

```
pedestrian1[_, _, _, 1, 0, _, _, _, _] := 1;
```

```
pedestrian1[_, _, _, _, 1, 0, _, _, _] := 0;
pedestrian1[_, _, _, _, 1, 1, _, _, _] := 1;
pedestrian1[_, _, _, _, 1, 2, _, 0 | 2, 0] := 0;
pedestrian1[_, _, _, _, 1, 2, _, 1, 0] := 1;
pedestrian1[_, _, _, _, 1, 2, 0 | 2, 0, 1] := 0;
pedestrian1[_, _, _, _, 1, 2, 1, 0, 1] := 1;
pedestrian1[_, _, _, _, 1, 2, _, 1 | 2, 1] := 1;
pedestrian1[_, _, _, _, 1, 2, 0 | 2, 0, 2] := 0;
pedestrian1[_, _, _, _, 1, 2, 1, 0, 2] := 1;
pedestrian1[_, _, _, _, 1, 2, _, 1 | 2, 2] := 1;
```

```
pedestrian1[0, 0, _, 0 | 2, 0, _, _, _, _] := 0;
pedestrian1[0, 1, 0 | 1, 0 | 2, 0, _, _, _, _] := 0;
pedestrian1[0, 1, 2, 0 | 2, 0, 0, _, _, _] := 0;
pedestrian1[0, 1, 2, 0 | 2, 0, 1 | 2, _, _, _] := 1;
pedestrian1[0, 2, _, 0 | 2, 0, _, _, _, _] := 0;
pedestrian1[1, 0, _, 0 | 2, 0, _, _, _, _] := 0;
pedestrian1[1, 1, 0 | 1, 0 | 2, 0, _, _, _, _] := 0;
pedestrian1[1, 1, 2, 0 | 2, 0, 0, _, _, _] := 0;
pedestrian1[1, 1, 2, 0 | 2, 0, 1 | 2, _, _, _] := 1;
pedestrian1[1, 2, _, 0 | 2, 0, _, _, _, _] := 1;
pedestrian1[2, 0 | 2, _, 0 | 2, 0, _, _, _, _] := 0;
pedestrian1[2, 1, 0 | 1, 0 | 2, 0, _, _, _, _] := 0;
pedestrian1[2, 1, 2, 0 | 2, 0, 0, _, _, _] := 0;
pedestrian1[2, 1, 2, 0 | 2, 0, 1 | 2, _, _, _] := 1;
pedestrian1[_, _, _, _, 2, _, _, _, _] := 2;
```

Because this rule is rather complicated, the risk of an error is not negligible. It is therefore wise to check conservation of the number of pedestrians.

```
Clear[init,iter]
init = initialMultiLaneConfig[234,347,10,1000];
iter = neighborhood[pedestrian1, init];
{Apply[Plus,DeleteCases[Flatten[init], 2]],
Apply[Plus,DeleteCases[Flatten[iter],2]],
Apply[Plus,DeleteCases[Flatten[init],1]] / 2,
Apply[Plus, DeleteCases[Flatten[iter],1]] / 2}
```

{234, 234, 347, 347}

36.4 Moving Rules for Type 2 Pedestrians

Similarly, we write moving rules for type 2 pedestrians.

```
pedestrian2[_, _, _, _, 1, _, _, _, _] := 1;
```

```
pedestrian2[_, _, _, _, 0, 0 | 1, _, 0, 0] := 0;
pedestrian2[_, _, _, _, 0, 0 | 1, 0 | 2, 2, 0] := 0;
pedestrian2[_, _, _, 0, 0, 0 | 1, 1, 2, 0] := 0;
pedestrian2[_, _, _, 1 | 2, 0, 0 | 1, 1, 2, 0] := 2;
pedestrian2[_, _, _, _, 0, 0 | 1, _, 1, 0] := 0;
pedestrian2[_, _, _, _, 0, 0 | 1, _, 0, 2] := 0;
pedestrian2[_, _, _, _, 0, 0 | 1, 0 | 2, 2, 2] := 0;
pedestrian2[_, _, _, 0, 0, 0 | 1, 1, 2, 2] := 0;
pedestrian2[_, _, _, 1 | 2, 0, 0 | 1, 1, 2, 2] := 2;
pedestrian2[_, _, _, _, 0, 0 | 1, _, 1, 2] := 2;
pedestrian2[_, _, _, _, 0, 0 | 1, _, 0 | 1, 1] := 0;
pedestrian2[_, _, _, _, 0, 0 | 1, 0 | 2, 2, 1] := 0;
pedestrian2[_, _, _, 0, 0, 0 | 1, 1, 2, 1] := 0;
pedestrian2[_, _, _, 1 | 2, 0, 0 | 1, 1, 2, 1] := 2;
```

```
pedestrian2[_, _, _, 0, 2, _, _, _, _] := 0;
pedestrian2[_, _, _, 2, 2, _, _, _, _] := 2;
pedestrian2[0, 2, _, 1, 2, _, _, _, _] := 2;
pedestrian2[2, 0, 2, 1, 2, _, _, _, _] := 2;
pedestrian2[1, 0, 2, 1, 2, _, _, _, _] := 2;
pedestrian2[2, 0, 0 | 1, 1, 2, _, _, _, _] := 0;
pedestrian2[0, 0 | 1, _, 1, 2, _, _, _, _] := 0;
pedestrian2[2, 1 | 2, _, 1, 2, _, _, _, _] := 2;
pedestrian2[1, 0, 0 | 1, 1, 2, _, _, _, _] := 0;
pedestrian2[1, 1 | 2, _, 1, 2, _, _, _, _] := 2;
pedestrian2[_, _, _, _, 0, 2, _, _, _] := 2;
```

```
Clear[init, iter]
init = initialMultiLaneConfig[269, 327,10,1000];
iter = neighborhood[pedestrian2, init];
{Apply[Plus,DeleteCases[Flatten[init], 2]],
Apply[Plus,DeleteCases[Flatten[iter], 2]],
Apply[Plus,DeleteCases[Flatten[init], 1]] / 2,
Apply[Plus,DeleteCases[Flatten[iter],1]] / 2}
```

{269, 269, 327, 327}

36.5 Evolution of Pedestrians of Both Types

To obtain the moving rules for both pedestrian types, we alternatively apply
type 1 pedestrian moving rules and type 2 pedestrian moving rules.

```
update1[configuration_] :=
MapThread[pedestrian1,
Map[RotateRight[configuration, #] &,
{{1, 1}, {1, 0}, {1, -1}, {0, 1},
{0, 0}, {0, -1}, {-1, 1}, {-1, 0}, {-1, -1}}], 2];
update2[configuration_] :=
MapThread[pedestrian2,
Map[RotateRight[configuration, #] &,
{{1, 1}, {1, 0}, {1, -1}, {0, 1},
{0, 0}, {0, -1}, {-1, 1}, {-1, 0}, {-1, -1}}], 2];
evolve[configuration_] := update2[update1[configuration]];
```

Checking number conservation for types 1 and 2 pedestrians when both types
move, we have

```
init = initialMultiLaneConfig[321, 279,10,1000];
iter =evolve[init];
{Apply[Plus,DeleteCases[Flatten[init],2]],
Apply[Plus, DeleteCases[Flatten[init],1]] / 2,
Apply[Plus,DeleteCases[Flatten[iter],2]],
Apply[Plus,DeleteCases[Flatten[iter],1]] / 2}
```

{321, 279, 321, 279}

36.6 Animation

We can now show an animation of the evolution of both types of pedestrians.
Considering a passageway of length 100, after 200 iterations the self-organized
spatiotemporal pattern is manifest: pedestrians walking in opposite directions
do form lanes along which walkers move in the same direction.

```
Clear[init, ca]
init = initialMultiLaneConfig[150, 150, 10, 100];
time = 200;
ca = NestList[evolve, init, time];
Map[Show[Graphics[RasterArray[# /.
{0 → RGBColor[1, 1, 0], 1 → RGBColor[0, 0, 1],
2 → RGBColor[1, 0, 0]}]],
AspectRatio → Automatic] &,
Table[Reverse[ca[[k]]], {k, 1, }]];
```

We do not display the 200 lattices. We just display the the first and the last
exhibiting the self-organized pattern.

Fig. 36.1. *Initial pedestrian configuration. Type 1 pedestrians (blue squares) move
to the right, and type 2 (red squares) move to the left.*

Fig. 36.2. *Final pedestrian configuration. Type 1 pedestrians (blue squares) move
to the right, and type 2 (red squares) move to the left.*

References

1. M. BARNSLEY, *Fractals Everywhere* (San Diego: Academic Press 1988).
2. G. BAUMANN, *Mathematica for Theoretical Physics: Classical Mechanics and Nonlinear Dynamics* (New York: Springer-Verlag 2005).
3. G. BAUMANN, *Mathematica for Theoretical Physics: Electrodynamics, Quantum Mechanics, General Relativity, and Fractals* (New York: Springer-Verlag 2005).
4. N. BOCCARA, *Functional Analysis: An Introduction for Physicists* (Boston: Academic Press, 1990).
5. N. BOCCARA, *Probabilités* (Paris: Ellipses, 1995).
6. N. BOCCARA, *Intégration* (Paris: Ellipses, 1995).
7. N. BOCCARA, *Fonctions analytiques* (Paris: Ellipses 1996).
8. N. BOCCARA, *Distributions* (Paris: Ellipses 1997).
9. N. BOCCARA, *Modeling Complex Systems* (New York: Springer-Verlag 2004).
10. G. CANTOR, *De la puissance des ensembles parfaits de points*, Acta Mathematica, **4** 381–392 (1884).
11. H. S. CARSLAW, *Introduction to the Theory of Fourier's Series and Integrals* (New York: Dover 1930).
12. P. COLLET and J. P. ECKMANN, *Iterated Maps on the Interval as Dynamical Systems* (Boston: Birkhäuser 1980).
13. DEVANEY R. L., *Chaotic Dynamical Systems* (Redwood City, CA: Addison-Wesley 1989).
14. M. J. FEIGENBAUM, *Quantitative universality for a class of transformations*, Journal of Statistical Physics **19** 25–52 (1978).
15. M. J. FEIGENBAUM, *Universal behavior in nonlinear systems*, Los Alamos Science **1** 4–27 (1980).
16. E. FLORES and T. J. OSLER, *The tautochrone under arbitrary potentials using fractional derivatives*, American Journal of Physics **67** 718–722 (1999).
17. M. FUKUI and Y. ISHIBASHI, *Self-organized phase transitions in cellular automaton models for pedestrians*, Journal of the Physical Society of Japan **68** 2861–2863 (1999)
18. M. FUKUI and Y. ISHIBASHI, *Jamming transition in cellular automaton models for pedestrians on passageway*, Journal of the Physical Society of Japan **68** 3738–3739 (1999)

530   References

19. R. J. GAYLORD and L. J. D'ANDRIA, *Simulating Society: A Mathematica Toolkit for Modeling Socioeconomic Behavior* (New York: Springer-Verlag 1998).

20. J. W. GIBBS, *Collected Works* (New York: Longmans, Green 1928), volume 2, pp 258–260.

21. W. J. GILBERT and R. J. GREEN, *Negative based number systems*, Mathematics Magazine, **52** 240–244 (1979).

22. W. J. GILBERT, *Fractal geometry derived from complex bases*, The Mathematical Intelligencer, **4** 78–86 (1982).

23. W. J. GILBERT, *Arithmetic in complex bases*, Mathematics Magazine, **57** 77–81 (1984).

24. R. L. GRAHAM, D. E. KNUTH, and O. PATASHNIK, *Concrete Mathematics: A Foundation of Computer Science* (Reading MA: Addison-Wesley 1989).

25. F. HAUSDORFF, *Dimension und äusseres Maß*, Mathematische Annalen, **79** 157–179 (1919).

26. G. JULIA, *Mémoire sur l'itération des fonctions rationnelles*, Journal de Mathématiques Pures et Appliquées, **8** 47-245 (1918).

27. C. KAMP, *Élémens d'arithmétique universelle* (Cologne, 1808).

28. H. VON KOCH, *Sur une courbe continue sans tangente, obtenue par une construction géométrique élémentaire*, Arkiv för Matematik, Astronomi och Fysik, **1** 681–702 (1904).

29. J. C. LAGARIAS, *The 3x + 1 problem and its generalizations*, American Mathematical Monthly, **92** 3–23 (1985).

30. A. LINDENMAYER, *Mathematical models for cellular interactions in development*, Journal of Theoretical Biology, **18** 280–300 (1968).

31. E. N. LORENZ, *Deterministic nonperiodic flow*, Journal of Atmospheric Sciences, **20** 130–141 (1963).

32. E. N. LORENZ, *The Essence of Chaos* (Seattle: University of Washington Press 1993).

33. J. LÜTZEN, *Heaviside operational calculus and the attempts to rigorise it*, Archive for the History of Exact Science, **21** 161–200 (1979).

34. B. B. MANDELBROT, *Fractals, Form, Chance and Dimension*, (San Francisco: W. H. Freeman 1977).

35. B. B. MANDELBROT, *The Fractal Geometry of Nature* (San Francisco: W. H. Freeman 1983).

36. A. MENEZES, P. VAN OORSCHOT, and S. VANSTONE, *Handbook of Applied Cryptography* (Boca Raton, FL: CRC Press 1997).

37. K. S. MILLER and B. ROSS, *An Introduction to the Fractional Calculus and Fractional Differential Equations* (New York: Wiley 1993).

38. P. M. MORSE, *Diatomic molecules according to the wave mechanics. II Vibrational levels*. Physical Review **34** 57–64 (1929).

39. P. J. MYRBERG, *Sur l'itération des polynômes quadratiques*, Journal de Mathématiques Pures et Appliquées **41** 339–351 (1962).

40. K. B. OLDHAM and J. SPANIER, *The Fractional Calculus; Theory and Applications of Differentiation and Integration to Arbitrary Order* (Mineola, NY: Dover 2006).

41. T. J. OSLER, *Fractional derivatives and Leibniz rule*, American Mathematical Monthly **78** 645–649 (1971).

A complete list of Osler's papers on fractional calculus can be found at http://www.rowan.edu/mars/depts/math/osler/my_papersl.htm

42. R. PEARL and L. J. REED, *On the rate of growth of the population of the United States since 1790 and its mathematical representation*, Proceedings of the National Academy of Sciences of the United States of America **21** 275–288 (1920).

43. H. POINCARÉ, *Science et Méthode* (Paris: Flammarion 1909). English translation by G. B. HALSTED, in *The Foundations of Science: Science and Hypothesis, The Value of Science, Science and Method* (Lancaster, PA: The Science Press 1946).

44. B. VAN DER POL, *Forced oscillations in a circuit with nonlinear resistance*, London Edinburgh and Dublin Philosophical Magazine **3** 65–80 (1927).

45. P. PRUSINKIEWICZ and J. HANAN, *Lindenmayer Systems, Fractals, and Plants* (New York: Springer-verlag, 1989).

46. P. PRUSINKIEWICZ and A. LINDENMAYER, *The Algorithmic Beauty of Plants* (New York: Springer-Verlag 1990).

47. R. RIVEST, A. SHAMIR, and L. ADLEMAN, *A method for obtaining digital signatures and public-key encryption*, Communications of the ACM, **21**, 2 120–126 (1978).

48. H. RUSKEEPÄÄ, *Mathematica Navigator* (Amsterdam: Elsevier. Academic Press, 2004)

49. H. SAGAN, *Space-Filling Curves* (New York: Springer-Verlag 1994).

50. D. SCHWALBE and S. WAGON, *VisualDSolve: Visualizing Differential Equations with Mathematica* (New York: Springer Telos 1996).

51. P. C. SHIELDS, *The Ergodic Theory of Discrete Sample Paths*, Graduate Studies in Mathematics Series, volume 13 (Providence, RI: American Mathematical Society 1996).

52. W. SIERPIŃSKI, *Œuvres choisies*, (Warszawa: Państowe wydawnictwo naukowe, 1975).

53. C. SPARROW, *The Lorenz Equations, Bifurcations, Chaos, and Strange Attractors* (New York: Springer-Verlag 1982).

54. P. ŠTEFAN, *A Theorem of Šarkovskii on the existence of periodic orbits of continuous endomorphisms of the real line*, Communications in Mathematical Physics **54** 237–248 (1977).

55. M. TROTT, *The Mathematica Guidebook for Graphics* (New York: Springer-Verlag 2004).

56. M. TROTT, *The Mathematica Guidebook for Programming* (New York: Springer-Verlag 2004).

57. M. TROTT, *The Mathematica Guidebook for Numerics* (New York: Springer-Verlag 2005).

58. M. TROTT, *The Mathematica Guidebook for Symbolics* (New York: Springer-Verlag 2005).

59. B. VAN DER POL, *Forced oscillations in a circuit with nonlinear resistance*, London, Edinburgh and Dublin Philosophical Magazine **3** 65–80 (1927).

60. B. VAN DER POL and J. VAN DER MARK, *The heartbeat considered as a relaxation oscillation, and an electrical model of the heart*, Philosophical Magazine Supplement **6** 763–775 (1928).

61. I. VARDI, *Computational Recreations in Mathematica* (Redwood City, CA: Addison-Wesley 1991).

62. P. F. VERHULST, *Notice sur la loi que la population suit dans son accroissement*, Correspondances Mathématiques et Physiques **10** 113–121 (1838).

63. D. VVEDENSKY, *Partial Differential Equations with Mathematica* (Reading, MA: Addison-Wesley 1992).
64. S. WAGON, *Mathematica in Action* (New York: W. H. Freeman, 1991).
65. J. V. WHITTAKER, *An analytical description of some simple cases of chaotic behavior*, American Mathematical Monthly **98** 489-504 (1991).
66. T. WICKHAM-JONES, *Mathematica Graphics*, 5th edition (New York: Springer-Verlag 1994).
67. H. WILBRAHAM, *On a certain periodic function*, Cambridge and Dublin Mathematical Journal **3** 198-201 (1848).
68. S. WOLFRAM, *The Mathematica Book*, 5th edition (Champaign, IL: Wolfram Media 2003).
69. E. ZECKENDORF, *Représentation des nombres naturels par une somme de nombres de Fibonacci ou de nombres de Lucas*, Bulletin de la Société Royale des Sciences de Liège, **5** 179–182 (1972).

Index

Printed in the United States
By Bookmasters